RADAR PRINCIPLES WITH APPLICATIONS TO TRACKING SYSTEMS

RADAR PRINCIPLES WITH APPLICATIONS TO TRACKING SYSTEMS

Philip L. Bogler
University of Missouri
Columbia, Missouri

A WILEY–INTERSCIENCE PUBLICATION

JOHN WILEY & SONS

New York · Chichester · Brisbane · Toronto · Singapore

Library of Congress Cataloging-in-Publication Data

Bogler, Philip L.

Radar principles with applications to tracking systems / Philip L. Bogler.

p. cm.

"A Wiley-Interscience publication."

Bibliography: p

Includes index.

ISBN 0-471-50192-1

1. Tracking radar. 2. Signal processing. I. Title.

TK6580.B64 1989

621.3848—dc20 89-34215

CIP

Printed in the United States of America

10 9 8 7 6 5 4 3 2 1

To Cheryl, Sam and my parents

PREFACE

This book provides a thorough introduction to radar-tracking theory and practice. It is written primarily for first-year graduate students and new engineers in the radar field. Although the emphasis is on radar tracking, necessary background material is presented on the radar sensor and signal-processing features that gather and initially process the target information. Furthermore, the tracking logic often interfaces with the radar's front end to optimize the total system performance, and so an understanding of these processes is considered essential.

The book is organized into four main parts. The first part, Chapters 1 through 5, covers the traditional aspects of radar theory including the sensor technology, detection theory, signal-processing algorithms, and waveform selection. Particular emphasis is placed on those algorithms and technology which are used in new radar system design; older technology is discussed only where historically significant. The second part, Chapters 6, 7, and 8, introduces the reader to those aspects of estimation theory which are used in radar tracking. Chapters 9 through 12 illustrate the practical use of the theory by means of examples drawn from the fields of single- and multiple-target tracking. The necessary correlation logic is also presented to allow the estimation algorithms to operate on a single target. The final two chapters concentrate on the evolving area of electronically scanned array tracking. The emphasis here is on presenting an orderly introduction to the subject of system optimization and resource allocation, rather than an exhaustive catalog of available technology.

A more detailed survey of the book follows. Chapter 1, "Introduction," provides a concise description of basic track definitions and applications.

Chapter 2, "The Radar Sensor," presents the background needed to understand the radar hardware as it pertains to tracking. In order to quickly distill the required technical knowledge, particular emphasis is placed on a modern fully-coherent digital monopulse radar. Chapter 3, "Signal Processing," reviews signal-processing features and resulting detection theory. Chapter 4, "Waveform Selection," discusses waveform parameters to be selected including the radar's pulse repetition frequency (PRF), pulse width (both compressed and uncompressed), and illumination time. Chapter 5, "Pulse Compression," briefly describes some of the codes commonly used including Chirp, Barker, ripple suppressed, PN, complementary, Frank, P3 and P4 codes.

Chapter 6, "Measurement Theory," derives maximum-likelihood estimators for monopulse angle, range, and Doppler. Important suboptimum estimators, such as split-gate, noncoherent, and power centroid estimators, are described, and the rationale for their use explained. Accuracy expressions are derived and related to tracking performance in single- and multiple-target environments. Chapter 7, "Kalman Filtering," reviews the one-dimensional Kalman filter and square-root filter. Chapter 8, "Adaptive Kalman Filtering," begins with simple ad hoc approaches and then proceeds to more sophisticated logic to describe the principles of adaptive Kalman filtering.

Chapter 9, "Coordinate Systems," presents several coordinate systems previously used in radar-tracking applications, including antenna, velocity, NED (north, east, down), covariance, measurement, site, stereographic projection, earth-centered, and range–velocity coordinates. Chapter 10, "A Representative STT System," discusses the philosophy and functioning of single-target tracking (STT) systems. The subjects of decoupling, ECCM, and passive ranging techniques are also discussed. Chapter 11, "Data Correlation Logic," postulates a generalized framework in which to quickly explain the major tenets of data correlation. Chapter 12, "A Representative TWS System," reviews choice of state variables and processing logic and discusses the subject of decoupling from the TWS perspective.

Chapter 13, "ESA Allocation Logic," discusses techniques to adaptively allocate the radar's time and energy and presents critical trade-offs. Chapter 14, "A Representative ESA Radar System," draws on the previously presented ideas to propose designs for future electronically scanned array (ESA) systems.

Jefferson City, Missouri PHILIP L. BOGLER

July 1989

ACKNOWLEDGMENTS

A number of people have contributed to this book. Sam Blackman provided continual encouragement and numerous comments on all aspects of this project. Sam deserves as much credit for this book as I, for without his help this book would not have been written. I am also indebted to Robert Fitzgerald and Alfonso Farina for taking the time to review and critique this work despite the difficulties of long distance communication. Howard Nussbaum and Belvin Freeman also suggested many improvements and were instrumental in establishing a high level of technical content. I would like to thank Belvin in particular for his advice, encouragement, and friendship throughout the course of this project.

R. T. Marloth reviewed and commented upon the entire manuscript, and Stephen Lipsky, Tom Zahm, Kan Jew, and Milton Crane carefully read portions of the rough draft and provided critical comments. Conversations with Ted Broida and Steve Cummings were very helpful and allowed me to improve Chapters 9 and 13. John Wittmond provided a basis for the initial glossary and extensive revisions to the final glossary. Charles Weber provided a basis for the examples in the first six chapters. Finally, the atmosphere of cooperative research I found while working at Hughes Aircraft was a primary component in shaping my professional career, and I would like to thank Jeff Hoffner, Tom Robinson, Gary Graham, and Tom Abbott for creating such an intellectual environment.

CONTENTS

CHAPTER 1

INTRODUCTION

The radar scene includes the radar itself, a target at range R, and the transmitted waveform that travels to the target and back, as illustrated in Figure 1.1. Information about the target's spatial position is first obtained by measuring the changes in the backscattered waveform relative to the transmitted waveform. The time shift provides information about the target's range, the frequency shift provides information about the target's radial velocity (through the Doppler effect), and the received voltage magnitude and phase provide information about the target's angle.

At the heart of tracking is the concept of measuring a target parameter in this manner, and then processing over time a sequence of these measurements to refine our knowledge of that parameter. A related goal is to estimate a parameter not directly measured, such as a higher-order derivative, and to predict future target behavior.

For the broad-brush purposes of this text, the various approaches to radar tracking will be grouped into two main categories: *single-target tracking* (*STT*) and *multiple-target tracking* (*MTT*). Multiple-target tracking is further subdivided into *track-while-scan* (TWS) and *electronically scanned antenna* (ESA) tracking (also called *phase array tracking*). This division is important only in the sense that each of these categories reflects a different priority in functional objectives, and it is the intent of this book to relate the functional objectives to the system design.

As shown in Figure 1.2, STT refers to a dedicated process of tracking a single target or target cluster closely spaced in angle. STT is also referred to as *closed-loop tracking*, *precision tracking*, or *fire control tracking*. The term *tracking radar*, when used without qualification, usually refers to this type of tracking system,

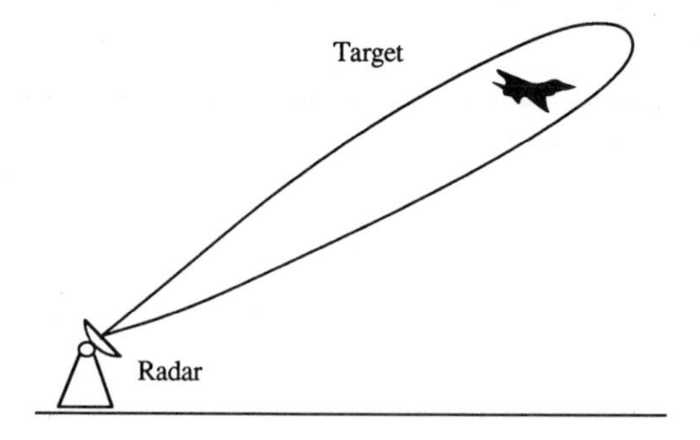

Figure 1.1 Typical radar operational environment.

and the term *tracking* is most closely identified with this type of closed-loop operation.

One of the earliest applications of STT was for antiaircraft fire control, first with guns and later with missiles. This fire control background is reflected in the terminology still used in the literature today, such as target (meaning object of interest), boresight (meaning center of the antenna beam), and range (meaning distance to the target). Today many civilian applications exist as well such as instrumentation radars, satellite-tracking radars, and navigation-aiding radars. As shown in Figure 1.3, modern STT radars may possess very sophisticated feedback logic to optimize their performance relative to the target being tracked. The chief characteristic of all STT radars, however, is that they are highly specialized for the function of tracking, and suffer the inability to track more than a single target (or target cluster) at any given time.

Track-while-scan (*TWS*), on the other hand, refers to the process of automatically tracking targets directly from data supplied by a search radar. The search radar continuously scans for targets in a regular predetermined search

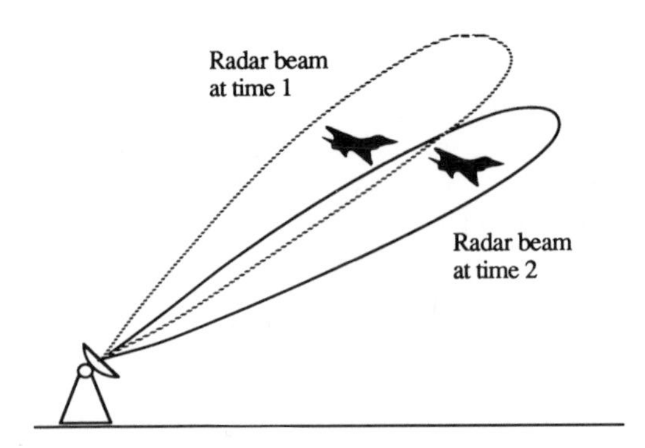

Figure 1.2 A "tracking radar," in its purest sense, is configured to continuously follow the target with the radar beam.

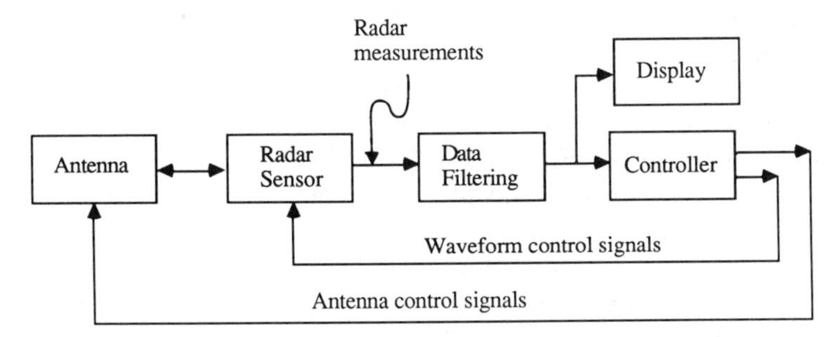

Figure 1.3 Functions that comprise an STT system.

pattern (open loop). Target detections are received at the regular scan rate. The TWS computer forms this raw data into meaningful target tracks.

The principal difference between a TWS system and a pure search radar is the TWS computer. The primary function of this computer is to correlate new radar measurements with previous target tracks to allow sequential measurements from the same target to be filtered as in STT. This correlation process is made complex due to the primary search function, which necessarily entails (1) a long time between target revisits and (2) a possible multiplicity of target declarations. The functions described—parameter measurement, data correlation and trackfile maintenance, and data filtering—are the functions that comprise a TWS loop. This loop is configured as shown in Figure 1.4.

The advantage of TWS over pure search is that tracking is automated and hence operator workload alleviated. The advantage of TWS over STT is that multiple targets can be simultaneously engaged. The advantage of STT over TWS is that high track accuracy can be achieved continuously. While, theoretically, a TWS system can achieve the same accuracy, it can only achieve it momentarily as it mechanically scans past the target.

TWS and STT systems often work together to jointly search and track. The function of the TWS system is to detect new targets, to perform initial target tracking, and to enable the STT radar to acquire the target. The function of the STT radar is to provide refined estimates of the target's kinematic parameters. Sometimes these two functions are performed by two physically separate radars,

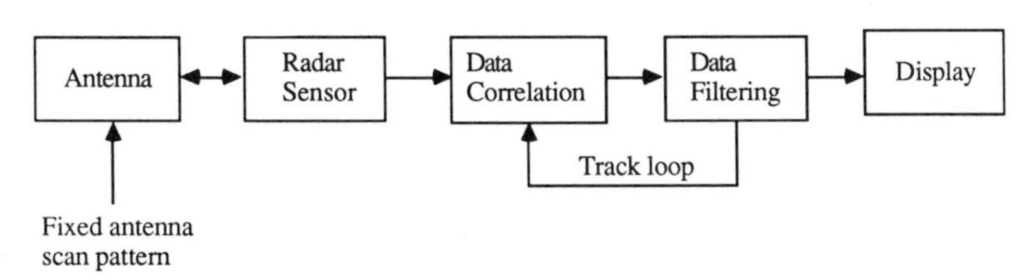

Figure 1.4 Functions that comprise a TWS system.

and sometimes a single radar interleaves its times between the two modes of operation in a time-shared fashion.

The unique characteristics of the electronically steered antenna (ESA) can be exploited to overcome the deficiencies apparent in the older STT and TWS systems (e.g., Wirth, 1978; Fleskes and Van Keuk, 1980). The ESA is capable of electronically steering its beam to virtually any position in space without suffering mechanical inertia. Thus, continuous STT-like track accuracy can be achieved on multiple selected targets while realizing a large TWS-like search coverage from a single aperture. In conclusion, the key feature that distinguishes this type of tracking system is the rapid rate with which it can transition between different modes of operation. A hypothetical ESA tracking loop can be configured as shown in Figure 1.5. As envisioned here, the ESA system incorporates both STT and TWS design attributes.

The purpose of this book is to introduce the basic tenets underlying modern STT and TWS systems and to extrapolate the theory to the design of ESA systems. Unfortunately, no standard approaches have been generally accepted in any case, and the radar technology is itself undergoing widespread and rapid evolution as it moves toward the twenty-first century. The approach taken here is to present representative architectures that illustrate the major drivers that go into new radar system design.

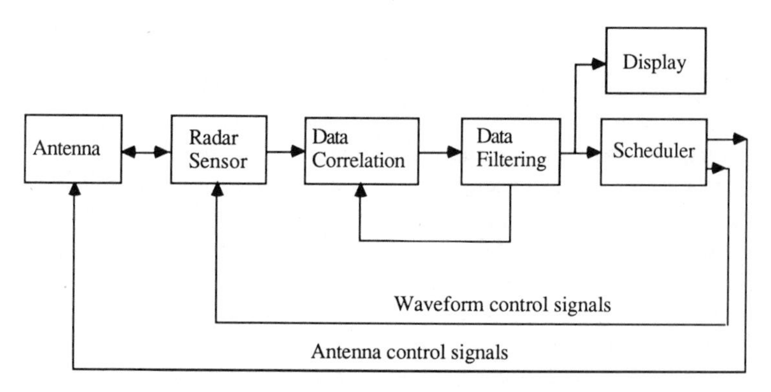

Figure 1.5 Functions that might comprise an ESA system.

CHAPTER 2

THE RADAR SENSOR

The major elements of a coherent radar sensor consist of the following: a transmitter (or transmitters in the case of an active array), an exciter (or local oscillator), an antenna(s), duplexer(s), a receiver tuned to the specific transmit frequency, analog signal processor, analog/digital (A/D) converter, digital signal processor, data processor, and finally a display for user interfacing. Figure 2.1 charts the interrelationship between these elements. Beginning with the transmitter, the functions performed by each of these elements are briefly outlined in turn.

2.1 TRANSMITTER

Radar transmitters provide the high-power signal needed for transmission. The transmitter output, as shown in Figure 2.2, is switched on and off to form pulses

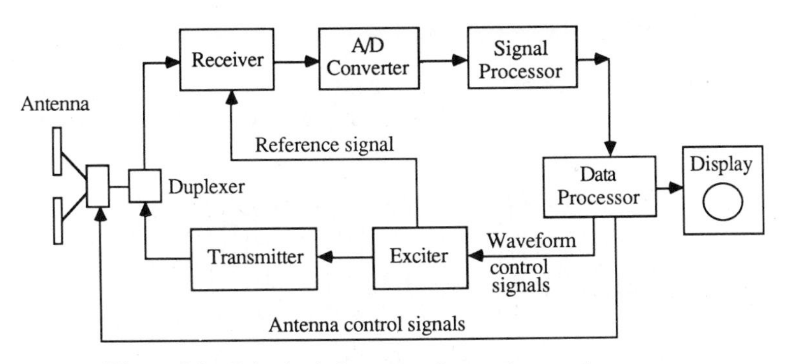

Figure 2.1 Principal elements of a modern radar sensor.

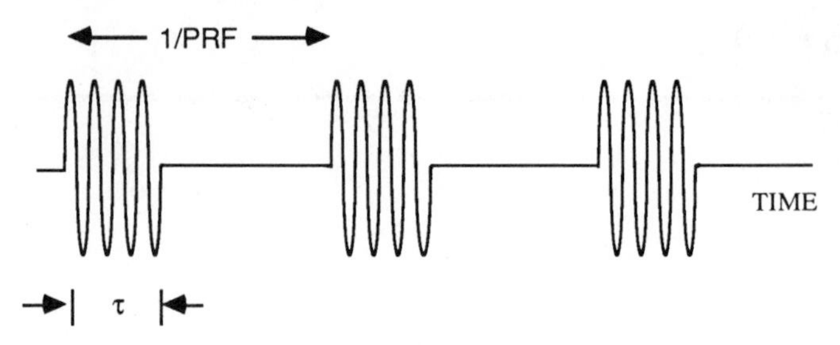

Figure 2.2 Example of transmitter output.

of the desired range resolution, τ. Pulses are repeated at a semiregular pulse repetition frequency (PRF) in order to increase the output signal-to-noise ratio and enhance target detection. In STT, both τ and PRF may be selected from prior track file analysis.

2.2 EXCITER

Coherent processing is often desired for clutter rejection. By measuring the Doppler shift in frequency of the received array of pulses relative to the transmitted array, the moving target of interest can be separated from nearby stationary objects (termed *clutter*).

Phase stability may be maintained by referencing the transmitted output to a continuous-wave (CW) highly stable low-power (several milliwatts) reference signal (see Figure 2.3). In those radar systems employing power amplification, such as with a traveling-wave tube or klystron, the transmitter amplifies segments of the CW reference signal. The source of this reference signal is referred to as an *exciter, master oscillator,*" or *radar master oscillator* (*RMO*) (Eaves and Reedy, 1987, Chapter 5; Skolnik, 1980, Chapter 6). In coherent-on-receive (COR) systems, the transmitted signal itself is not coherent, but a local oscillator locks the received pulse in phase to the transmitted pulse during one of the receiver mixing stages.

2.3 DUPLEXER

To conserve space and minimize cost, most pulsed radars employ only one antenna. The antenna is then time-shared between the functions of transmission

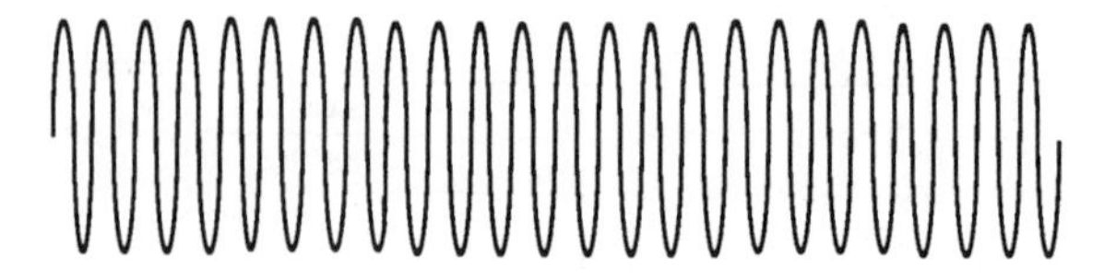

Figure 2.3 Example of exciter output.

and reception. To avoid leakage of transmitted energy into the receiver and causing harm to the equipment, a duplexer (or circulator) is typically provided to divert energy away from the receiver during the transmission period (Eaves and Reedy, 1987, pp. 198–201; Skolnik, 1970, pp. 8-31). For example, a transmit/receive (TR) tube is a common type of duplexer. The total time that the receiver is shut down is equal to the pulse length time τ plus the duplexer recovery time τ_R. During this shut-down time $(\tau + \tau_R)$, any target return that arrives back will suffer an "eclipsing" loss due to partial obscuration of the signal.

2.4 ANTENNA

The functions of the antenna are: (1) to provide sufficient gain to detect targets, (2) to separate targets from clutter and from other targets at different angular positions, and (3) to provide the required angular resolution and measurement accuracy to the target. Because of the importance of these tasks, the antenna will be discussed in some detail.

2.4.1 Two Baseline Antennas

In order to focus our discussion, only two types of antenna will be described: the parabolic reflector antenna and the array antenna. These two antenna types deserve special attention because they are so often utilized in tracking applications.

The parabolic antenna, shown in Figure 2.4, is an example of a "reflector" type of antenna (Eaves and Reedy, 1987, p. 165). A central radiating element is used to radiate energy, which is reflected by a parabolic dish. Due to the geometric shape of the parabolic dish, the energy exiting the antenna face has a uniform phase front and is, therefore, focused into a directive beam.

The array antenna, shown in Figure 2.5, consists of an array of radiating elements, usually mounted on a flat plate. The radiating elements may be $\lambda/2$ dipoles, slots, or active modules (active array). The elements are interconnected

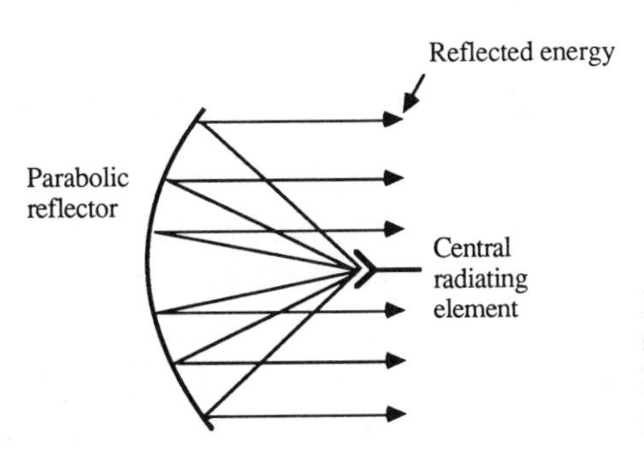

Figure 2.4 The parabolic antenna consists of a central radiating element located at the focus of a parabolic reflector dish.

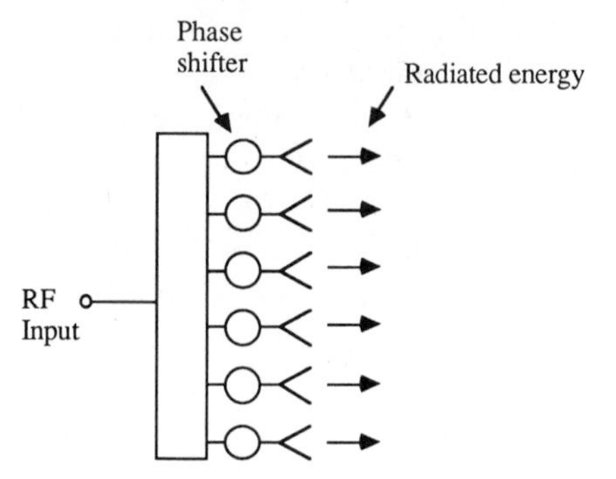

Figure 2.5 The array antenna consists of an RF input located behing an array of radiating elements.

by transmission lines, waveguides, or strip line feeders.* Each radiating element has associated with it a phase shifter (or fixed length of transmission line) whose function is to compensate for any path length differences encountered. This phase shifter ensures that the energy exiting the antenna has a uniform phase front and thus is focused into a directive beam.

The relative merits of the parabolic antenna versus the array antenna have been debated for decades. At the risk of oversimplifying, it may be stated that the parabolic antenna is simpler to fabricate and, hence, is less of a technological risk. Also, in terms of *existing* technology, it is capable of achieving a larger percentage bandwidth (multioctave) by virtue of its geometric shape.

The primary advantage of the array antenna relates to its ability to achieve a wider angular scan volume for given size antenna dimensions. In particular, the array antenna occupies less physical depth and hence requires less volume in which to mechanically scan, and is capable of electronic scanning over a wider angular scan volume. Other advantages often cited include: (1) no spillover, (2) better control of the sidelobes in the sum pattern, and (3) better purity in the direction or "sense" of the polarization.

Scanning options include:

1. Mechanically scanned antenna (MSA).
2. Electronically scanned antenna (ESA).
3. One-dimensional electronically scanned antenna augmented with mechanical scanning (1-D ESA).
4. Two-dimensional electronically scanned antenna augmented with mechanical scanning (mechanically scanned 2-D ESA).

As depicted in Figures 2.6a and 2.6b, electronic scanning can be accomplished either by electronically varying the phase or the frequency in an array

*As used here, the term *array antenna* refers to an antenna with constrained feeds. An array antenna with a space feed acts essentially like a lens or reflector antenna and may be classified as such for this discussion.

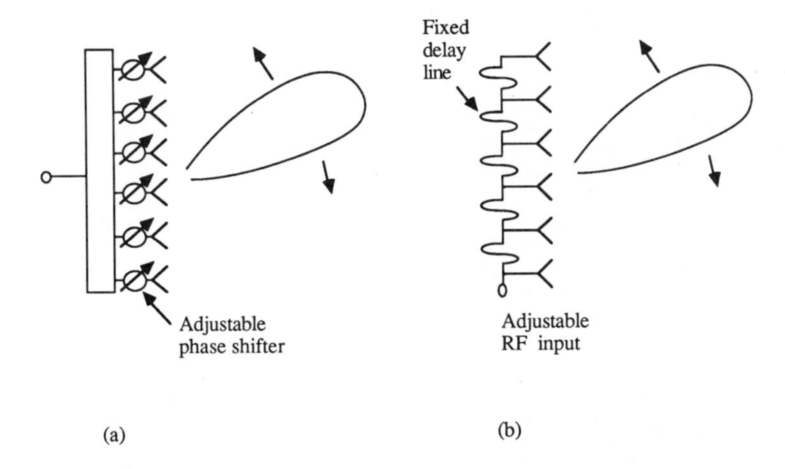

Figure 2.6 Electronic steering can be accomplished either by (a) adjusting discrete phase shifters or (b) by adjusting the RF input frequency in conjunction with fixed delay lines.

antenna (or both). In either case the objective is to electronically induce a linear phase shift over the antenna aperture. With electronic phase control (called *phase scan*), the linear phase shift is induced by adjusting discrete phase shifters spaced over the array face. With electronic frequency control (called *frequency scan*), the phase shift is induced by fixed delay lines in conjunction with varying the input radio frequency (RF). Frequency scanning is usually one dimensional due to implementation difficulties with two-dimensional scanning.

Reflector antennas have also been built to scan electronically, for example, by electronically steering the central radiating feed. This type of operation is used in those applications where the small instantaneous ESA scan volume poses no difficulty (Brookner and Mahoney, 1983, p. 472).

2.4.2 Monopulse Processing

The shape of the antenna beam is characterized by a central lobe in the main scan direction (mainlobe) and low response in all other directions (sidelobes). The mainlobe width (width between half-power points) is given by $\theta_{HP} = k\lambda/D$, where λ is the operating RF wavelength, D is the antenna diameter, and k is a constant varying between 0.9 and 1.5 (the exact value depending on antenna weighting). For most tracking applications θ_{HP} is on the order of 1° to 4°.

The foregoing should not be taken to imply that the accuracy with which a radar can measure angle to the target is limited to the antenna mainlobe width. Tracking systems are capable of measuring the target's angle to much greater accuracy. Better angle accuracy can be achieved by comparing the outputs from two or more beam positions displaced in angle. In the simplest configuration, as shown in Figure 2.7, a single beam is swept past the target. Since the amplitude

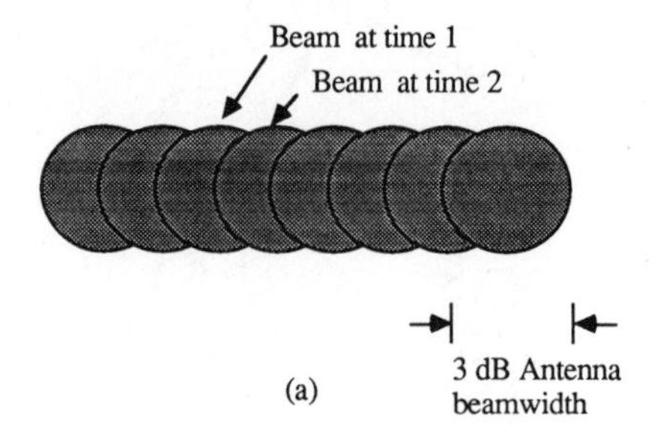

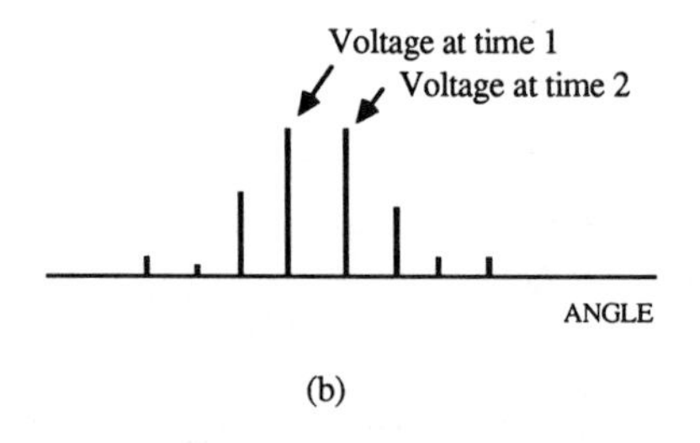

Figure 2.7 Pattern of voltage hits received by the radar follow the antenna scanning: (a) Typical scan pattern. (b) Associated voltage pattern.

of the received returns varies as the beam sweeps over the target, the target's direction can be determined to within a fraction of the beamwidth.

The configuration shown in Figure 2.7 is used in some TWS systems and has the advantage of simplicity. However, it suffers several limitations. Primary among these include (1) the need to scan over the target in sufficiently small steps to accurately measure angle and (2) the presence of shoft-term changes in the strength of the target returns—caused by amplitude scintillation or electronic countermeasures (ECM)—which introduce spurious noise into the angle measurement process. Both of these problems are compounded if both azimuth and elevation (or height) information are sought simultaneously.

To overcome these deficiencies, the antenna may be designed to produce several displaced beams simultaneously. The term *monopulse* is used to denote this procedure since theoretically all of the angular information can be obtained from a single reflected pulse. The name is somewhat of a misnomer from a system viewpoint, however, as multiple pulses are often required for detection purposes.

2.4.3 Amplitude Comparison Monopulse

There are two generic types of monopulse. The first type, called *amplitude comparison monopulse*, is illustrated in Figure 2.8. Amplitude comparison monopulse compares the outputs of at least two beams that are slightly displaced in angle (i.e., they point at slightly different angular directions). Thus, the amplitude of the received returns differ, with the difference proportional to

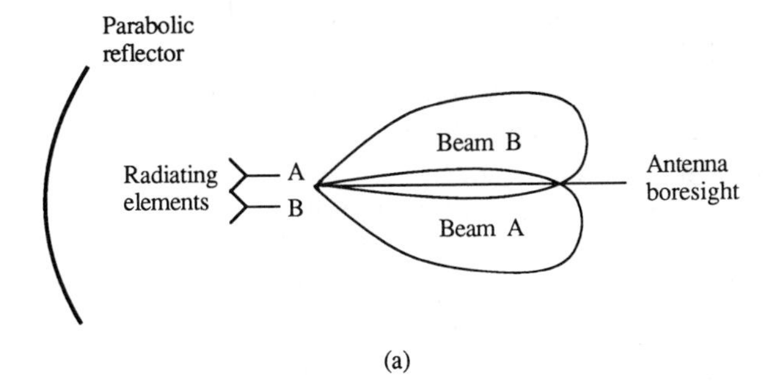

(a)

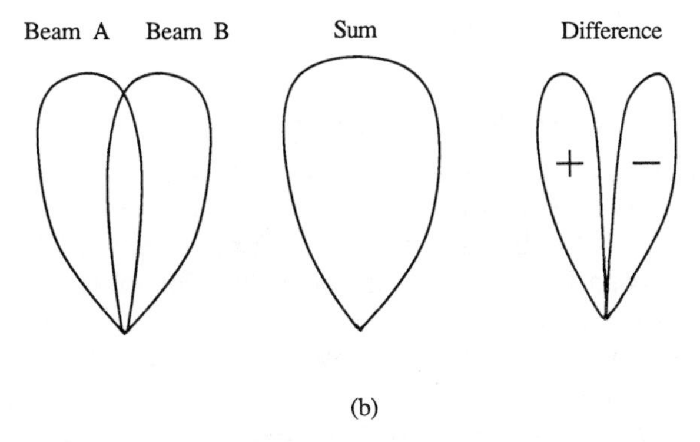

(b)

Figure 2.8 In amplitude comparison monopulse two or more beams are formed that are displaced in angle. (a) Typical (**A**, **B**) feed structure. (b) Resulting sum ($\mathbf{S} = \mathbf{A} + \mathbf{B}$) and difference ($\mathbf{D} = \mathbf{A} - \mathbf{B}$) antenna beam patterns. Angular error is proportional to difference signal after normalization by sum signal ($\mathbf{D}/\mathbf{S}$).

the target's angular displacement from the antenna's boresight (Page, 1955; Barton, 1964).

In the one-dimensional case, either elevation or azimuth, amplitude comparison monopulse typically results in the following received voltage signals:

$$\mathbf{S} \triangleq \mathbf{A} + \mathbf{B} \tag{2.1a}$$

$$\mathbf{D} \triangleq \mathbf{B} - \mathbf{A} \tag{2.1b}$$

where **A** and **B** represent the input voltages arriving at elements A and B, respectively, and boldface denotes complex-valued. Theoretically, the signals that emerge from the antenna may be in any format. In practice, however, either (**S**, **D**) format or (**A**, **B**) format seems to be conventional. The (**S**, **D**) format is slightly easier to implement since the transmitter need access only the sum port to transmit. In order to avoid having to treat both cases, (**S**, **D**) processing is assumed in this text.

To generalize this procedure to two angles, a third or fourth beam is sometimes added, and separate comparisons made. There are also 5- and even 12-beam monopulse systems. Adding more beams permits better comparisons between adjacent beams. In any TWS or search application, adding more beams also permits a wider scan volume to be swept in a shorter amount of time (Skolnik, 1980, p. 310; Eaves and Reedy, 1987, p. 172).

It may be asserted that amplitude comparison monopulse is more compatible with reflector-type antennas, such as the parabolic reflector described earlier. The additional beams are easily formed by placing additional elements in the parabolic feed focus. By displacing the elements, the resulting beams are displaced. It is also possible to implement amplitude comparison monopulse by using an array antenna, namely by interlacing two or more array feeds. However, this last procedure is relatively uncommon.

2.4.4 Phase Comparison Monopulse

The other major type of monopulse operation is referred to as *phase comparison monopulse*. In this type of operation the individual radiating elements are separated by a large distance, and the resultant beams are parallel. In contrast, in amplitude comparison monopulse the individual radiating elements are located close together, and the resultant beams are displaced (squinted).

Figure 2.9 depicts a simple example of phase comparison monopulse operation. An array antenna is divided into two halves (called *ports*) with the beams produced by both ports pointing in the same direction. Consequently, the return received through one port, **A**, has roughly the same amplitude as that received through the other port, **B**. However, if the target is located off boresight, the phases of the two returns differ due to differences in the distance to each antenna port. Referring to Figure 2.9, the distance differential is equal to ζd, for d the antenna phase center separation and ζ the direction cosine of the angle to the

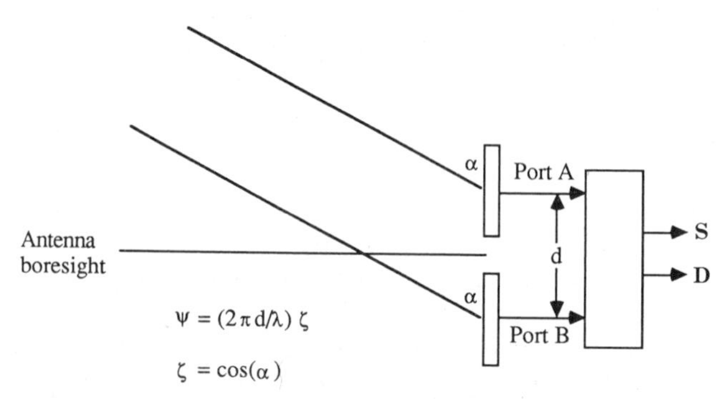

Figure 2.9 In phase comparison monopulse the amplitude received in each port is equal but the phase differs by an angle proportional to angular error, α.

target. Thus, the resulting phase shift, ψ, is

$$\psi = \frac{2\pi\, d\zeta}{\lambda}$$

where λ is the transmitted wavelength.

As Figure 2.10 illustrates, a difference monopulse signal, **D**, may be formed by introducing a 180° phase shift in the output of port A and summing the two outputs. This procedure is usually done in a microwave device called the *comparator*, located immediately behind the antenna ports. A sum signal **S** is also provided by combining the two outputs without the 180° phase shift. In the one-dimensional case the above process results in

$$\mathbf{S} \triangleq \mathbf{A} + \mathbf{B} \tag{2.2a}$$

$$\mathbf{D} \triangleq j(\mathbf{B} - \mathbf{A}) \tag{2.2b}$$

where **A** and **B** represent the inputs arriving at ports A and B, respectively. The (**S**, **D**) signals are then sent to the receiver.

Since the **A** and **B** signals are equal in amplitude but differ in phase, (**S**, **D**) processing results in

$$\begin{aligned} \mathbf{S} &\triangleq \mathbf{A} + \mathbf{B} \\ &= \mathbf{A}(1 + e^{j\psi}) \\ &= 2|\mathbf{A}|\cos(\psi/2) \end{aligned} \tag{2.3a}$$

For notational simplicity, (2.3a) drops a constant phase term and assumes equal (**A**, **B**) weighting. Likewise,

$$\begin{aligned} \mathbf{D} &\triangleq j(\mathbf{B} - \mathbf{A}) \\ &= j\mathbf{A}(1 - e^{j\psi}) \\ &= 2|\mathbf{A}|\sin(\psi/2) \\ &= \mathbf{S}\tan(\psi/2) \end{aligned} \tag{2.3c}$$

where

$$\frac{\psi}{2} = \frac{\pi\, d\zeta}{\lambda}.$$

Thus, the output ratio Re(**D**/**S**) is proportional to the direction cosine ζ to the target's angle off boresight. The constant of proportionality is approximately given by $\frac{\pi d}{\lambda}$ (on boresight).

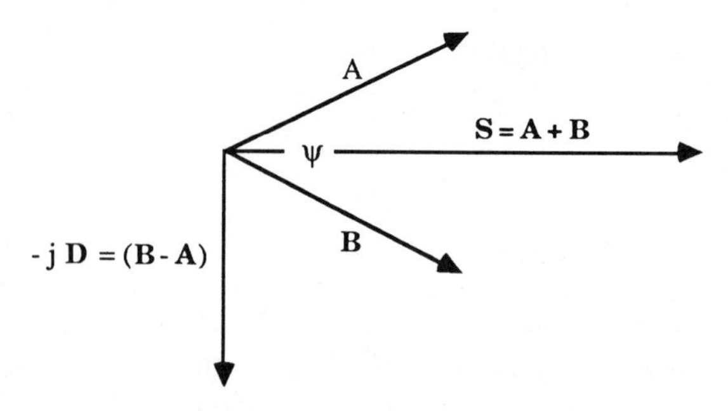

Figure 2.10 Phase relationship between signals **A** and **B** at the output of the antenna ports.

If the antenna is mechanically positioned such that the difference signal **D** is zero, then the *angle* to the antenna boresight axis gives a direct indication of the target's angular position. This method is commonly used by contemporary STT systems.

In the case of ESA steering the beam is electronically steered without moving the antenna itself. One way to visualize this is to insert a variable phase shift in one port relative to the other. As a consequence of the relationship expressed by (2.3c), any phase shift so introduced will result in a proportional shift in the *direction cosine* to the target. Thus, an ESA tracking system of this type measures direction cosine data, rather than angle data per se. The ramifications of this fact will be discussed in Chapter 14.

Finally, it may be asserted that phase comparison monopulse is more compatible with the use of an array antenna. The array antenna possesses distributed radiating elements and so phase differences are more readily sensed. It is technically possible to implement phase-comparison monopulse by employing two or more adjacent parabolic antennas (e.g., Sherman, 1984, p. 91), but this procedure is rarely employed today.

Two-Dimensional Phase Comparison Monopulse. By dividing the antenna into quadrants, as shown in Figure 2.11, the monopulse error signal **D** can be produced in both the azimuth and elevation angles, $\mathbf{D}_{EL}$ and $\mathbf{D}_{AZ}$. This

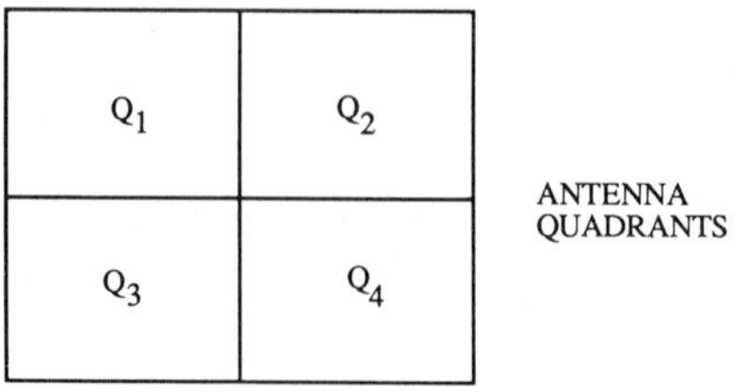

Q_1	Q_2
Q_3	Q_4

ANTENNA QUADRANTS

Figure 2.11 Two-dimensional antenna quadrants. For elevation monopulse tracking, $\mathbf{A} = \mathbf{Q}_1 + \mathbf{Q}_2$ and $\mathbf{B} = \mathbf{Q}_3 + \mathbf{Q}_4$. For azimuth monopulse tracking $\mathbf{A} = \mathbf{Q}_1 + \mathbf{Q}_3$ and $\mathbf{B} = \mathbf{Q}_2 + \mathbf{Q}_4$.

four-quadrant structure is analogous to the four-element structure discussed earlier in regard to parabolic antennas.

The previous "quadrant" method is a straightforward means of generating the sum and difference patterns. Its main drawback is that the antenna weighting cannot be independently specified for both the sum and difference patterns, resulting in relatively large sidelobes in the difference pattern. Low sidelobes in the difference pattern are often required to suppress multipath and hostile interferences entering through the one-way sidelobes (receive sidelobes). This objection is conventionally overcome by interlacing the dipole density in the transition region, as shown in Figure 2.12, thereby smoothing the dipole density near the edges. Interlacing the dipole density has no effect on the sum pattern but does affect the difference pattern since it changes the relative distribution of dipoles that contribute to the ports A and B.

An alternate procedure is to employ *independent* feed networks for the sum **S** and difference patterns ($\mathbf{D}_{EL}, \mathbf{D}_{AZ}$). This is the optimum monopulse solution since the difference patterns can then be specified completely independent of the

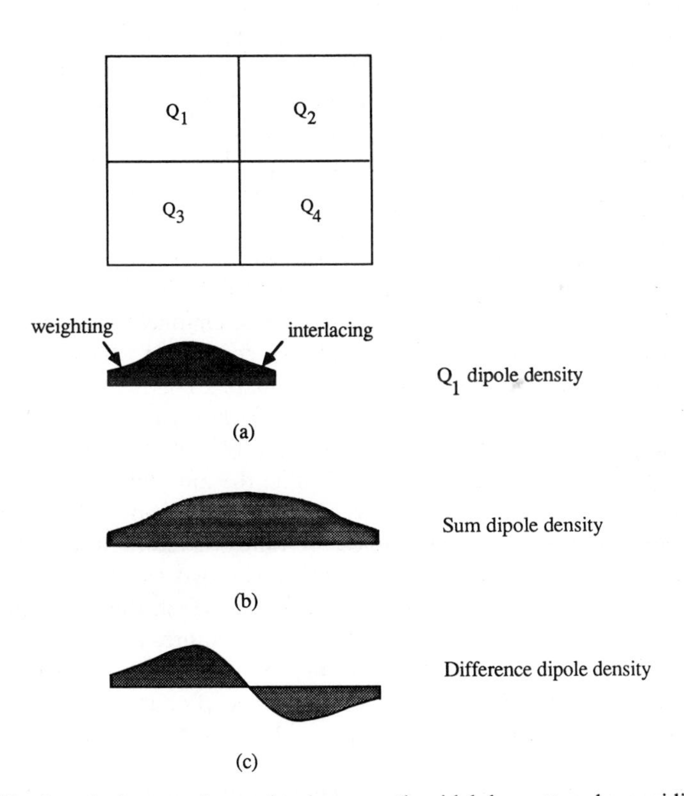

Figure 2.12 Interlacing may be used to improve the sidelobe pattern by avoiding sharp discontinuities: (a) Dipole density over Q_1 quadrant after interlacing; (b) Sum dipole density; (c) difference dipole density.

sum pattern. The drawback is the requirement for triple the number of feed connections in the antenna, and the associated antenna weight. When this type of monopulse solution is employed, the distinction between phase comparison monopulse and amplitude comparison monopulse becomes somewhat vague, and perhaps a more apt description is simply "monopulse with independent feeds."

2.4.5 Required Number of Channels

To form a complete two-dimensional monopulse picture, three separate monopulse signals are required—one for the sum signal S, and one each for the two difference signals $\mathbf{D}_{EL}$ and $\mathbf{D}_{AZ}$. In most current tracking radars employing monopulse, however, it is difficult to accommodate all three signals simultaneously since doing so requires at least three separate receivers and associated channels. To reduce the radar's weight (and cost), it sometimes makes more sense to time share the three signals through two channels (Sherman, 1984, pp. 186–195) or even through just one channel (Skolnik, 1970, Section 21-3).

One practical approach to time sharing follows. The azimuth and elevation difference signals are alternately received on a time-shared basis. That is, two independent transmissions are processed, one to collect the elevation monopulse information, and another to collect the azimuth information. Between these two transmissions, the radar's front-end hardware is switched among the appropriate antenna quadrants to time share the single difference channel between the dual tasks of azimuth and elevation difference signals. The other receiver is dedicated to the sum signal.

This procedure can be extended to using just one receiver channel. Examples include the so-called conical scan radar and the COSRO radar (Conical Scan on Receive Only). The conical scan radar is a single-channel system used in some civilian applications; COSRO is its military equivalent, designed to be less sensitive to deceptive jamming (Schleher, 1986, p. 239).

Time sharing relaxes both hardware and software requirements. However, it imposes a penalty in terms of performance. In the two-channel approach described previously, the penalty is to double the amount of time required to form a complete angle measurement. A second penalty is that the elevation and azimuth measurements are decoupled in time, therefore implying partially decorrelated elevation and azimuth signals. These two sacrifices can usually be tolerated in most STT applications. For TWS, however, the reduced data rate restriction makes this technique infeasible. TWS requires at least three separate channels to achieve two-dimensional monopulse operation while simultaneously maintaining a reasonable scan rate. This, of course, leads to increased cost, as well as additional gain and phase balancing to equalize the various channels.

For these reasons TWS systems have traditionally not employed two-dimensional monopulse. Rather, various suboptimum approaches have been employed instead. These include (1) using monopulse for only one angle (either elevation or azimuth) and then relying upon a narrow beamwidth and interpolation for the other angle, (2) relying on a height-referenced transponder for the other angle, or (3) stopping the scan and staring at the target. Suffice it to say that three channel monopulse is the preferred solution if it is feasible and if a high scan rate is desired.

ESA systems deserve some special comments in this regard. Since an ESA radar system does not require monopulse while in the search mode, nonmonopulse operation can be used for search, and time-shared monopulse can be used for track. This is one way to achieve monopulse accuracy for high-priority targets while simultaneously not degrading the search timeline. If three channels are available, then the ESA system could be configured as fully monopulse in both search and track. However, there may be other advantages to time sharing in this case. For one, a provision for channel time sharing relaxes hardware requirements and provides for channel redundancy. It also frees the third receiver to be utilized as a sidelobe canceler, thus allowing the sidelobes to be operationally matched to the local environment for high-priority targets (e.g., Farina, 1977).

2.4.6 Front-end Signals

The transmitted RF signal is of the form

$$\mathbf{S} = e^{j\omega_{\mathrm{RF}}t}p(t)$$

where $\omega_{\mathbf{RF}}$ is the RF angular frequency, t is time, and complex notation is used for convenience; $p(t)$ is the pulse-train function as illustrated by Figure 2.2. The pulse-train function $p(t)$ models the transmitted signal as nonzero only inside the pulse width τ and with the pulse repeated at a regular repetition frequency (PRF).

The signal that arrives at the front end of the antenna is a shifted version of this transmitted pulse:

$$\mathbf{S} = \mathbf{V}e^{j(\omega_{\mathrm{RF}} - \omega_T)t}p(t - 2R/c) \tag{2.4}$$

where $2R/C$ = time delay between transmission and reception
$\mathbf{V}$ = random complex voltage strength
c = velocity of light
ω_T = target's Doppler shift

The signals that emerge from the antenna front end depend on the design of the antenna. To avoid having to treat all cases separately, we generally represent the signals as

$$\mathbf{S} = \mathbf{V}e^{j(\omega_{\mathrm{RF}} - \omega_T)t}p(t - 2R/c) \tag{2.5a}$$

$$\mathbf{D} = \mathbf{V}e^{j(\omega_{\mathrm{RF}} - \omega_T)t}F(\Delta\psi)p(t - 2R/c) \tag{2.5b}$$

In general, $F(\Delta\psi)$ corresponds to some function of angle off boresight.

2.5 RECEIVER

The radar receiver translates the signals from the high carrier frequency to a lower frequency at which it can be more conveniently processed. Frequency translation begins by mixing the received signal at the target's frequency, $\omega_{RF} - \omega_T$, with a reference signal supplied by the radar's exciter as shown in Figure 2.13. In the particular case of the superheterodyne receiver, this reference signal is offset from the RF signal by a precise amount, $\omega_{RF} \pm \omega_{IF}$, where ω_{IF} is the *intermediate frequency* (or IF). Next, the signals are added together and then passed through a nonlinear device such as the following simple squarer (Eaves and Reedy, 1987, p. 201):

$$\left| e^{j(\omega_{RF} - \omega_T)t} + e^{j(\omega_{RF} - \omega_{IF})t} \right|^2$$

This process is referred to as *mixing*. After mixing, the term of interest is the intermediate cross-product term, represented here as $e^{j(\omega_{IF} - \omega_T)t}$. The other terms are removed via IF bandpass filtering. The outputs of the first IF stage are denoted here as $\mathbf{S}'$ and $\mathbf{D}'$, respectively:

$$\mathbf{S}' = \mathbf{V} e^{j(\omega_{IF} - \omega_T)t} p(t - 2R/c) \tag{2.6a}$$

$$\mathbf{D}' = \mathbf{V} e^{j(\omega_{IF} - \omega_T)t} F(\Delta\psi) p(t - 2R/c) \tag{2.6b}$$

The advantage of the superheterodyne receiver is that the subsequent filters need not be tunable. Rather, tuning is accomplished by varying the IF supplied by the exciter. Since the subsequent filters need not be tunable, sharper filters can be built. A second advantage is that IF amplification is easier to mechanize than RF amplification. The drawbacks of the superheterodyne receiver is the increased cost and complexity of the multiple-frequency conversion. Also, additional spurious noise components are introduced after mixing that may fall within the passband of subsequent receiver stages.

2.5.1 Second Intermediate Frequency (2nd IF)

A second mixing frequency is sometimes added to avoid potential problems with image frequencies. This process begins by mixing the signal with a second (2nd) IF as depicted in Figure 2.14.

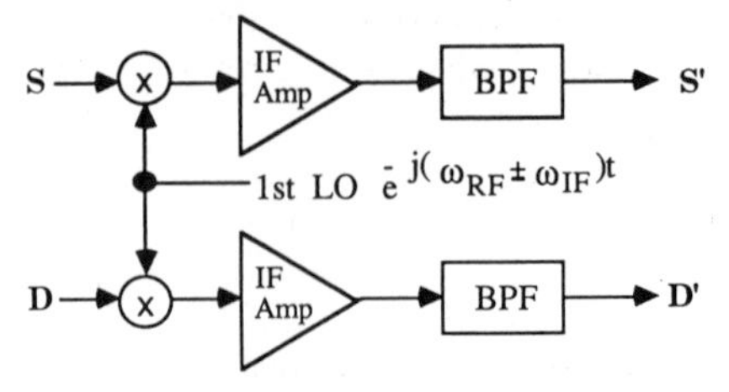

Figure 2.13 The superheterodyne receiver first "mixes" the signals, that is, translates the received microwave signals to a lower frequency (IF). Then the signals are amplified and bandpass filtered (BPF) to produce an output proportional to the input at a lower frequency.

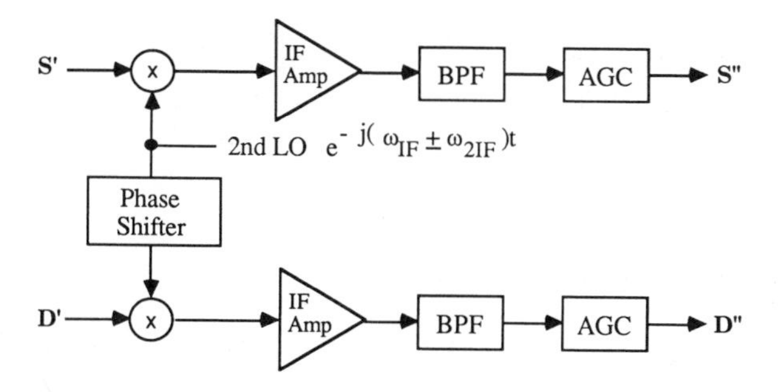

Figure 2.14 A second mixing frequency is sometimes added to avoid problems with image frequencies.

To understand the image phenomenon, notice that there are at least two frequencies that pass unattenuated through the IF stage. If the mixer frequency is set at $\omega_{\mathbf{RF}} - \omega_{\mathrm{IF}}$, then one frequency is centered at $\omega_{\mathbf{RF}}$ and its image is centered at $\omega_{\mathbf{RF}} - 2\omega_{\mathrm{IF}}$ (Ziemer and Tranter, 1976, p. 109). Assuming it is necessary to reject this image frequency, as well as any potential cross product of the image frequencies (called two-tone interference), then the passband of the first IF needs to be narrower than the center frequency of the second IF. With available near-term technology, this necessarily implies that each IF stage can undergo at most a fixed percentage down-conversion.

A typical value for the first IF may be around 1 GHz. The choice of the second IF is governed by the need to be several multiples of the radar's operating bandwidth. Second IF values used in the past lie in the range from 3–300 MHz.

Next, the signals are amplified and bandpass filtered to remove unwanted high-frequency components (Eaves and Reedy, 1987; p. 216). The sum and difference signals at the 2nd IF output are labeled $\mathbf{S}''$ and $\mathbf{D}''$, respectively, where

$$\mathbf{S}'' = \mathbf{V}e^{j(\omega_{2\mathrm{IF}} - \omega_T)t}p(t - 2R/c) \tag{2.7a}$$

$$\mathbf{D}'' = \mathbf{V}e^{j(\omega_{2\mathrm{IF}} - \omega_T)t}F(\Delta\psi)p(t - 2R/c) \tag{2.7b}$$

Automatic gain control (AGC) may be applied near this point. The AGC controls the gain of the radar to protect the receiver from strong targets or clutter. This involves attenuating short-range signals to prevent receiver saturation, and passing long-range signals unattenuated in order to be detected above thermal noise.

The required AGC dynamic range depends on the specific application. If the ratio of maximum to minimum detection range is 20:1, for instance, then the dynamic range required to compensate for the R^4 propagation effect is 52 dB [52 dB = 10 log(20^4)]. The target cross section (RCS) can conceivably contribute another 20–40 dB of variation. Thus, the required dynamic range lies somewhere between 70 and 90 dB (Skolnik, 1980, p. 158).

To provide the minimum amount of AGC necessary, the AGC attenuation may be adjusted in a closed-loop fashion. In the case of ground-based radars, the AGC level may be calculated from data contained in previous beam positions pointed at the same angular direction. In the case of radars on moving platforms, where the clutter is dynamic, the required AGC level may be calculated from one or two pilot pulses that proceed the dwell or inferred from a prior beam position.

2.5.2 *I/Q* Detection

The signal at this stage consists of a sinusoidal that possesses an arbitrary phase relationship with respect to the radar's phase reference. This can be thought of as a rotating phasor possessing both real and imaginary components. For coherent Doppler and monopulse processing, it is necessary to convert both components into digital signals.

As depicted in Figure 2.15, real and imaginary components are obtained by mixing each channel with two reference signals that are 90° out of phase with each other (sine and cosine). Afterward, the A/D converter independently samples both components of data. In radar terminology, the terms *in phase*, or *I*, and *quadrature phase*, or *Q*, are used to describe the resultant real and imaginary components. The terminology arises because the two components are in phase (0°) and quadrature (90°) with the radar's phase reference, respectively.

The output of the *I/Q* detector is

$$\mathbf{S}''' \triangleq \mathbf{S}_I''' + j\mathbf{S}_Q''' \tag{2.8a}$$

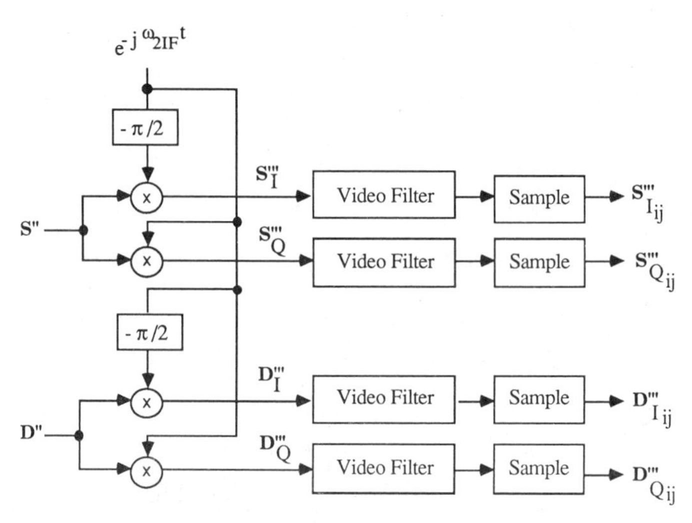

Figure 2.15 In *I/Q* detection the signals are first separated into *I* and *Q* components, literally in-phase (real) and quadrature-phase (imaginary) components.

for

$$\mathbf{S}_I''' = \mathrm{Re}[\mathbf{V}e^{j\omega_T t}]p(t - 2R/c)$$
$$\mathbf{S}_Q''' = \mathrm{Im}[\mathbf{V}e^{j\omega_T t}]p(t - 2R/c)$$

and

$$\mathbf{D}''' \triangleq \mathbf{D}_I''' + j\mathbf{D}_Q''' \tag{2.8b}$$

for

$$\mathbf{D}_I''' = F(\Delta\psi)\,\mathrm{Re}[\mathbf{V}e^{j\omega_T t}]p(t - 2R/c)$$
$$\mathbf{D}_Q''' = F(\Delta\psi)\,\mathrm{Im}[\mathbf{V}e^{j\omega_T t}]p(t - 2R/c)$$

where Re[ ] denotes "the real part of," Im[ ] denotes "the imaginary part of," and $F(\Delta\psi)$ is assumed to be approximately real.

2.5.3 Amplitude/Phase Detection

An alternative approach is to sense the amplitude and phase of the signal (Skolnik, 1970, pp. 5-41). Amplitude/phase detection results in some hardware simplifications since the phase can be more coarsely quantized and the amplitude can be processed nonlinearly.

The disadvantage of any kind of nonlinear processing is the increased Doppler smearing that results (Davenport and Root, 1958; Ward, 1969; Zeoli, 1971a; Nathanson, 1969, pp. 119–130). Contrary to popular belief, a target's Doppler frequency can be measured in the presence of an amplitude nonlinearity in the receiver. However, the process of *separating* signals on the basis of sensed Doppler is based on the principle of linearity. Therefore, linearity is needed up to that point where the target's signal is the dominant signal in the passband or alternatively the nonlinear effects must be removed in postprocessing. A common solution to this dilemma is to first separate signals by means of a bank of analog filters prior to nonlinear amplitude/phase detection, and then followed by digital Doppler filtering.*

If digital Doppler filtering is not required, then some additional hardware simplifications can be introduced. In this case it is only necessary to detect that portion of the signal that carries the monopulse information (e.g., Sherman, 1984, pp. 159–164; Leonov and Fomichev, 1986, Chapter 7). These simplifications are routinely implemented in noncoherent systems.

2.5.4 $S \pm jD$, $S \pm D$ Detection

The previous methods are straightforward approaches to the problem of A/D conversion. However, for hardware reasons it is sometimes more advantageous

*A similar situation occurs in FM communications theory, wherein two signals may interfere with each other after nonlinear FM processing. This phenomenon is known as the *capture effect*.

to detect the rotating phasors $\mathbf{S} \pm j\mathbf{D}$ or $\mathbf{S} \pm \mathbf{D}$ instead (Kirkpatrick, 1952). To avoid having to treat both cases, only $\mathbf{S} \pm j\mathbf{D}$ detection is discussed here.

$\mathbf{S} \pm j\mathbf{D}$ detection is implemented by inserting an IF combiner prior to the A/D converter and amplification. The purpose of this combiner is to convert the $(\mathbf{S}, \mathbf{D})$ signals into $(\mathbf{S} + j\mathbf{D}, \mathbf{S} - j\mathbf{D})$ signals. Afterward, the $(\mathbf{S} + j\mathbf{D}, \mathbf{S} - j\mathbf{D})$ signals are sent to the A/D converter. The A/D converter may involve either I/Q detection or amplitude/phase detection, as previously discussed. The A/D converter is preceded by an IF limiter, with the limit set so that A/D saturation cannot occur under any possible circumstance. After A/D conversion, the signals are reconverted into their original $(\mathbf{S}, \mathbf{D})$ format for further digital processing.

The advantage of $(\mathbf{S}, \mathbf{D})$ detection is that the radar's dynamic range is considerably improved. An important property of the signals in $\mathbf{S} \pm j\mathbf{D}$ format is that the angle information is decoupled from the amplitude information. That is, all the angle information is contained in the ratio $(\mathbf{S} + j\mathbf{D})/(\mathbf{S} - j\mathbf{D})$, as seen by examining Figure 2.10, whereas the amplitudes $|\mathbf{S} + j\mathbf{D}|$, $|\mathbf{S} - j\mathbf{D}|$ carry (relatively) little angle information. Therefore, it follows that valid angle information can be extracted even after significant amplitude nonlinearities have occurred.

Of course, this assumes that the amplitude information is not required as part of the coherent reception process, which is rarely true in general. Also, the quadrature component of the difference signal may be lost if IF limiting occurs, which implies that the channels must be perfectly aligned up to the IF stage. In general, therefore, $\mathbf{S} \pm j\mathbf{D}$ processing is not a panacea to the problem of dynamic range. But it generally remains true that $\mathbf{S} \pm j\mathbf{D}$ processing (or $\mathbf{S} \pm \mathbf{D}$ processing) helps reduce the adverse effect of amplitude nonlinearities in the presence of a strong close-in signal or a jammer in the mainlobe (Di Lazzaro et al, 1983). For weak (clutter-limited) signals the processing may revert back to the linear part of the processing region to better separate the target from clutter.

2.5.5 Video Filtering

Video filtering refers to filtering the signal with a lowpass filter whose characteristics are "matched" to the transmitted pulse waveform. The nomenclature "video" arises because the signals are at a low frequency and, hence, resemble those signals found in a TV receiver.*

Normally the transmitted and received waveform is approximately a square pulse of width τ. Hence, the impulse response of the matched filter is also a square pulse of width τ. After video filtering, as illustrated in Figure 2.16, the resultant pulse is roughly triangular in shape of width 2τ. The actual design of the video filter is perhaps best described in Lathi (1983, p. 507), Schleher (1986, p. 381), and Levanon (1988, p. 106).

*Historically the nomenclature "video" refers to those signals that are sent to the video display. In general, a video signal is one that has been translated down to baseband for further filtering and processing.

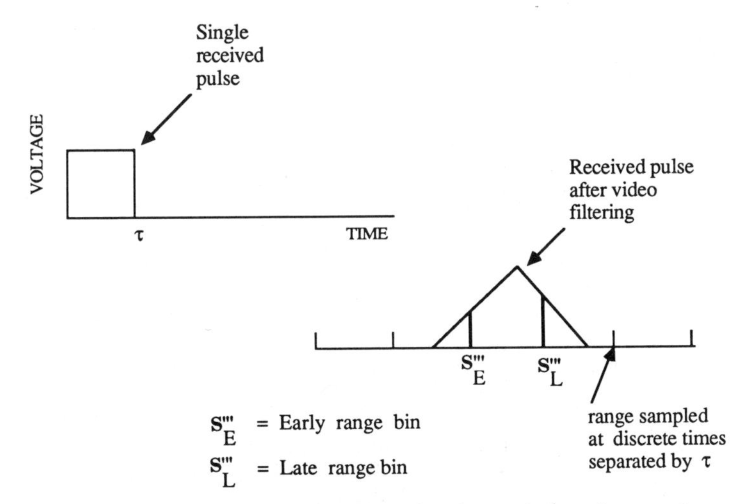

Figure 2.16 The signal is video filtered and sampled at discrete times.

2.6 A/D CONVERTER

After video filtering, the signal is sampled at discrete time intervals. Each sample point is referred to as a *range sample* or *range bin*. Because of the triangular shape of the video filter response, typically at least two range samples are received per target return.

At this stage of the processing, the radar return consists of a digital stream of data, $\mathbf{S}'''_{ij}$, $\mathbf{D}'''_{ij}$, for the jth range sample and ith pulse transmission interval. The A/D sampler quantizes this $\mathbf{S}'''_{ij}$, $\mathbf{D}'''_{ij}$ data into discrete numbers or "words" as shown in Figure 2.17 (Schmidt, 1970; Stark et al., 1988, p. 173; Skolnik, 1970, pp. 5–46).

The A/D sampling process introduces an extra noise term referred to as *quantization noise*. By definition, quantization noise is the difference between the analog signal and its digitized equivalent. To control the level of quantization noise, it is important that adequate A/D word size be provided. For most radar-tracking applications, word sizes vary from 1–12 bits (2–13 bits including sign $\pm$). On the basis of 6 dB of dynamic range per bit [i.e., 6 dB = 20 log(2)], a 12-bit A/D word size provides 72 dB of dynamic range. If one adds on the additional 11 dB due to the $\frac{1}{12}$ bit-to-sigma reduction ratio (see Problem 2.8), and subtracts out the 2:1 increase in quantization error due to the two channels (I/Q) of data being processed, then a 12-bit A/D word size corresponds to an A/D dynamic range of 80 dB.

2.6.1 Required Level of A/D Dynamic Range

By definition, the dynamic range is the ratio of signals that can be processed in the same manner. For radar receivers the upper bound is established by the receiver saturation level, and the lower bound is established by the lowest signal

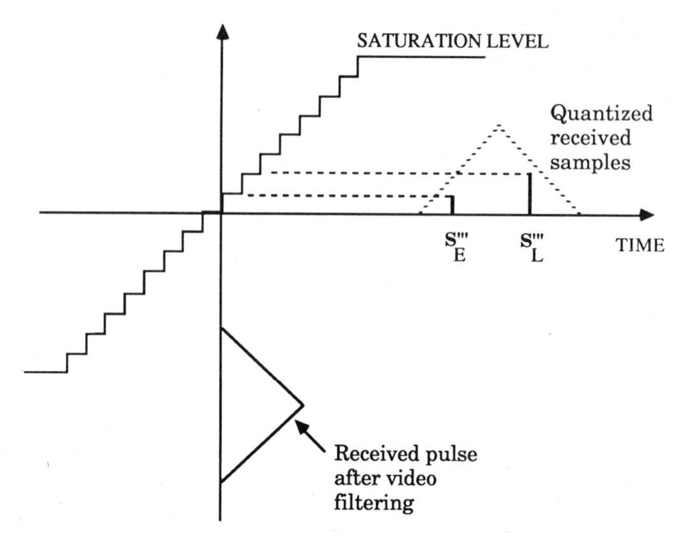

Figure 2.17 Transfer characteristic of the A/D sampler.

level that can be detected above noise. The total dynamic range is further broken up into the *AGC dynamic range* and the *instantaneous dynamic range*. For digital radar receivers the instantaneous dynamic range is equivalent to the A/D dynamic range.

Providing adequate A/D dynamic range is perhaps the most important consideration in the design of any digital receiver. The following example illustrates some of the more important tradeoffs involved. Let us consider the case of trying to detect a target with a radar cross section (RCS) of $0.1\,m^2$. Conceivably, unwanted *clutter* returns from nearby stationary objects can have a much larger RCS. A common traffic stop sign, for instance, acts like a mirror-like specular reflector when viewed perfectly straight on. In fact, it is not unusual for discrete specular reflections from small metallic objects (called discretes) to generate a clutter RCS of up to $10^4\,m^2$ or larger!

It is the function of the AGC to provide the minimum amount of attenuation necessary to avoid A/D saturation. Let us assume that the AGC references the A/D's saturation level to $10^6\,m^2$. The extra 10–20 dB of headroom is inserted to ensure a low probability of the unknown clutter RCS randomly "scintillating up" to the maximum saturation level. In airborne TWS an additional 5–10 dB of headroom may be inserted to account for any scanning delay in the AGC loop.

If the A/D's word size is 12 bits, for example, then the total A/D dynamic range is 80 dB. Thus, the quantization noise is down -80 dB or is equal to -20 dBsm. Later, this quantization noise is further reduced by (1) the ratio of the Doppler filter size to the PRF and (2) the ratio of pulse compression. Thus, relative to a $0.1\,m^2$ target, a 12-bit word size reduces the quantization noise to negligible values relative to that needed for accurate monopulse angle measurement.

At the other extreme a single bit (1-bit) A/D is sometimes employed if digital Doppler processing and monopulse operation are not required. This illustrates the potential range of A/D values involved.

2.7 CLUTTER FILTER

A logical next step is to get rid of the bulk of the clutter. The target signal at this stage is a slowly varying voltage with frequency determined by the target's specific Doppler frequency shift, ω_T. Added to this is the ground clutter return with zero mean frequency (i.e., $\omega_C = 0$ after IF compensation for ownship motion). This signal-plus-clutter arrangement is illustrated in Figure 2.18a. By notching out those frequencies containing clutter, the target return is enhanced relative to clutter. The resulting output of uncanceled target return plus clutter residue is depicted in Figure 2.18b.

The advantages of clutter filtering are fourfold. First, it immediately gets rid of the bulk of the clutter. More importantly, it reduces the dynamic range required in subsequent processing. Third, it reduces the clutter rejection requirements on subsequent signal processing. Finally, it effectively "whitens" the clutter, implying that any residual clutter left over is best removed by matched filtering (i.e., matched to white Gaussian noise). Due to this latter role, the clutter filter is sometimes referred to as a "prewhitening" filter.

The disadvantage of the clutter filter is that it is inherently suboptimum. It is impossible to physically separate the whitening process from the Doppler filter bank without inducing some performance loss (Capon, 1964; Hansen and

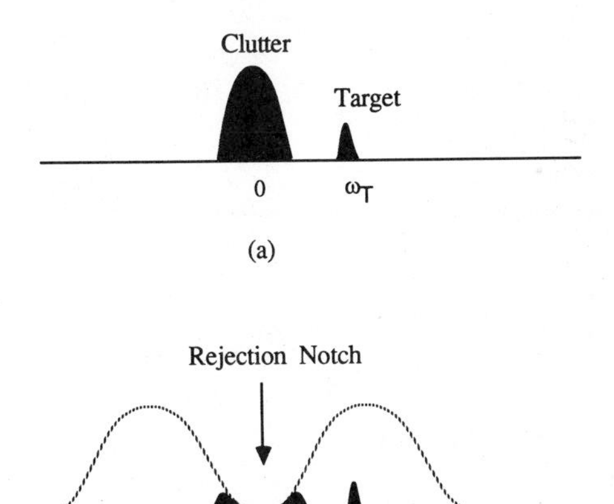

Figure 2.18 (a) Spectrum of clutter and target before clutter filtering. (b) Spectrum of residual clutter and target after clutter filtering.

Michelson, 1980; D'Addio and Galati, 1985). To be more precise, any pulses used to form the clutter filter are unavailable to form the subsequent Doppler filter bank, and the Doppler filter bank is more efficient at trading off clutter rejection and signal-to-noise ratio (SNR) improvement. This may help explain why, for any application where Doppler filtering is performed and the number of pulses available is limited, it is typical to use the minimum number of pulses to form the clutter filter subject to dynamic range constraints.

2.7.1 Design of the Clutter Filter

For comparative purposes clutter filters can be classified as follows: (1) Digital–analog; (2) Analog–analog; (3) Digital–digital.

The class of filters implemented with analog delay lines may be classified as *digital–analog filters*. The discrete nature of time is manifest in the transfer function:

$$\mathbf{S}_{ij}'''' = \sum_{m=0}^{N} a_m \mathbf{S}_{i-m,j}''' \tag{2.9}$$

where a_m = the filter coefficients, and N = the number of pulses employed by the clutter filter. N is also the maximum number of pulses unavailable to form the subsequent doppler filter bank. In summary, digital–analog filters are formed from a finite number of delays, where each delay is implemented with an analog storage element.

The simplest and most popular kind of digital–analog clutter filter is the single delay-line canceler, expressed as (see Figure 2.19)

$$\mathbf{S}_{ij}'''' = \mathbf{S}_{ij}''' - S_{i-1,j}''' \tag{2.10}$$

or, in terms of our prior notation,

$$a_0 = 1 \qquad \text{and} \qquad a_1 = -1$$

This kind of clutter filter has been extensively discussed in the literature (Nathanson, 1969, p. 331; Toomay, 1982, pp. 112–115; Eaves and Reedy, 1987, pp. 454–459; and Levanon, 1988, p. 240).

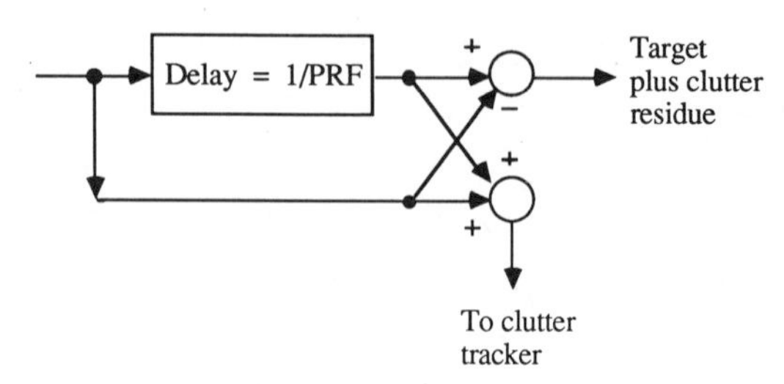

Figure 2.19 Single delay-line canceler.

In the case of ground-based radar systems, it is normally desired to have zero response at zero frequency. This restriction is mathematically written as

$$\sum_{m=0}^{N} a_m = 0 \tag{2.11}$$

Filter synthesis consists of selecting the a_m coefficients to give the required spectrum characteristics subject to (2.11).

Analog–analog clutter filters are filters that operates on an analog signal and produces an analog response; a good example is bulk quartz crystal filters. Crystal filters are used predominantly in those applications where the signal is continuous wave (CW), that is, where the pulsed modulation has been removed (stripped off) from the waveform prior to clutter filtering.

Fully digital filters consist of those filters that are implemented after A/D conversion. Digital filters can be implemented using any type of configuration and using either feedforward or feedback connections. Digital filters have become increasingly widespread as a result of recent technology advances, and most contemporary range-gated radar systems use this class of clutter filter.

Fully digital clutter filters offer the advantage of flexibility and convenience (Skolnik, 1980; Farina and Galati, 1985). In addition, the physical size of a digital clutter filter may be smaller, especially if the number of range bins is large (since all analog devices require a separate storage element for each range bin used). Finally, digital filters are more compatible with three-channel monopulse operation since they are easier to balance among multiple channels. Conversely, the primary drawback of all fully digital filters is the required dynamic range (both A/D and internal). This drawback becomes more pronounced as the level of clutter rejection needed increases.

2.8 DIGITAL SIGNAL PROCESSOR

The digital part of the signal processor performs the coherent integration process, target detection, false alarm control, noncoherent integration, and, depending on the radar's partitioning, may do some of the preliminary steps in measurement formation. Digital signal processing is described in Chapter 3.

2.9 DATA PROCESSOR

The function of the data processor is to evaluate the target's location through various estimation strategies, including maximum-likelihood estimation, data filtering, data prediction, clutter estimation, adaptive threshold selection, data correlation, adaptive waveform selection, and adaptive antenna control. Also, the data processor monitors the overall radar hardware integrity via built-in-test (BIT) and calibration procedures.

2.10 SUMMARY

The discussion in this chapter was mainly tutorial. Enough radar technology was presented for the reader to understand subsequent chapters. The monopulse principle was reviewed and issues related to channel selection and dynamic range were brought out. Special attention was given to the A/D converter since this device is often the limiting component in modern radars.

PROBLEMS

2.1 The simplest radar receiver consists of an RF bandpass filter followed by an envelope detector. What is the primary drawback of this simple arrangement?

2.2 Monostatic radar systems (i.e., single-site radar systems) conventionally employ pulsed modulation as outlined in the text. Conversely, bistatic radar systems often transmit a continuous-wave (CW) waveform. List at least two hardware advantages to using a CW waveform in the bistatic case.

2.3 What is the primary reason for using single-channel monopulse systems? Is there any additional reason present in the case of an airborne radar?

2.4 In the simplest case, $\mathbf{S} \pm j\mathbf{D}$ detection can be accomplished with a single envelope detector, a phase comparator, and two RF bandpass filters. Sketch a possible arrangement.

2.5 Any nonlinear device can serve as a mixer. Sketch a possible configuration using a single diode, a single resistor, bandpass filter, and a source of IF reference signal.

2.6 Let us model the antenna as an RF bandpass filter, with passband between 9 and 11 GHz. Select a first IF value such that the first image frequency always falls outside the antenna's passband and so is rejected.

2.7 Suggest one reason for inserting an IF limiter prior to the A/D during the process of $\mathbf{S} \pm j\mathbf{D}$ detection.

2.8 Assume the probability of quantization error q is uniformly distributed between $-\Delta/2$ and $\Delta/2$:

$$P(q) = \begin{cases} 1/\Delta & -\Delta/2 < q < \Delta/2 \\ 0 & \text{otherwise} \end{cases}$$

Find the expected value of q^2 in terms of Δ. The coefficient describing this relationship is referred to as the *bit-to-sigma reduction ratio.*

2.9 Consider an input signal with SNR = 30 dB, $\tau = 1\,\mu$sec, and PRF = 1000 Hz. Assume no clutter. It is required that the quantization noise be 10 dB below the thermal noise level in the sum channel. What is the corresponding value of A/D word size and the A/D sampling rate?

2.10 If a simple RC circuit is used to form the video filter, show that for $\tau/RC = 1.2$ the ratio of the peak amplitude to the amplitude delayed by τ from the peak is 10.4 dB.

2.11 Find the output response of a cascade of two delay-line cancelers (i.e., a double canceler) to an input signal in the jth range bin as given by

$$\mathbf{S}_j''' = \mathbf{V}e^{j\omega_T t}$$

2.12 How could one generalize a double-delay line canceler to notch out bimodal clutter? Bimodal clutter refers to clutter at two different frequencies, such as ground clutter and chaff clutter. Assume the second clutter is chaff located at mid PRF, and the primary clutter is ground backscatter at zero frequency. Leave your answer in terms of a_m.

CHAPTER 3

SIGNAL PROCESSING

Digital signal processing (DSP) and its performance are discussed in this chapter. The basic functions performed by the DSP are depicted in the flow diagram in Figure 3.1. The input consists of digitized data from the A/D converter containing target signal plus clutter and noise. The functions performed on the data include enhancing the signal power relative to the clutter and noise power, performing initial detection decisions, and, depending on the partitioning of the radar, further data reduction. The output consists of target declarations and associated information. Errors in signal processing generate false alarms that appear in subsequent data processing.

3.1 RANGE/DOPPLER GRID FORMATION

The returns are first sorted by range (assuming conventional range gating). Range sorting is repeated for all the M pulse transmissions in the coherent dwell. As depicted in Figure 3.2, this results in an array of M data samples for each of the L range bins. An optional next step is to convert this array of M samples into

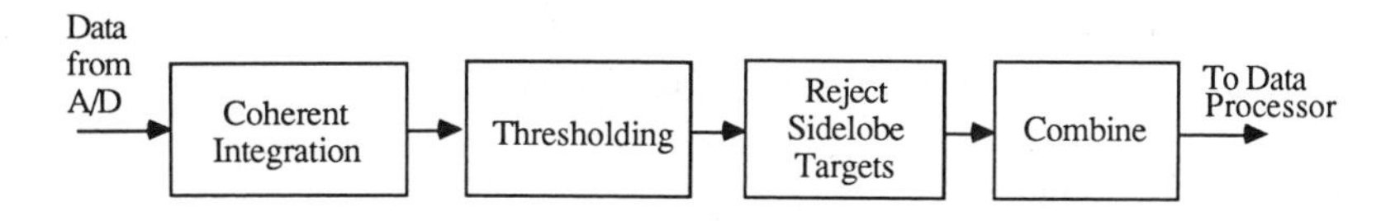

Figure 3.1 The signal processor does all the background tasks needed to isolate the target returns from noise.

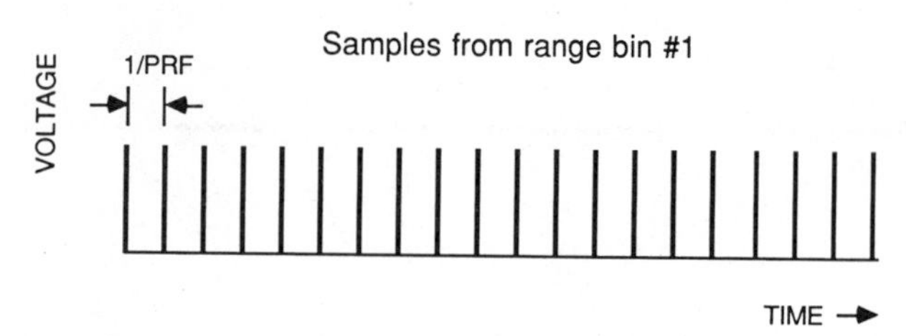

Figure 3.2 The signal processor first sorts the returns according to range. Over the coherent dwell time T, this results in an array of M samples for each range bin, $M = \text{PRF} \times T$, where here $M = 20$.

the frequency domain, thereby forming a "bank" of Doppler filters for each range bin. The end result is a "grid" of range/Doppler cells.

The output of the Doppler filter bank is labeled $\mathbf{S}_{kj}$ and $\mathbf{D}_{kj}$ respectively:

$$\mathbf{S}_{kj} = \sum_{i=0}^{M-1} \mathbf{w}_{ki} \mathbf{S}_{ij}^{''''} \tag{3.1a}$$

$$\mathbf{D}_{kj} = \sum_{i=0}^{M-1} \mathbf{w}_{ki} \mathbf{D}_{ij}^{''''} \tag{3.1b}$$

where j is the range bin index, i is the pulse transmission index, k is the filter index, and the filter weights are labeled $\mathbf{w}_{ki}$, $i = 0, \ldots, M - 1$. In the particular case of uniform amplitude/phase weighting, the weights would be given by

$$\mathbf{w}_{ki} = e\{j\omega_k i/\text{PRF}\}$$

for ω_k the angular frequency at which the kth filter is tuned.

A taper is typically applied to the filter weights $\mathbf{w}_{ki}$ to control the sidelobe level in the frequency domain, as depicted in Figure 3.3. The price paid for this taper is an effective increase in filter bandwidth and, hence, an associated loss in signal-to-noise performance.

3.2 COHERENT INTEGRATION GAIN

The performance of the signal processing can be evaluated by means of the gain attributed to the Doppler filter bank. Without loss of generality, assume a signal in the jth range bin as given by

$$\mathbf{S}_{ij}^{''''} = \mathbf{V} e\{-j\omega_T i/\text{PRF}\}$$

In this case the signal output is

$$\mathbf{S}_{kj} = \mathbf{V} \sum_{i=0}^{M-1} \mathbf{w}_{ki} e\{-j\omega_T i/\text{PRF}\} \tag{3.2}$$

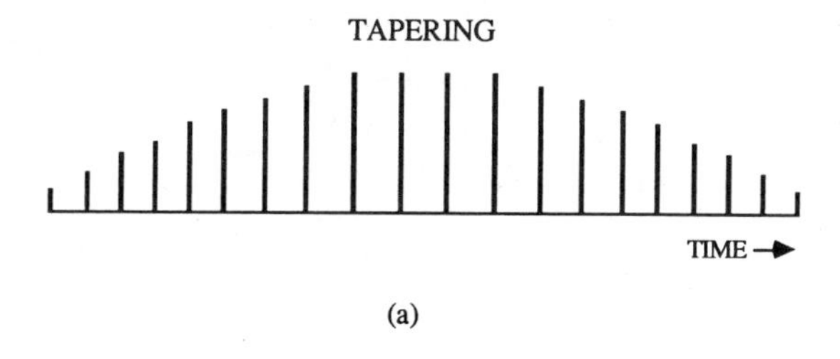

(a)

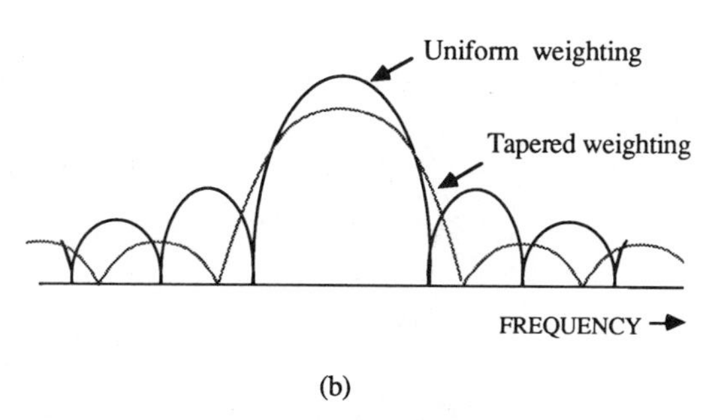

(b)

Figure 3.3 Weighting may be applied for sidelobe reduction. (a) Example of weighting (tapering). (b) The weighting reduces the sidelobe levels after filter formation, but this reduction is accompanied by an increase in the 3 dB filter bandwidth and associated decrease in SNR gain.

Thus, the sum-channel signal experiences a peak gain of

$$\text{Signal gain} = \left| \sum_{i=0}^{M-1} \mathbf{w}_{ki} e\{-j\omega_T i/\text{PRF}\} \right|^2 \tag{3.3}$$

where M is the number of data samples and $\mathbf{w}_i$, $i = 0, \ldots, M-1$, are the weights. If noise is present, its power gain is given by

$$\text{Noise gain} = \left| \sum_{i=0}^{M-1} \sum_{l=0}^{M-1} \mathbf{w}_{ki} \mathbf{w}_{kl}^* R(l-i)/R(0) \right| \tag{3.4}$$

where $R(l-i)$ denotes the noise autocorrelation function. The net signal-to-noise gain of the kth filter is thus

$$\text{Gain} \triangleq \frac{R(0) \left| \sum_{i=0}^{M-1} \mathbf{w}_{ki} e\{-j\omega_T i/\text{PRF}\} \right|^2}{\left| \sum_{i=0}^{M-1} \sum_{l=0}^{M-1} \mathbf{w}_{ki} \mathbf{w}_{kl}^* R(l-i) \right|} \tag{3.5}$$

It is instructive to convert this equation into the frequency domain. Define the Fourier transform of the weighting function $\mathbf{w}_{ki}$ to be

$$\mathbf{W}_k(\omega) \triangleq (1/\text{PRF}) \sum_{i=0}^{M-1} \mathbf{w}_{ki} e\{-j\omega i/\text{PRF}\} \tag{3.6a}$$

with inverse Fourier transform relationship:

$$\mathbf{w}_{ki} = \int_0^{\text{PRF}} \mathbf{W}_k(\omega) e\{j\omega i/\text{PRF}\}\, d\omega/2\pi \tag{3.6b}$$

In the frequency domain the convolution summation in the denominator of (3.5) becomes a simple product, and so

$$\text{Gain} = \frac{\left[\int R(\omega)\, d\omega/2\pi\right] |\mathbf{W}_k(\omega = \omega_T)|^2}{\int |\mathbf{W}_k(\omega)|^2 R(\omega)\, d\omega/2\pi} \tag{3.7}$$

where $R(\omega)$ is the Fourier transform of the noise autocorrelation function.

At this stage the equation can be generalized to include the cascaded effect of the clutter filter a_i and the Doppler filter $\mathbf{w}_{ki}$. Recognizing the multiplicative nature of transfer functions in the frequency domain, the net gain becomes

$$\text{Net gain} = \frac{\left[\int R(\omega)\, d\omega/2\pi\right] |\mathbf{H}_k(\omega = \omega_T)|^2}{\int |\mathbf{H}_k(\omega)|^2 R(\omega)\, d\omega/2\pi} \tag{3.8}$$

where $\mathbf{H}_k(\omega) \triangleq \mathbf{W}_k(\omega)\mathbf{A}(\omega)$ and $R(\omega)$ refers to the Fourier transform of the noise autocorrelation function at the input to the clutter filter.

The noise autocorrelation function, $R(\omega)$, is generally written as the sum of two parts. One part consists of "white" noise, N_0, and a second part consists of narrowband "colored" noise $C(\omega)$. This is denoted by

$$R(\omega) = C(\omega) + N_0 \tag{3.9}$$

White noise N_0 includes all wideband noise components contained in the PRF interval, such as thermal noise, A/D quantization noise, receiver roundoff errors and other types of processing noise, sidelobe clutter on an airborne radar, and broadband jamming noise. *Colored noise* $C(\omega)$ refers to all sources of narrowband interference. For instance, mainlobe ground clutter is often modeled as Gaussian (Nathanson, 1969, p. 88):

$$C(\omega) = C_0 e\{-\omega^2/(2\text{K}^2)\} \qquad \text{K} = 2\pi\sigma_C.$$

for σ_C the spectral bandwidth of mainlobe clutter. Other types of colored noise may also be present including rain clutter, chaff, neighboring targets, sidelobe clutter,* jamming, and so on.

The net signal-to-clutter ratio (SCR) gain (also called *clutter attenuation*) is given by

$$G_{\text{SCR}} \triangleq \frac{|\mathbf{H}_k(\omega = \omega_T)|^2 \int C(\omega)\, d\omega/2\pi}{\int |\mathbf{H}_k(\omega)|^2 C(\omega)\, d\omega/2\pi} \tag{3.10}$$

Figure 3.4 illustrates the relationship between the clutter spectrum $C(\omega)$, target spectrum, and system response $\mathbf{H}_k(\omega)$ as it appears at the various signal-processing stages.

Likewise, the net signal-to-noise ratio (SNR) gain is

$$G_{\text{SNR}} = \frac{\text{PRF}\, |\mathbf{H}_k(\omega = \omega_T)|^2}{\int |\mathbf{H}_k(\omega)|^2\, d\omega/2\pi}$$

If it can be assumed that the Doppler filter is much narrower than the clutter filter, then $\mathbf{H}_k(\omega) \cong \mathbf{W}_k(\omega)$ for ω values centered around ω_k, and so

$$G_{\text{SNR}} \cong \frac{\text{PRF}\, |\mathbf{W}_k(\omega = \omega_T)|^2}{\int |\mathbf{W}_k(\omega)|^2\, d\omega/2\pi} \tag{3.11}$$

This is just an illustration of the well-known fact that the clutter filter has a negligible effect on the output SNR assuming the Doppler filters are narrowband.

The integral in (3.11) can be identified as the equivalent noise bandwidth (ENB):

$$\text{ENB} \triangleq \frac{\int |\mathbf{W}_k(\omega)|^2\, d\omega/2\pi}{|\mathbf{W}_k(\omega = \omega_k)|^2}$$

such that G_{SNR} is of the general form

$$G_{\text{SNR}} = L_{\text{DS}}(\text{PRF}/\text{ENB})$$

*Sidelobe clutter may be classified as white noise or colored noise depending on the context and detection region. For example, a MPRF airborne radar detecting a receding target may classify sidelobe clutter as essentially white over the detection bandwidth of interest. The same radar when detecting a closing target in HPRF may view the sidelobe clutter as colored. This may seem confusing, but one must remember that the concept of "whiteness" is a relative term.

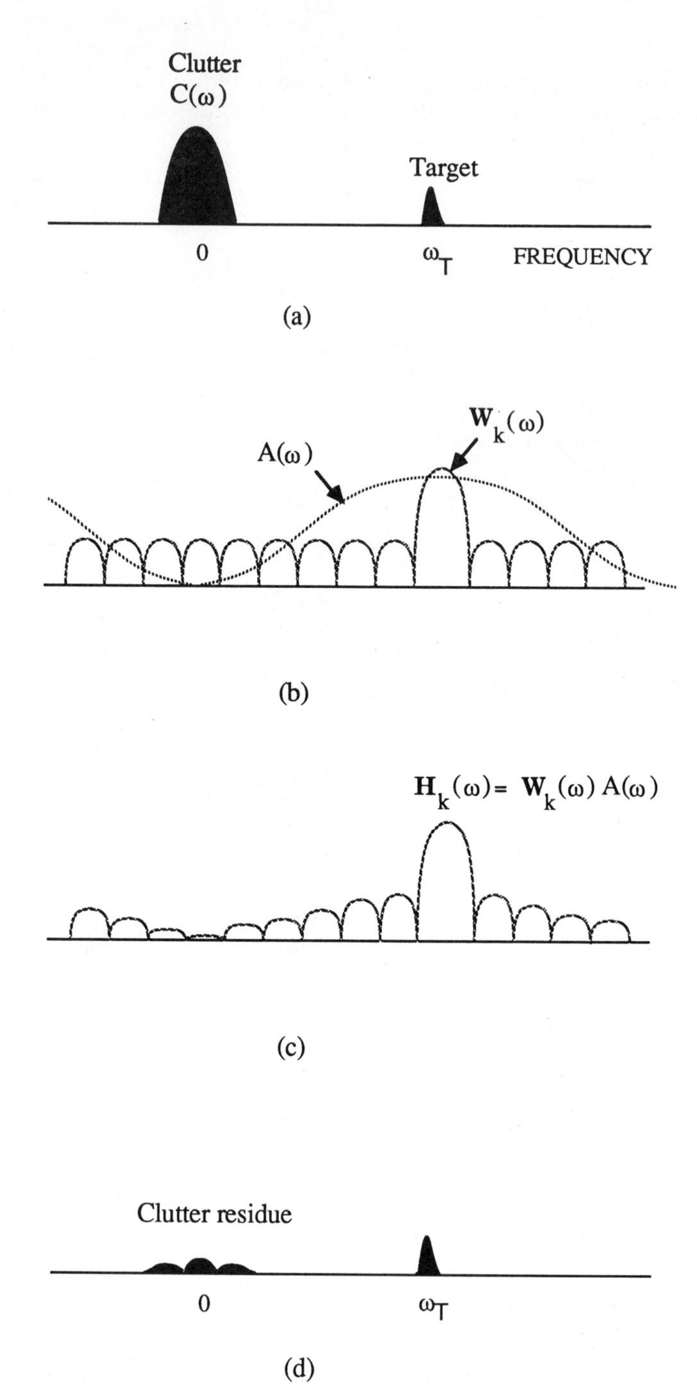

Figure 3.4 The overall system response. (a) Spectrum of clutter and target before filtering. (b) Individual spectrums of the Doppler filter $\mathbf{W}_k$ and clutter filter $\mathbf{A}$. (c) Spectrum of the cascaded system response $\mathbf{H}$. (d) Spectrum of clutter and target after filtering.

This equation emphasizes the fact that the gain is inversely proportional to the filter's equivalent bandwidth ENB. The term

$$L_{\mathrm{DS}} \triangleq \frac{|\mathbf{W}_k(\omega_T)|^2}{|\mathbf{W}_k(\omega_k)|^2}$$

represents the loss in gain attributed to the target's Doppler frequency ω_T not being situated in the center of the passband (Doppler filter straddling loss) and is normally included as part of the loss budget.

Amplitude weighting loss (AWL) represents another way to evaluate weighting performance. Reconverting (3.11) back into the time domain,

$$G_{\mathrm{SNR}} \cong \frac{\left|\sum_{i=0}^{M-1} \mathbf{w}_{ki} e\{-j\omega_T i/\mathrm{PRF}\}\right|^2}{\sum_{i=0}^{M-1} |\mathbf{w}_{ki}|^2}$$

where the denominator conversion follows from Parseval's theorem. From the Schwartz inequality, this equation has a maximum value of M, which is attained whenever $\mathbf{w}_{ki} = e\{j\omega_k i/\mathrm{PRF}\}$ and $\omega_T = \omega_k$. Thus, it is reasonable to separate G_{SNR} into two parts: one part describing the "maximum" value of G_{SNR} corresponding to a filter matched to white noise, M, and another part that represents any losses due to mismatched processing (i.e., not matching the receive processing to the transmitted processing). Correspondingly, let us define the AWL to be the loss relative to a matched filter:

$$\mathrm{AWL} \triangleq \frac{\left|\sum_{i=0}^{M-1} \mathbf{w}_{ki} e\{-j\omega_k i/\mathrm{PRF}\}\right|^2}{M \sum_{i=0}^{M-1} |\mathbf{w}_{ki}|^2} \tag{3.12}$$

such that $G_{\mathrm{SNR}} = M L_{\mathrm{DS}}\,\mathrm{AWL}$. The more conventional approach is to describe the SNR performance in terms of AWL and L_{DS}, and to describe SCR performance in terms of some kind of gain function. However, no single procedure is universally accepted in any case.

In theory, the weights $\mathbf{w}_{ki}$ should be optimally selected to balance the various noise outputs. In practice, it is often desired to place the clutter residue far below the receiver noise level at the output of the range/Doppler grid. The idea is that, unlike receiver noise, the clutter statistics and probability distribution are only partially known and randomly vary as a function of range, ownship altitude, look angle, season, and time of day. Perhaps more importantly, mainlobe clutter does not fully decorrelate from dwell to dwell, whereas noise decorrelates and so tends to average out when processed over multiple dwells. For these reasons overspecification of the clutter rejection level is often desired at this stage to prevent later desensitization.

Assuming noise-limited operation can be achieved, then the clutter-plus-noise at the output of the range/Doppler grid obeys the relationship:

$$\text{Clutter}_{\text{out}} \ll \text{noise}_{\text{out}}$$

which implies that

$$G_{\text{SCR}} \gg \text{CNR}_{\text{in}} G_{\text{SNR}} \tag{3.13}$$

where CNR_{in} is the input clutter-to-noise ratio, and G_{SCR} is the total signal-to-clutter gain. Equation (3.13) needs to be independently satisfied for each filter in the radar's passband for optimal performance.

3.3 IMPLEMENTATION ISSUES

There are many ways to accomplish the task of forming the Doppler filter bank. If the number of Doppler bins M is not too large, then the direct approach is best. This involves direct evaluation of each term in (3.1).

The advantage of direct evaluation is that the filter weights $\mathbf{w}_{ki}$ can be optimally matched to the clutter environment. For example, in some ground-based systems, the weights $\mathbf{w}_{ki}$ are selected to place a deep null at zero frequency for every filter in the filter bank. A related approach is to derive the optimum "mismatched" Doppler filter using the theory of filter optimization; more discussion on this topic may be found in the literature (Capon, 1964; Hansen et al., 1973; Hsiao, 1974; Hansen and Michelson, 1980; Schleher, 1982; D'Addio and Galati, 1985; Monzingo and Miller, 1980, p. 96). In any case independent control of the filter weights $\mathbf{w}_{ki}$ is desired if filter performance is the primary criteria of interest.

For large values of M the fast Fourier transform (FFT) algorithm becomes more appropriate (Oppenheim and Schafer, 1975). When using the direct approach, the evaluation of each term (k, j) requires $4M$ complex multiplications. Evaluation of the complete spectrum requires $4M^2$ complex multiplications. The FFT reduces the number of computations to $2M \log_2 M$.

The drawback of the FFT is that the weights $\mathbf{w}_{ki}$ and filter frequency ω_k cannot be independently specified for each Doppler filter (Taylor, 1982). The FFT sacrifices this ability in order to achieve the stated computational efficiency. In particular, the filter frequencies must be uniformly spaced over the PRF interval as given by $\omega_k = 2\pi k\ \text{PRF}/N$, for N a radix 2 number. The weights $\mathbf{w}_{ki}$ must be the same for all filters in the filter bank as given by $\mathbf{w}_{ki} = \mathbf{w}_i e\{j\omega_k i/\text{PRF}\}$. Thus, FFT implementation makes no claim to optimality.

3.3.1 FFT (Precomputed) Weights

Dolph–Chebyshev weights (or Tschebyscheff weights) are the most frequently used FFT weights. Dolph–Chebyshev weights possess the desirable property of minimizing the peak sidelobe level for a given Doppler filter bandwidth (Dolph,

1946; Ward, 1973; Hamming, 1983). In addition, Dolph–Chebyshev weights result in constant, or *equiripple*, sidelobes. Equiripple sidelobes make it easier to predict the sidelobe performance after filter formation, and also performance is conveniently independent of filter index number. Dolph–Chebyshev weights can be generated using the subroutine provided in Figure 3.5.

Hamming weights are "cosine on a pedestal" weights. The exact expression for the amplitude is given by (Hamming, 1983)

$$\mathbf{w}_i = 0.54 + 0.46\cos[(2\pi/M)(i - M/2)] \tag{3.14}$$

for $\mathbf{w}_{ki} = \mathbf{w}_i e\{j\omega_k i/\text{PRF}\}$. Hanning weights are raised cosine weights:

$$\mathbf{w}_i = \cos^{\alpha}[(\pi/M)(i - M/2)] \qquad \alpha = 1, 2, 3 \ldots \tag{3.15}$$

Kaiser weights are zero-order Bessel functions of the first kind (Harris, 1978, p. 73).

```
      SUBROUTINE DOLPH (N, SLL, w)

c     Inputs: N  = number of weights
c             SLL = filter sidelobe level relative to peak (dB)
c     Output:  w = array of N weights

      DIMENSION w(N)

      S     = 10.0**(SLL/20.0)
      BP    = 2.0*ALOG(S + SQRT(S**2-1.0)) / (N-1)
      BP    = EXP(BP)
      B     = (BP-1.0)**2 / (BP+1.0)**2
      M     = (N+1)/2
      w(1)  = 1.0

      Do k=2,M
         A = (N-k)*B
         jj = k-1
         ww = A

         j = 2
         Do WHILE ( (j .LE. jj) .AND. (ww .GE. 1.0E10*A) )
            jp = k-j
            A = A*JP*(N-2*k+1+JP)*B/((k-jp-2)*(k-jp))
            ww = ww + A
            j = 2
         END DO

         w(k) = (N-1)**ww / (N-k)
         w(N-k+1) = w(k)
      END DO

      ww  = w(M)
      Do i = 1,N
         w(i) = w(i) / ww
      END DO

      IF (MOD(N, 2) .EQ. 1) THEN
          w(N/2 + 1) = 1.0
      END IF

      RETURN
      END
```

Figure 3.5 Computer program to calculate Dolph–Chebyshev weights.

Weights with decaying sidelobes, such as Hamming, Hanning, and Kaiser weights, also have several advantages. Primary among these include the enhanced detection performance that occurs in the critical mid-PRF region. Moreover, any sidelobe leakage that does occur is likely to have come from the close-in sidelobe region, and so the relative impact on measurement error is reduced. Finally, decaying sidelobes are usually simpler to generate. Both Hamming and Hanning weights can be simply formed after an unweighted FFT (Rabiner and Gold, 1975; Cartledge and O'Donnell, 1977, p. 403) since a cosine in the time domain corresponds to an impulse in the frequency domain. Similarly, Kaiser weights are said to be easier to generate than an equivalent set of Dolph–Chebyshev weights (Harris, 1978, p. 82) and can be easily parameterized.

3.3.2 Performance of Some Precomputed Weights

Table 3.1 charts the relationship between amplitude weighting loss (AWL) and sidelobe level for various weights. The performance of Dolph–Chebyshev weights, in particular, are intuitively easy to understand since the sidelobe level is uniform and, hence, not a function of filter-to-clutter separation. Thus, Table 3.1 may be construed as charting the relationship between AWL and G_{SCR} in the particular case of Dolph-Chebyshev weights.

TABLE 3.1 Filter weights and performances.

Filter weighting		Peak sidelobe level.	Sidelobe falloff (dB/oct)	Amplitude weighing loss (dB)	Signal loss (dB)
Uniform		-13	-6	0.00	0.00
Dolph-Chebyshev	M = 8	-20	0	0.30	2.07
	16	-30	0	0.65	3.70
	32	-40	0	1.10	4.81
	64	-50	0	1.49	5.70
	64	-60	0	1.82	6.40
	128	-70	0	2.16	7.12
	512	-90	0	2.66	8.18
Hamming		-42.8	-6	1.34	5.35
Hanning $\cos^{\alpha}(x)$	$\alpha = 1.0$	-23	-12	0.90	3.97
	$\alpha = 2.0$	-32	-18	1.76	6.02
	$\alpha = 3.0$	-39	-24	2.39	7.55
	$\alpha = 4.0$	-47	-30	2.88	8.42
Kaiser-Bessel	$\alpha = 2.0$	-46	-6	1.76	6.20
	$\alpha = 2.5$	-57	-6	2.17	7.12
	$\alpha = 3.0$	-69	-6	2.55	7.86
	$\alpha = 3.5$	-82	-6	2.85	8.63

3.3.3 Data Turning

Data turning may be applied prior to FFT formation. The total number of samples M is given by

$$M = \mathrm{PRF} * T \qquad \text{for } T = \text{coherent dwell time} \tag{3.16}$$

Assuming FFT computation, this number should be a radix 2 number, that is, $M = 2^k$. In many applications, however, the choice for PRF and dwell time are dictated by factors that preclude M from being a radix 2 number. In this case equivalent performance may be attained by "turning" the data into a radix 2 number (Rivers, 1977; Levanon, 1988, p. 219).

Data turning, as shown in Figure 3.6, is performed by decomposing the data into overlapping segments of length N, $N = 2^k$, and then adding the appropriate segments before FFT formation. Since the filter weights have not been changed, the AWL and sidelobe levels remain unchanged after data turning. However, the filter-to-filter spacing is changed.

Data turning frees M from the radix 2 restriction imposed by the need for efficient FFT computation. Data turning also has the side benefit in that the

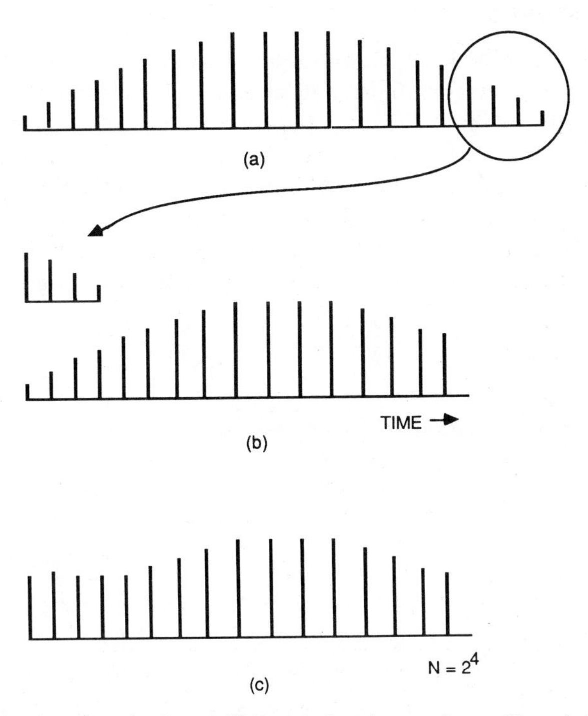

Figure 3.6 The data may be "turned" by overlapping sections of length N. (a) Original data. (b) Adding two sections. (c) Resulting overlapped data.

computational loading can be reduced since the FFT size can be deliberately chosen smaller than the array size. The drawback of data turning is that the filter-to-filter spacing is increased, for a fixed M, and consequently Doppler straddling loss is increased. Zero padding, on the other hand, may be used to reduce the straddling loss (at the expense of more computational loading).

3.4 DETECTION

Thresholding is applied next to isolate the target return from any residual noise and clutter present on the range/Doppler grid. As depicted in Figure 3.7, thresholding consists of comparing the voltage magnitude, $|\mathbf{S}_{kj}|$, to a detection threshold T_h,

$$|\mathbf{S}_{kj}| \underset{\text{no target}}{\overset{\text{target}}{\gtrless}} T_h \tag{3.17}$$

If the signal plus noise exceed T_h, a target is declared to be present at that range/Doppler location.

3.4.1 Threshold Selection in Search

The *false alarm rate*, or FAR, is an important criteria by which to judge any continuous search operation. In fact, the approach most often taken is to assign a constant value to the FAR and then accept the resulting probability of detection and probability of false alarm.

As depicted in Figure 3.8, the probability of false alarm, P_{fa}, is equal to the probability of noise exceeding the threshold ($|\mathbf{N}| > T_h$) and thus inducing a false target declaration. To relate the FAR to the P_{fa}, one must multiply the P_{fa} by the total number of opportunities for a false alarm to occur in a given amount of

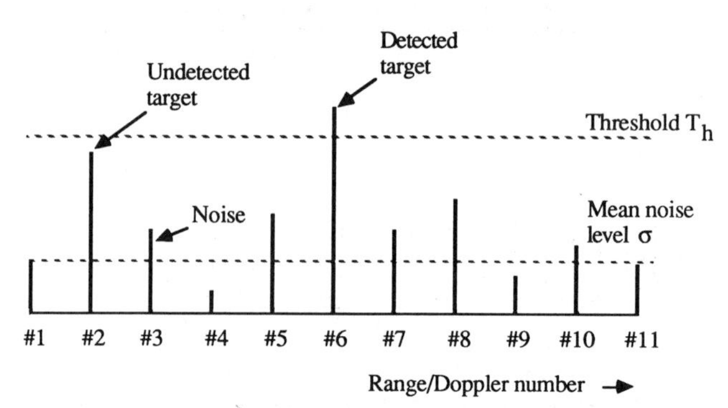

Figure 3.7 A target is declared if the voltage magnitude exceeds a noise threshold T_h.

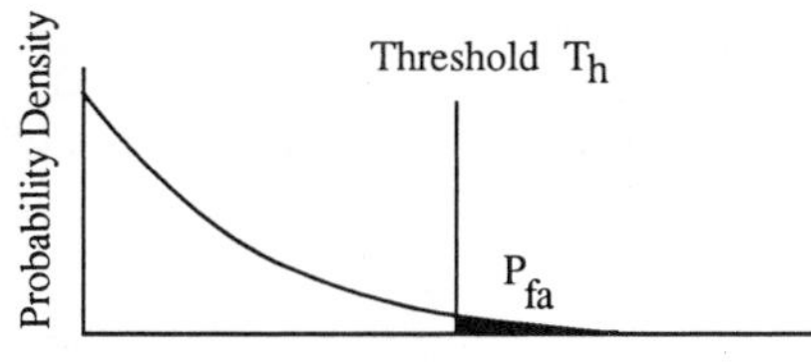

Figure 3.8 Probability of false-alarm P_{fa}.

time:

$$\text{FAR} \cong \frac{\text{number of opportunities}}{\text{processing time}} \times P_{fa} \tag{3.18}$$

where "processing time" refers to the time between detection decisions (dwell time plus dead time). For a given FAR, the corresponding P_{fa} is a direct function of the "number of opportunities," and so maintaining a constant FAR is equivalent to a Neyman Pearson test with constant P_{fa}. The threshold T_h is determined from the calculated P_{fa} and the assumed value of noise variance σ^2.*

If a detection decision is made on the basis of a single coherent dwell, then the number of opportunities is equal to the number of range bins examined, N_{RB}, multiplied by the number of Doppler filter bins examined, N_{DB}. In this special case the FAR expression approximately reduces to

$$\text{FAR} \cong \frac{N_{RB} N_{DB}}{\text{processing time}} \times P_{fa}$$

$$\simeq \frac{1}{\tau} \times P_{fa}$$

where $N_{RB} N_{DB} \tau$ = processing time (assuming negligible dead time). In general, however, a detection decision may be based on the outcome of multiple dwells, as the information gathered in a single dwell is usually insufficient. In this case the number of opportunities refers to the number of opportunity paths that can lead to a target declaration.

3.4.2 Threshold Selection in Track

Historically, STT has used only the simplest thresholding rules, if thresholding is even applied at all. A common procedure is to send those voltages that straddle the center of the gate region to the measurement algorithm for processing. However, recent concerns with rejection of electronic countermeasures (ECM)

*Equation (3.18) is based on the approximation that P_{fa} is small enough that $1 - (1 - P_{fa})^n \cong nP_{fa}$, for n the number of opportunities.

artifacts can lead to a larger gate region. Thresholding is applied in this case the same as in search.

The designer of a TWS system has, in the past, been limited in his choices of detection thresholds. In other words, usually the detection threshold was established early in the design based on the required search FAR. However, an idea that has been growing in popularity is to select the threshold T_h to optimize individual track accuracy (Kurniawan and McLane, 1985). That is, since track accuracy is itself a function of P_d and P_{fa}, the threshold in TWS should be selected on this basis. This idea is interesting and may prove useful in those situations where a separate threshold is desired in track, independent of search and false alarm considerations.

3.4.3 CFAR Thresholding

Constant FAR (CFAR) algorithms attempt to maintain a constant FAR value independent of the fluctuating noise environment (Skolnik, 1980; Eaves and Reedy, 1987, Chapter 12). The most popular form of CFAR algorithm involves estimating the noise level σ in real time and using this estimate to calculate the noise threshold T_h, for example, $T_h = k_S\sigma$, for σ the estimated noise level and k_S predetermined to maintain a constant FAR value. Real-time estimation is desired since σ dynamically changes as a function of local clutter environment, pulse compression code, Doppler filter sidelobe level, thermal noise, spurious signals created in the mixing process, hardware instabilities, neighboring target environment, spurious Doppler spectra, changing receiver gains, birds, sidelobe clutter (in an airborne radar), meteorological phenonenon, and intentional and unintentional RF interference.

As shown in Figure 3.9, σ may be estimated by averaging the outputs of an ensemble of J cells on either side of the one in question. Since most of these filter outputs can be safely assumed to contain only noise, the average should be an indication of the mean noise level. The averaging may be carried out in range,

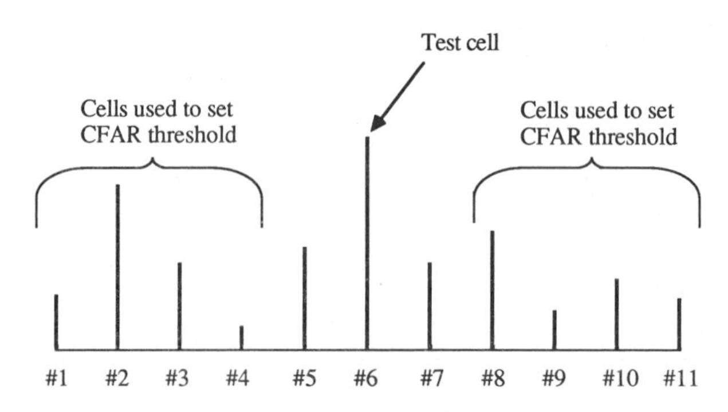

Figure 3.9 The most popular form of CFAR algorithm involves averaging the output of neighboring cells to obtain an estimate of the noise level σ, and choosing the threshold T_h to be a constant multiplier of σ.

Doppler, or time (scan to scan). The averaging should be carried out in that dimension where the noise exhibits the most homogeneous (or stationary) statistics.

The drawback of CFAR averaging is that an additional loss phenomenon results called *CFAR loss*. CFAR loss occurs because the threshold multiplier k_S must be increased to allow for any uncertainty in estimating σ (Blake, 1986, p. 336; Mitchell and Walker, 1971).

It can be shown that the cell-averaging CFAR procedure is statistically optimum provided the background noise is independent from cell to cell, homogeneous, and Gaussian distributed. In order to guarantee the validity of this hypothesis, normally some additional steps are taken. The most straightforward step is to limit the number of cells used. Another technique is to censor the K largests cells before averaging (Dillard, 1967; Richard and Dillard, 1977) or to use multiple passes to identify cells containing nonstationary noise spikes. A related technique is the "greatest of" method (Hansen and Sawyers, 1980; Levanon, 1988, p. 254).

Limiting the number of cells used, however, results in increased CFAR loss. This illustrates the basic CFAR tradeoff: a larger average is desired for a low-variance noise estimate, whereas a smaller average is desired to ensure homogeneity and to avoid desensitization by nonstationary noise (Ritcey, 1986).

As a final check on the CFAR algorithm, a feedback provision may be employed. If an excessive number of false alarms seem to be occurring, in either the tracking algorithm or the range resolver, then this information is fed back to change the CFAR parameters. This may include changing either k_S, J, or K.

3.4.4 Sidelobe Blanking

All radars are vulnerable to unwanted returns entering through the radar's sidelobes and generating a false target declaration. This includes the radar's sidelobes in angle (antenna sidelobes), Doppler (filter sidelobes), and range (pulse compression sidelobes). In response to this dilemma, *sidelobe blanking* procedures have been developed. Sidelobe blanking refers to any signal-processing procedure that detects those sidelobe returns strong enough to appear above the CFAR threshold and suppress them by local desensitization of the radar.

In range and Doppler, the actual cause of the disturbance may be *inside* the scan volume and hence detected. Thus, since both the parent and the sidelobe pattern are known, the magnitude of any sidelobe returns can be estimated as a function of their relative offset from the cause. Sidelobe hits are identified as such by their relative SNR and discarded.

3.4.5 Angle Sidelobe Blanking

Identification of returns entering through the antenna's sidelobes is difficult since the actual cause of the disturbance may itself be outside the scan volume and, hence, itself undetected. One way to identify the presence of this type of false sidelobe targets is to provide the radar with a *guard* antenna and associated receiver channel (Maisel, 1968; Arancibia, 1979).

The guard antenna is essentially a broad-beam antenna as depicted in Figure 3.10. The width of the guard's mainlobe is sufficient to encompass the entire region illuminated by the antenna's principal sidelobes. Consequently, by comparing the signal level in the guard receiver $\mathbf{G}_{kj}$ to the signal level in the radar's sum-channel receiver $\mathbf{S}_{kj}$, sidelobe targets can be identified. The guard antenna is also referred to as a control antenna, sidelobe blanker, or sidelobe inhibitor.

Sum/guard processing is mathematically written as

$$|\mathbf{S}_{kj}|/|\mathbf{G}_{kj}| \underset{\text{no mainlobe target}}{\overset{\text{mainlobe target}}{\gtrless}} k_{S/G}. \tag{3.19}$$

In the case of radars on moving platforms, the guard antenna plays a crucial role. Since the radar itself is moving, any large RCS structure on the ground illuminated by the antenna's sidelobes appears with a shifted Doppler. Given that the radar processes that portion of the spectrum occupied by sidelobe clutter, then these false sidelobe targets will be detected no differently than an aircraft in the mainlobe. The guard antenna serves to flag the presence of any false detections of this type and eliminate them.

When used on ground-based radars, the function of the guard antenna is to detect strong targets or interference entering through the antenna's sidelobes. The radar is then desensitized in those regions containing interference, or, to avoid desensitization, the detection may activate a procedure that adaptively nulls out the unwanted return.

Ideally, the guard pattern is always much larger than the sum pattern everywhere within the principal sidelobe region. Realistically, however, the guard pattern may deviate from this ideal. In this case a common procedure is to artificially raise the guard pattern above the sum pattern as follows:

$$|\mathbf{S}_{kj}|/|\mathbf{G}_{kj}| \underset{\text{sidelobe target}}{\overset{\text{mainlobe target}}{\gtrless}} k_{S/G}. \qquad \text{if } |\mathbf{G}_{kj}| > T_h \tag{3.20}$$

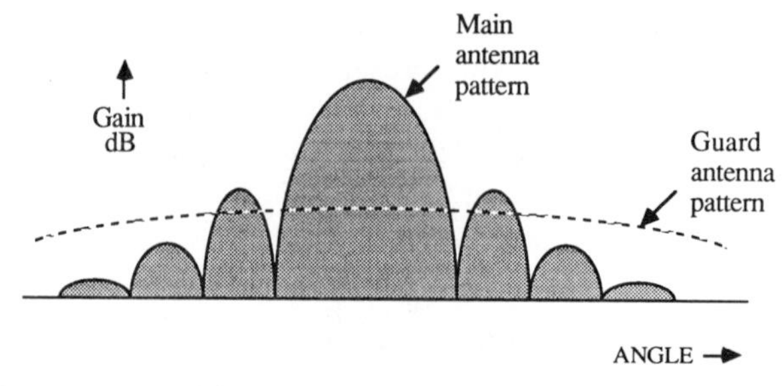

Figure 3.10 A guard antenna is an broadbeam antenna used to detect the presence of unwanted radiation entering through the sidelobes of the radar's main antenna.

where $k_{S/G} > k_S$. If $|\mathbf{G}_{kj}| < T_h$ and $|\mathbf{S}_{kj}| > T_h$, then a valid mainlobe detection is also declared.

The disadvantage of raising $k_{S/G}$ is that noise in the guard channel may trigger a false alarm. This possibility must be reduced by thresholding the guard channel as well as the sum channel. Unfortunately, the probabilities after guard thresholding are now supoptimum since a sidelobe target may falsely result in the condition $|\mathbf{G}_{kj}| < T_h$ and $|\mathbf{S}_{kj}| > T_h$. This is the price one pays for using a nonideal guard pattern.

3.4.6 Angle Ratio Thresholding

An approximation to the guard test is often mechanized by thresholding the ratio between the sum and difference signal magnitudes, that is, $|\mathbf{D}|/|\mathbf{S}| > k_{D/S}$, as illustrated in Figure 3.11. If this ratio is greater than a constant, typically between 0.8 and 1.2, the received signal is rejected. This procedure is called angle ratio thresholding (ART).

ART processing is less effective than guard processing since the difference pattern does not form a perfect omnidirectional pattern over the principle sum-channel sidelobe region. Nevertheless, ART is frequently used, mainly because it can be so easily mechanized. The chief difficulty with guard processing is that a separate guard receiver is required, or alternatively the guard signal may be received in the difference channel (at the expense of more time to collect all the data).

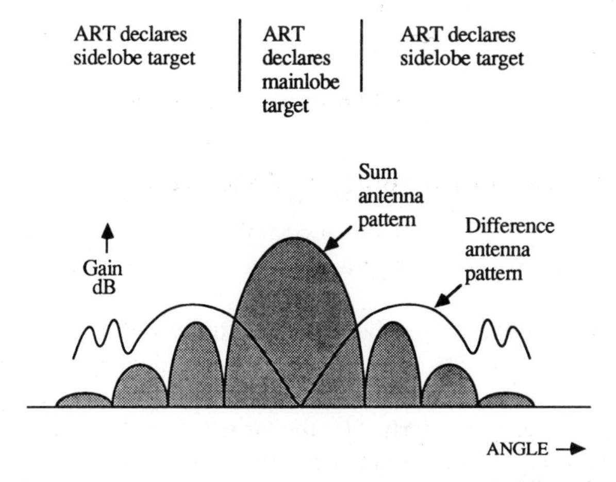

Figure 3.11 Sidelobe targets can also be rejected by thresholding the angular ratio between the sum and difference signals.

3.5 SINGLE-DWELL PROBABILITIES

3.5.1 Probability of False Alarm

With few exceptions the noise probability density at the output of each range/Doppler cell can be assumed to be Gaussian (from the central limit theorem). To be precise, both the real and imaginary components are jointly Gaussian and independent. Since the magnitude of two Gaussian random variables is Rayleigh distributed and the magnitude squared is exponentially distributed, the probability density of interest is

$$P(Z) = e^{-Z} \tag{3.21}$$

where $Z \triangleq |\mathbf{N}|^2/(\sigma^2)$
$= (N_I^2 + N_Q^2)/(\sigma^2)$
N_I = noise in I component of the sum channel **S**
N_Q = noise in Q component of the sum channel **S**
σ^2 = noise power = $E(N_I^2 + N_Q^2)$*

Thus, the probability of noise exceeding the threshold, $|\mathbf{N}| > T_h$, is

$$P_{\text{fa}} = \int_{T_h^2/\sigma^2}^{\infty} P(Z)\,dZ$$
$$= e^{-T_h^2/\sigma^2} \tag{3.22}$$

The quantity P_{fa} is referred to as the *single-dwell probability of false alarm.* To illustrate the use of this expression, to achieve a single-dwell P_{fa} of 10^{-8},

$$10^{-8} = e^{-T_h^2/\sigma^2}$$
$$T_h = 4.3\sigma$$
$$= 12.6\,\text{dB above noise power}$$

3.5.2 Single-Dwell Probability of Detection

As shown in Figure 3.12, P_d is the probability of a target return plus noise exceeding the threshold T_h. In this section a first-cut expression for P_d is calculated.

Unfortunately, unlike noise, the signal fluctuation does not have a simple universal probability density. Rather, it varies considerably from one target to the next and from one operational situation to another. Gaussian statistics are sometimes justifiable, particularly if (1) the target's extent is large compared to a wavelength and (2) many individual scattering centers contribute to the overall

*In some texts the noise power appears as $2\sigma^2$.

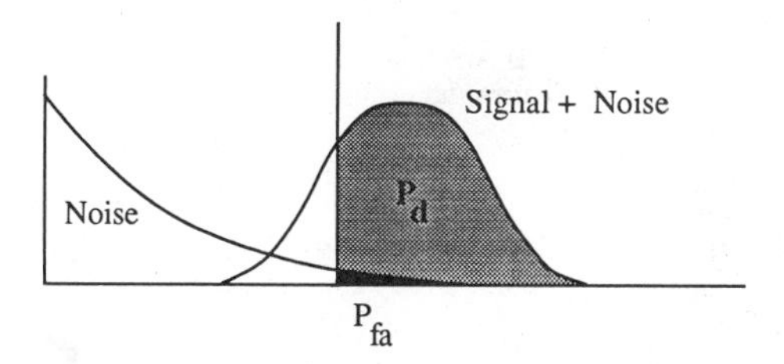

Figure 3.12 Probability of detection P_d and false alarm P_{fa}.

signal level. Such targets are referred to as Swerling types I or II, depending on the rate of variation of the statistics (Swerling, 1957).

Assuming Gaussian statistics, the single-dwell P_d is then given by

$$P_d = \int_{T_h^2/\sigma^2}^{\infty} P(Z)\,dZ$$
$$= e^{-T_h^2/[\sigma^2(\mathrm{SNR}+1)]} \tag{3.23}$$

where SNR is the target's average signal-to-noise ratio. Note that, since the signal plus noise is zero-mean Gaussian, the probability P_d is completely specified by the signal power, σ^2SNR, the noise power, σ^2, and the threshold T_h.

3.6 POSTDETECTION INTEGRATION

Typically multiple dwells are transmitted as the beam sweeps over the target. Figure 3.13 illustrates a typical scenario where the time on target comprises three separate dwells. To integrate these multiple dwells into a single detection solution, a common scheme is to add their powers and apply the sum to the detection threshold. This process of noncoherent integration and thresholding is called *postdetection integration* (PDI).

The PDI process is described by

$$X = \frac{1}{N} \sum_{i=1}^{N} (S_{I_i}^2 + S_{Q_i}^2) \tag{3.24}$$

where N is the total number of integrated dwells. A common implementation is to group N returns together by summing each new power return with the preceding $N - 1$ returns (sliding window). A target is declared whenever the integrated output X exceeds a threshold:

$$X \underset{\text{no target}}{\overset{\text{target}}{\gtrless}} T_H^2$$

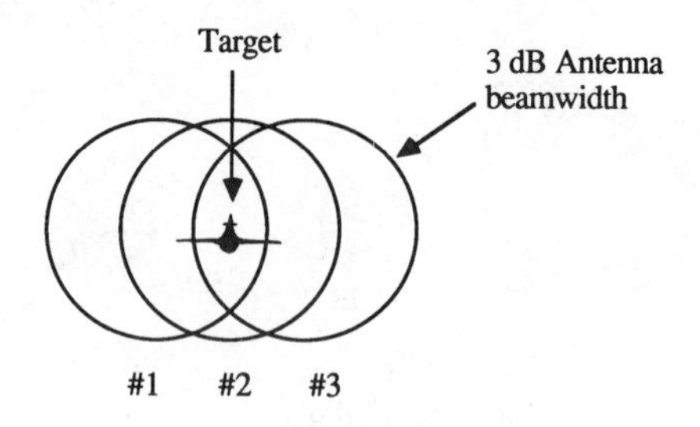

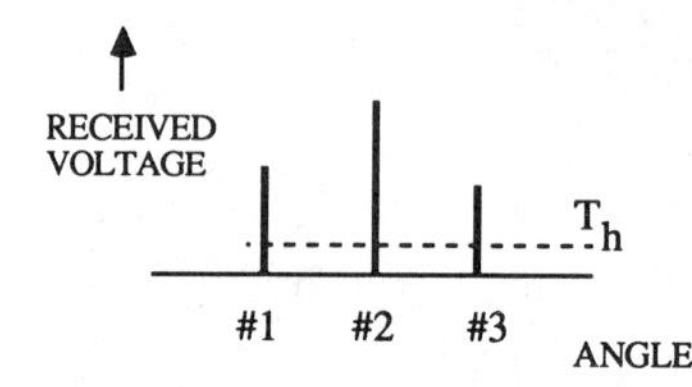

Figure 3.13 In any search application, a target is typically viewed at least N times per time on target, where $N = 3$ here.

There are several approaches to PDI thresholding (Cantrell, 1984). In the double-threshold approach only those range/Doppler bins where the power crossed the initial threshold T_h are used in the noncoherent integration process. In the single-threshold approach, powers are integrated along all possible range/Doppler walks and then each summation is tested against the PDI threshold T_H.

3.6.1 Swerling I and III Target Models

The signal $\mathbf{S} = (S_I, S_Q)$ is modeled as a signal $\mathbf{V}$ embedded in Gaussian noise $\mathbf{N} = (N_I, N_Q)$, that is, $\mathbf{S} = \mathbf{V} + \mathbf{N}$. Hence, the conditional probability density of $\mathbf{S}$ given $\mathbf{V}$ is Gaussian:

$$P(S_{I_i}, S_{Q_i}|\mathbf{V}) = \frac{1}{\pi\sigma^2} e^{-|S_{I_i}+jS_{Q_i}-\mathbf{V}|^2/\sigma^2} \tag{3.25}$$

Next, it is convenient to introduce the following notation:

$$X_i \triangleq S_{I_i}^2 + S_{Q_i}^2$$

where the probability density of X_i is well known to be Rician (Lathi, 1983, p. 439; Ziemer and Tranter, 1976, p. 247):

$$P(X_i|\mathbf{V}) = (1/\sigma^2)e^{-(X_i+|\mathbf{V}|^2)/\sigma^2} I_0(2|\mathbf{V}|\sqrt{X_i}/\sigma^2) \tag{3.26}$$

for $I_0(\,)$ the zero-order modified Bessel function.

Since $X = (1/N)\,\Sigma\, X_i$, the probability density of X is given by convolving the

X_i probability density with itself N times (Ziemer and Tranter, 1976, p. 195). The resulting convolution integral has been evaluated (DiFranco and Rubin, 1968, p. 346) to be

$$P(X|\mathbf{V}, N) = (N/\sigma^2)(X/|\mathbf{V}|^2)^{(N-1)/2} e^{-N(X+|\mathbf{V}|^2)/\sigma^2} I_{N-1}(2N|\mathbf{V}|\sqrt{X}/\sigma^2) \tag{3.27}$$

where N = number of samples PDIed
X = normalized signal power
$\mathbf{V}$ = instantaneous signal strength
$I_{N-1}(x)$ = $N-1$ -order modified Bessel function

The probability of X exceeding the threshold is thus

$$P_D(\mathbf{V}, N) = \int_{T_H^2}^{\infty} P(X|\mathbf{V}, N)\, dX \tag{3.28}$$

Using (3.27) and the following series for $I_n(X)$ (CRC, 1970, p. 521),

$$I_n(X) = (X/2)^n \sum_{k=0}^{\infty} \frac{(X/2)^{2k}}{k!(n+k)!} \tag{3.29}$$

and, after interchanging the order of summation and integration, (3.28) becomes

$$P_D(\mathbf{V}, N) = e^{-N|\mathbf{V}|^2/\sigma^2} \sum_{k=0}^{\infty} \frac{N^k(|\mathbf{V}|/\sigma)^{2k}}{k!} \sum_{m=0}^{N-1+k} e^{-NT_H^2/\sigma^2} \frac{(NT_H^2/\sigma^2)^m}{m!} \tag{3.30}$$

The resulting P_{FA} is, taking $\mathbf{V} \to 0$,

$$\begin{aligned} P_{FA}(N) &= P_D(\mathbf{V} = 0, N) \\ &= e^{-NT_H^2/\sigma^2} \sum_{m=0}^{N-1} \frac{(NT_H^2/\sigma^2)^m}{m!} \end{aligned} \tag{3.31}$$

which agrees with the tabulated results of Pachares (1958).

Consider next the following normalized version of the random target statistics:

$$Z = \frac{|\mathbf{V}|^2}{\sigma^2 \text{ SNR}} \tag{3.32}$$

where $|\mathbf{V}|^2/\sigma^2$ = instantaneous signal-to-noise ratio
SNR = average signal-to-noise ratio
Z = normalized signal power

such that (3.30) becomes

$$P_D(Z, N) = e^{-N\,\mathrm{SNR}\,Z} \sum_{k=0}^{\infty} \frac{(N\,\mathrm{SNR}\,Z)^k}{k!} \sum_{m=0}^{N-1+k} e^{-NT_H^2/\sigma^2} \frac{(NT_H^2/\sigma^2)^m}{m!} \tag{3.33}$$

If the target RCS is Gaussian (Swerling I), then the random fluctuations Z is described by exponential (or Gaussian squared) statistics:

$$P(Z) = e^{-Z} \tag{3.34a}$$

If the target RCS comprises one large reflector and several small ones, then Z is more aptly described by the Swerling III probability density:

$$P(Z) = 4Ze^{-2Z} \tag{3.34b}$$

This latter model is sometimes used to describe ship, missile, or satellite statistics.

The *single-look* P_D is then

$$P_D = \int_0^{\infty} P(Z)P_D(Z, N)\,dZ \tag{3.35}$$

where $P_D(Z, N)$ is given by (3.33).

In the case of Swerling I, Swerling has evaluated the preceding integral to be (DiFranco and Rubin, 1968, p. 390)

$$P_D = P(N-1, NT_H^2/\sigma^2) + \left(\frac{N\,\mathrm{SNR}+1}{N\,\mathrm{SNR}}\right)^{N-1} e^{-NT_H^2/[\sigma^2(N\,\mathrm{SNR}+1)]}\left[1 - P\left(N-1, \frac{N^2\,\mathrm{SNR}\,T_H^2/\sigma^2}{N\,\mathrm{SNR}+1}\right)\right]$$

where

$$P(N, Y) \triangleq \sum_{m=0}^{N-1} e^{-Y} \frac{Y^m}{m!} \tag{3.36}$$

Similar results are attained for Swerling III (Eaves and Reedy, 1987, p. 361; Schleher, 1986, p. 394). Efficient numerical methods for evaluating these integrals are found in Shnidman (1975).

In conclusion, PDI averaging enhances both the P_{FA} and the P_D. The threshold T_H can be set closer to the mean noise level σ without increasing the P_{FA}. Detection performance P_D is also enhanced since the probability of a target going undetected because of destructive interferences between the signal and thermal noise is reduced. In the case of Swerling I theoretical studies by Marcum have shown that the "effective SNR" to produce a target detection is reduced by a factor $N^{0.84}$ for N small, assuming $P_{FA} = 10^{-6}$ and $P_D = 0.5$, declining to $N^{0.69}$ for N equal to 100 (DiFranco and Rubin, 1968; Blake, 1986, p. 43). A linear curve fit is sometimes employed of the form $N^{0.84 - 0.076 * \log(N)}$.

The disadvantage of PDI is that it is only weakly effective in enhancing the target-to-clutter statistics. Normally the clutter is correlated from look to look and so does not average out during the PDI process.

3.6.2 Swerling II and IV Target Models

Equation (3.35) models the signal as having a constant value over the PDI averaging process, and thus the signal scintillation Z is incorporated outside the PDI averaging loop. This assumption is valid provided all N dwells are completed in less than the target's scintillation time, and the transmitted RF is not switched from dwell to dwell (no frequency agility).

If the RF is switched between dwells in steps greater than 30 MHz, however, then the signal will decorrelate from dwell to dwell [assuming a target dimension of at least 5 m (Skolnik, 1980, p. 171)]. In the case of Swerling II (Gaussian signals in Gaussian noise), the resulting P_D is given by (3.31) with σ^2 replaced by $\sigma^2(\text{SNR} + 1)$. The Swerling IV case is treated in the literature (Eaves and Reedy, 1987, p. 361; Schleher, 1986, p. 395).

Frequency agility reduces the probability of not detecting a strong target due to destructive fading or "fluctuation loss." For weak signals, on the other hand, frequency agility hurts detection performance since the averaging tends to eliminate the probabilistic outliers. For this reason frequency agility is most often used to enhance the detection of targets whenever a high P_D is required ($P_D > 30\%$). For applications where a low P_D is sufficient (e.g., as in ESA search), then the averaging inherent in frequency agility is not always a desirable property.

3.7 *M*-OUT-OF-*N* DETECTION

Another way to integrate multiple waveforms is to require M threshold crossings per N consecutive dwells for successful target declaration. This procedure has been variously denoted as *M-out-of-N detection, coincidence detection, binary integration, binary detection, double-threshold detection, dual-threshold detection,* and *nonparametric detection.*

3.7.1 Binomial Transformation

The probability of detection after M out of N is given by the familiar binomial distribution (Lathi, 1983, p. 362; Ziemer and Tranter, 1976),

$$P_D = \int P(Z)\, dZ \sum_{i=M}^{N} \frac{N!}{M!(N-M)!} P_d^i(Z)[1 - P_d(Z)]^{N-i} \qquad (3.37a)$$

for Swerling I and III or

$$P_D = \sum_{i=M}^{N} \frac{N!}{M!(N-M)!} P_d^i (1 - P_d)^{N-i} \tag{3.37b}$$

for Swerling II and IV, where P_d is the single-dwell probability of detection and P_D is the single-look probability of detection.

False alarms cannot be modeled in the same way since false alarms must be modeled as stochastic (i.e., continuously generated at discrete process times). Therefore, the N trials must be modeled as a sliding window over a series of experiments extending from negative infinity to positive infinity. Detection decisions based on overlapping (sliding window) data are made by declaring a new target whenever M target declarations occur in the present detection window, and $M - 1$ target declarations occurred one step in the past. Therefore, a false alarm occurs whenever the following sequence is observed: a "no" declaration with single-dwell probability $(1 - P_{\text{fa}})$, followed by exactly $M - 1$ false declarations out of $N - 1$ tries with probability:

$$\frac{(N-1)!}{(M-1)!(N-M)!} P_{\text{fa}}^{M-1} (1 - P_{\text{fa}})^{N-M}$$

followed by a single false alarm with probability P_{fa} (Levanon, 1988, p. 62). The resulting probability of a new false alarm is a multiple of these three probabilities:

$$P_{\text{FA}} = \frac{(N-1)!}{(M-1)!(N-M)!} P_{\text{fa}}^{M} (1 - P_{\text{fa}})^{N-M+1} \tag{3.38}$$

and

$$\text{FAR} = \frac{\text{Number of opportunities}}{\text{Processing time}} \times P_{\text{FA}}$$

where the number of opportunities refers to the number of independent range/Doppler paths examined for detection purposes, and processing time refers to the time between detection declarations.

3.7.2 Limitations of the Binomial Distribution

The binomial distribution gives a convenient theoretical expression for counting the numbers of ways that a detection and false alarm can occur, respectively. It is based on the assumption that the sequence of trials are stationary, statistically independent, and mutually exclusive.

In practice, counting the number of ways that a detection or false alarm can occur can be quite complicated. Normally the target returns are permitted to move between adjacent range/Doppler bins between adjacent dwell transmissions. Therefore, since not all range/Doppler bins can equally declare detections due to the changing clutter rejection notch and receiver dead zones, the statistics are *not* stationary. Furthermore, an additional complication arises in that some of the opportunity paths can share range/Doppler bins with other opportunity paths (nonmutually exclusive events). In general, although the binomial distribution provides a convenient starting point for the analysis, enumerating all possible opportunity paths and correctly modeling their interdependence is much more complicated than that implied by the binomial distribution.

In the case of P_D, it is often simpler just to write a computer program to calculate P_D after M-out-of-N detection. A computer simulation is also desired since the clutter statistics can be more accurately modeled and antenna modulation effects can be more easily incorporated. The speed and flexibility of modern digital computers are compelling reasons for taking this approach.

Figure 3.14 illustrates an example of an iterative computer program to calculate P_D. The program calculates P_D as a function of target range and

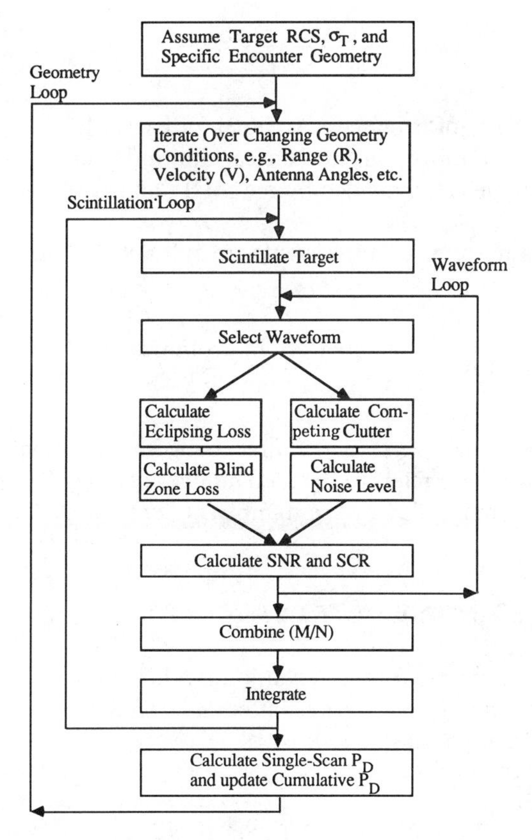

Figure 3.14 Flow diagram of a computer simulation to calculate P_D.

velocity since range and velocity determine the blind-zone losses for a given PRF. The output may be displayed as a function of target range/velocity (e.g., Stimson, 1983, p. 448) or averaged over an interval of target values.

Computer evaluation of the P_{FA} is usually not feasible unless convergence can be guaranteed. A simple analytical model for the P_{FA} may be used instead of the form:

$$P_{FA} = \sum_{K=M}^{N} \frac{(K-1)!}{(M-1)!(K-M)!} p_1(K) p_2(K) P_{fa}^M (1 - P_{fa})^{K-M+1} \tag{3.39}$$

The function $p_1(K)$ models the probability of K clear detection cells (per opportunity path) out of the last N samples after accounting for the changing clutter rejection notch and receiver dead zones. The function $p_2(K)$ models the probability that the current sample is clear whenever K samples are clear among a total of N; it can be approximated as (K/N). The function p_1 is more complicated, but by assuming independence among the shifting blind zones, the approximate expression results:

$$p_1(K) = \frac{N!}{K!(N-K)!} V^K (1 - V)^{N-K} \tag{3.40}$$

where $V =$ probability of being clear (visible) for each PRF,
$\cong$ number of range bins in the clear multiplied by number of Doppler bins in the clear, divided by the total number of range and Doppler bins.

Making the assumption that the events are mutually exclusive, then the FAR is bounded by:

$$\text{FAR} \leqslant \frac{\text{Number of opportunities}}{\text{processing time}} \times P_{FA} \tag{3.41}$$

where processing time refers to the time between detection decisions, and Number of opportunities refers to the total number of range/Doppler cells in the useful clear region multiplied by the number of permissible correlation paths.

3.8 SIGNAL-TO-NOISE RATIO CALCULATIONS

This section derives the radar range equation in a format that is applicable to all waveforms. To begin, start at the transmitter and define the peak radar power as P_p. The power density at distance R is

$$\text{Power density at } R = \frac{P_p G}{4\pi R^2}$$

for G the antenna gain. The target cross section (σ_T) indicates how much power is intercepted and reradiated back toward the radar. The power density at the radar antenna is therefore

$$\text{Power density at receiver} = \frac{P_p G \sigma_T}{(4\pi)^2 R^4} \tag{3.42}$$

The power captured by the antenna is dependent on the effective antenna aperture (A_e),

$$\text{Received power} = \text{power density} * A_e$$

From electromagnetics, it is known that

$$A_e = \frac{G\lambda^2}{4\pi} \tag{3.43}$$

where λ is the transmitted wavelength. This equation implicitly assumes monostatic operation ($G_R = G_T = G$). Substituting in (3.43) yields

$$\text{Received power} = \frac{P_p G^2 \sigma_T \lambda^2}{(4\pi)^3 R^4} \tag{3.44}$$

Equation (3.44) represents the power at the output of the antenna. Recall that the signal at this stage has been previously written as

$$\mathbf{S}''' = \mathbf{V}p(t - 2R/c)$$

where $p(t)$ is the pulse-train function of width τ and $\mathbf{V}$ is the instantaneous voltage strength. In terms of this notation, the received power is

$$\text{Received power} = E(|\mathbf{V}|^2) \tag{3.45}$$

Assuming, as it is usually done in preliminary radar analysis, that the pulse is rectangular, then the voltage actually captured by the jth range bin is

$$\begin{aligned} S''''_{ij} &= \int_{(j-1)\tau_V}^{j\tau_V} \mathbf{S}'''\,dt \\ &= \int_{(j-1)\tau_V}^{j\tau_V} \mathbf{V}p(t - 2R/c)\,dt \end{aligned} \tag{3.46}$$

where τ_V is the video filter integration time and $\mathbf{S}''''_{ij}$ is the sampled voltage. Note that $\tau_V = \tau$ assuming a perfectly matched filter; however, not all systems use a matched video filter.

If the receive pulse width τ is smaller than τ_V, $\tau < \tau_V$, then (3.46) is proportional to τ. Otherwise, it equals τ_V. Representing the two cases as

$$\mathbf{S}_{ij}'''' = \begin{cases} \mathbf{V}\tau_V & \tau \geqslant \tau_V \\ \mathbf{V}\tau & \tau < \tau_V \end{cases}$$
$$= \mathbf{V} \min(\tau_V, \tau)$$

then the received power captured by the jth range bin is

$$E(|\mathbf{S}_{ij}''''|^2) = \frac{P_p G^2 \sigma_T \lambda^2}{(4\pi)^3 R^4} \min(\tau_V, \tau)^2 \tag{3.47}$$

Receiver noise constitutes another component of the sum-channel voltage. The power spectral density (PSD) of receiver thermal noise is equal to Boltzmann's constant k multiplied by the receiver noise future:

$$\text{Receiver noise spectral density} = F_N k T_0$$

where k = Boltzmann's constant = 1.38×10^{-23} watt-second/K
T_0 = standard temperature (equal to 290 K by IEEE convention)
F_N = receiver noise figure.

F_N is a measure of how noisy the receiver is relative to T_0; T_0 is equal to 290 K (by definition) which is close to room temperature and conveniently makes kT_0 a round number (4×10^{-21} watt-second). Typical values for F_N lie between 2 and 7 dB (Eaves and Reedy, 1987, pp. 190–194).

The system noise figure F_S is an extension of the receiver noise figure concept to include both external environmental noise and internal (receiver) noise. The total noise is thus written

$$\text{Total noise spectral density} = F_S k T_0$$

The sampled noise is

$$\mathbf{N}_{ij} = \int_{(j-1)\tau_V}^{j\tau_V} \mathbf{N}(t)\, dt \tag{3.48}$$

so that the noise power captured by the jth range bin is written as

$$E(|\mathbf{N}_{ij}|^2) = \int_{(j-1)\tau_V}^{j\tau_V} dt_1 \int dt_2 E[\mathbf{N}(t_1)\mathbf{N}(t_2)]$$
$$= \int_{(j-1)\tau_V}^{j\tau_V} dt_1 \int dt_2 F_S k T_0 \delta(t_1 - t_2)$$
$$= F_S k T_0 \tau_V \tag{3.49}$$

where $\delta(\,)$ is the Dirac delta function. Defining the signal-to-noise ratio (SNR) to be

$$\mathrm{SNR} = \frac{\text{signal power captured by the range bin}}{\text{noise power captured by the range bin}} \tag{3.50}$$

and substituting in,

$$\mathrm{SNR} = \frac{P_p G^2 \sigma_T \lambda^2}{(4\pi)^3 \mathrm{R}^4 \mathrm{F}_S \mathrm{k} \mathrm{T}_0} \frac{\min(\tau_V, \tau)^2}{\tau_V} \tag{3.51}$$

3.8.1 Loss Budget

Equation (3.51) gives the ideal SNR assuming no losses. A typical loss budget contains contributions from the following sources:

- Range straddle loss.
- Atmospheric attenuation loss (two way).
- Blind-zone losses.
- Eclipsing losses.
- Beamshape loss (both elevation and azimuth).
- RF front-end losses (both transmit and receive).
- Radome loss.
- Field degradation loss ($\cong$1–3 dB).
- CFAR thresholding loss.

Including the loss term L, the net SNR is

$$\mathrm{SNR} = \frac{P_p G^2 \sigma_T \lambda^2 L}{(4\pi)^3 R^4 F_S k T_0} \frac{\min(\tau_V, \tau)^2}{\tau_V} \tag{3.52}$$

3.8.2 Coherent Doppler Filtering

If Doppler processing is employed, the SNR gain is equal to $G_{\mathrm{SNR}} = M$ AWL $= T$ PRF AWL, and the net SNR is thus

$$\mathrm{SNR} = \frac{P_p G^2 \sigma_T \lambda^2 L T \,\mathrm{PRF}}{(4\pi)^3 R^4 F_S k T_0} \frac{\mathrm{AWL}\, \min(\tau_V, \tau)^2}{\tau_V} \tag{3.53}$$

3.8.3 Alternate Forms

Next it is customary to introduce the concept of *average power* P_{avg}. The average power is equal to the peak transmitted power multiplied by the ratio of on time for which the radar operates:

$$\begin{aligned} P_{\mathrm{avg}} &\triangleq P_p \tau \,\mathrm{PRF} \\ &\triangleq P_p d_T \end{aligned} \tag{3.54}$$

where d_T is the transmit duty factor. Inserting this, we obtain for our final answer

$$\text{SNR} = \frac{P_{\text{avg}} T G^2 \sigma_T \lambda^2 L L_m}{(4\pi)^3 R^4 F_S k T_0} \tag{3.55}$$

where $L_m = \text{AWL}\,|\min(\tau_V,\ \tau)|^2/(\tau_V \tau)$ refers to the receiver mismatch loss. Sometimes the term $|\min(\tau_V,\ \tau)|^2/(\tau_V \tau)$ appears in the literature as $d_{T/R}^2/(d_T d_R)$, for $d_{T/R}$ the minimum of the transmit duty factor d_T and receive duty factor d_R, respectively. In other texts the term L_m is included in the loss budget L.

3.8.4 Conclusions

In words, (3.55) can be interpreted as follows. The average SNR is equal to the average transmitted power, P_{avg}, multiplied by the coherent dwell time, T, plus a loss term L_m due to the receive processing not being matched to the transmitted waveform, plus various other factors that account for two-way propagation effects. Note that the output SNR does not depend on the *waveform* of the transmitted signal, hence, waveform selection is important only in the manner that it influences the clutter level after signal processing.

From (3.55), and knowledge of σ_T and P_{avg}, the target's detection range can be calculated. The detection range is denoted by R followed by a subscript indicating the probability. For instance, R_{50} is commonly used to indicate the range for which the probability of detection is 50 percent on a single look.

Typical target cross sections σ_T in meters squared are as follows:

Jeep	small Jet	DC 10, 747	Large Bomber	Missile	Man	Aircraft carrier
5–100	0.1–10	15–20	2–200	0.5–0.05	1	20,000

3.9 SUMMARY

The basic features of a radar signal processor were outlined, including data weighting, Doppler filter formation, and rudimentary detection logic. The performance of the Doppler filter can be evaluated by means of the signal-to-clutter gain (3.10), the amplitude weighting loss (3.12), and the equivalent noise bandwidth. Further discussion focused on Dolph–Chebyshev weights and their performance. More optimal weights can be found for specific applications, but Dolph–Chebyshev weights are easy to analyze and are representative of the types of weights found in practice. Next data turning was described and the trade-off of filter overlapping versus zero padding described.

Thresholding rules were briefly discussed for both search and track. Thresholding against noise was discussed first, followed by thresholding against sidelobe targets. The probability of detection after thresholding was derived in a simple tutorial fashion, beginning with single-dwell (or single-pulse) probabilities and proceeding to more complicated single-look probabilities. Finally, derivation of the radar range equation was included at the end of the chapter and related to mismatched filtering.

PROBLEMS

3.1 Write an expression for the G_{SCR} of a single delay-line canceler given the Gaussian model for ground clutter. Assume the target frequency is uniformly distributed between (0, PRF), i.e., find the average value of $|H_k(\omega = \omega_T)|^2$ over (0, PRF). Leave your answer in terms of σ_C and PRF. Repeat for G_{SNR} and compare.

3.2 Consider the following set of FFT weights:

$$|w_0| = 0.1, \ |w_1| = 0.4, \ |w_2| = 1, \ |w_3| = 0.4, \ |w_4| = 0.1$$

Find the corresponding value of ENB and AWL assuming PRF = 1000 Hz.

3.3 Repeat 3.2 for the case of uniform and triangular weighting.

3.4 Consider a search radar with a PRF of 1000 Hz and a pulse length of 1 μsec. Assuming a P_{fa} of 10^{-6} per threshold decision, calculate the corresponding value of FAR.

3.5 On a single look and assuming a noise standard deviation σ, a given threshold T_h gives a P_{fa} of 10^{-6}. Calculate the new P_{fa} if the actual value of σ is doubled. Comment on the implication of this sensitivity.

3.6 Assuming $P(X_i|\mathbf{V})$ is known, where $X_i \triangleq \mathbf{S}_{Ii}^2 + \mathbf{S}_{Qi}^2$, derive an expression for the probability of detection after PDIing assuming $\mathbf{V}$ and T_H are known.

3.7 Given a search radar with a beamwidth of 2°, a scan rate of 50° per second, a pulse length of 1 μsec, and a PRF of 100 Hz, calculate:
- **(a)** The number of pulses per time on target.
- **(b)** Improvement in SNR if all the pulses per time on target are coherently integrated.
- **(c)** Improvement in SNR if all the pulses are noncoherently integrated (without RF agility).
- **(d)** The loss in decibels of (c) relative to (b).

3.8 M-out-of-N integration is employed with $N = 3$, $P_d = 0.9$, and $P_{\text{fa}} = 10^{-4}$. Calculate the P_D and P_{FA} after M out of N for the cases $M = 1$, 2, and 3.

3.9 Evaluate (3.31) for $T_H^2/\sigma^2 = 13.8$ and $N = 2$. Assuming Swerling II and SNR = 10 dB, calculate the corresponding P_D. Repeat this exercise with $N = 1$ and comment on the resulting change in probabilities.

3.10 Suppose we observe a target three times, with each observation being independent of the others and having a probability P_d of producing a positive declaration. We wish to calculate the probability of exactly two positive declarations out of three tries. An additional complication arises in that the individual detection probabilities are not uniform as given by

$$P_{d_1} = \tfrac{1}{2} \qquad P_{d_2} = \tfrac{1}{4} \qquad P_{d_3} = \tfrac{1}{8}$$

Question: Calculate the desired probability of exactly two detections out of the three tries. Is the probability binomially distributed?

3.11 Write down the radar range equation (SNR versus range) and identify every variable.

3.12 For the value of threshold that gives $P_{\text{fa}} = 10^{-6}$ on a single decision, calculate the value of SNR corresponding to R_{50}.

3.13 Determine the average range straddling loss if the return pulse is arbitrarily placed. Assume equal pulse widths, matched filtering $\tau_V = \tau$, and sampling at the pulse rate $1/\tau$.

CHAPTER 4

WAVEFORM SELECTION

This chapter discusses issues pertaining to waveform selection. Although a full discussion of waveform selection is outside the scope of this book, some introductory remarks can be stated simply. For a more in-depth treatment of the subject, the reader is referred to the references given.

4.1 BASIC MOVING TARGET WAVEFORM

The basic pulse-Doppler waveform consists of a single coherent block of pulses as depicted in Figure 4.1. This block of pulses is referred to as a coherent *dwell*, an *array* of pulses, a *burst*, or as a *coherent processing interval* (CPI). Each pulse is of duration τ. The duration of the entire dwell is denoted T.

When it is feasible, a constant and regular pulse repetition frequency (PRF) is employed. Pulse-to-pulse variations in the PRF (called *pulse staggering*) result in ambiguous clutter returns that do not fall on the same range bin (i.e., spread in range) and hence do not undergo the same coherent clutter rejection process as the unambiguous clutter returns. Pulse staggering can only be tolerated in those situations and/or those beam positions where the ambiguous clutter returns can be guaranteed to be negligible after range smearing.

In its simplest form information is extracted from the target as follows. Range is indicated by the observed time delay of a received pulse relative to a transmitted pulse. Doppler is indicated by the observed shift in the received frequency (assuming coherent processing of signals). By sensing this shift, the radar is able to separate the moving targets from clutter.

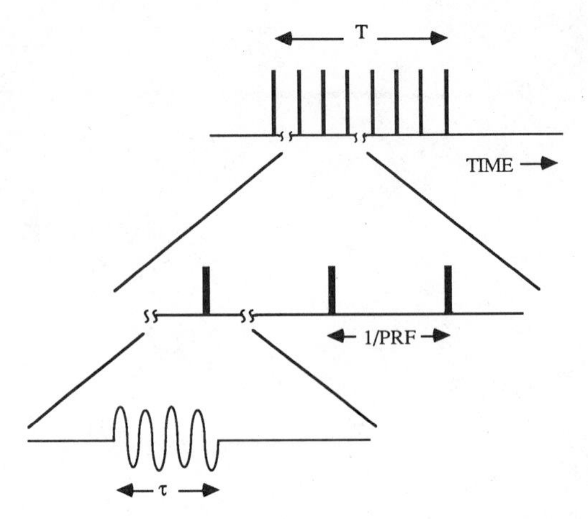

Figure 4.1 Example Pulse–Doppler waveform.

The Doppler frequency shift f_T is given by

$$f_T = \frac{2\dot{R}}{\lambda} \tag{4.1}$$

where $f_T = \omega_T/2\pi$ = Doppler frequency shift
$\dot{R}$ = radial component of relative velocity
λ = RF (radio frequency) transmitted wavelength

This shift occurs because, for every half wavelength that a target's range decreases (or increases), the phase of the received echo advances (or withdraws) by one cycle 2π.

4.2 TRACK WAVEFORM SELECTION

During track, the waveform is typically matched to the target characteristics (Barton, 1964). The compressed pulse width τ is matched to the range extent of the target if detection is limited by clutter and chaff. The target's range extent may be assumed *a priori* or measured as part of the initial STT processing. A larger pulse width may be commanded, on the other hand, if the target return is limited by receiver noise or target range walk rather than clutter.

The dwell time T is typically matched to the spectral characteristics of the target and/or transmitter/exciter characteristics. A large dwell time is desired to suppress both noise and clutter. Realistically, however, the dwell time is limited by the finite spectral bandwidth of the target and the radar. Given that the designer is free to specify the performance of the radar hardware, which is a safe assumption in the case of new system design, then the coherent dwell time is ultimately limited by the target characteristics.

All moving targets display a finite bandwidth due to vibrations alone (random yaw, pitch, roll and flexure), as Figure 4.2 demonstrates in the case of a jeep target. In order to capture all of the target's power, the Doppler filter width must be great enough to pass the full spectrum of target frequencies. From elementary Fourier transform theory, the Doppler filter width is approximately the inverse of the dwell time T. Therefore, the criterion that must be met is

$$\frac{1}{T} > \frac{2\Delta V}{\lambda} \tag{4.2}$$

Assuming X-band operation ($\lambda = 3\,\text{cm}$) and a jeep target ($\Delta V \cong 0.15$ m/sec from Figure 4.2), then T must be smaller than 100 ms. The value $T = 100$ msec corresponds to a Doppler filter width of roughly 10 Hz ($\cong 1/0.100$). The situation for other targets varies, but roughly the same spectral values apply. In the case of aircraft targets, for example, it is well known that the spectral width lies somewhere between 0.1 and 100 Hz, depending on the aircraft's exact size, speed, construction, and aspect angle (Nathanson, 1969; Hynes and Gardner, 1967; Riggs, 1975).

The previous remarks assume mean target maneuvers can be ignored. In general, it is necessary to accommodate the possibility of a large pilot-induced maneuver and/or aspect change. This is particularly true if steady tracking through maneuvers is a requirement and Doppler "tuning" is not applied. Doppler tuning can remove some of the effects of a target maneuver by moving the Doppler bin (motion compensating the bin) at the same rate as the mean acceleration.

Denote the unknown target acceleration as a_T. As illustrated in Figure 4.3, this acceleration produces a shift in velocity over the dwell time T as given by $\pm a_T T$. To prevent this shift from causing the target's energy to spill into adjacent Doppler filters, the criterion that must be met is

$$\frac{1}{T} > \frac{4a_T T}{\lambda} \tag{4.3}$$

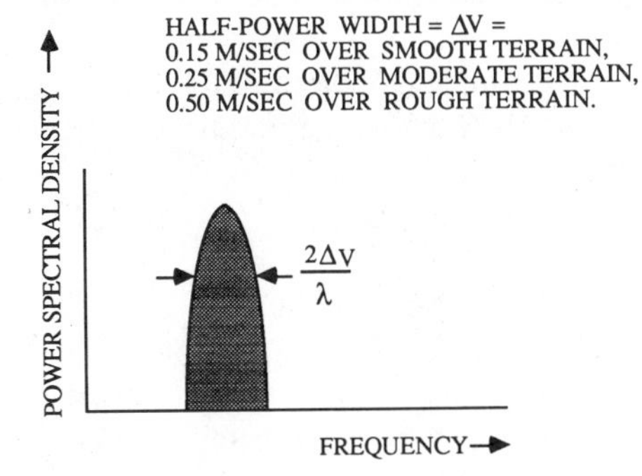

Figure 4.2 Spectral characteristics of a jeep target. Most targets of practical interest do not scatter monochromatically when illuminated by an energy source.

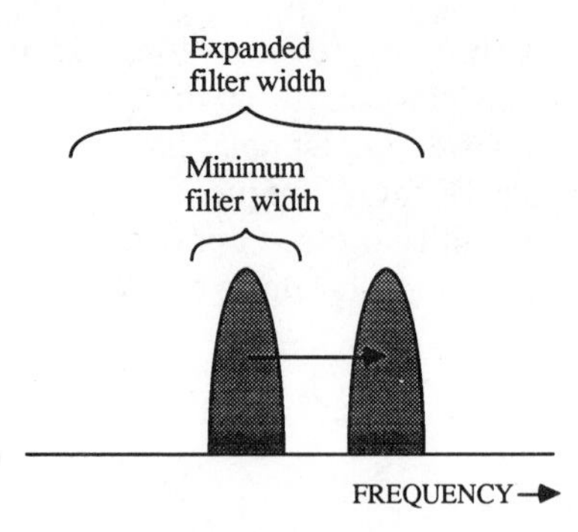

Figure 4.3 To avoid filter tuning, the Doppler filter passband may be made wide enough to accommodate mean target accelerations.

where a_T, the target's acceleration, can range anywhere between 0 and $10g$. For X-band operation ($\lambda \cong 3\,\text{cm}$) and assuming a tactical aircraft ($a_{\text{MAX}} > 5g$), T must be less than 12.5 msec. The corresponding Doppler filter size is roughly 80 Hz ($\cong 1/0.0125$). Assuming a mildly maneuvering aircraft ($a_T = 1g = 9.8\,\text{m/sec}^2$), then T must be less than 30 msec. These numbers may be scaled according to the application.

A turning maneuver also spreads the target's spectral width. To understand this, visualize the target as turning with rate ω_T. This turning causes the wingtips of the target to possess different radial velocities relative to each other. The resulting spectral width is proportional to the target's turn rate ω_T and length L. Therefore, the criterion that must be met is (Nathanson, 1969, p. 174)

$$\frac{1}{T} > \frac{2\omega_T L}{\lambda} \tag{4.4}$$

to prevent target energy from spilling over into adjacent Doppler filters. A turn rate ω_T can be physically caused by:

1. Target acceleration perpendicular to the velocity vector $\omega_T = a_T/v_T$.
2. Rotation of the line-of-sight vector $\omega_T = v_T/r$.

The first effect is dominant at long range, and the second effect is dominant at short range.

For X-band operation and assuming $\omega_T = 0.1$ rad/sec and $L = 8\,\text{m}$, the corresponding value for T is 20 msec. This type of Doppler smearing is potentially more serious than the previous type, since Doppler tuning cannot be applied even if ω_T is known. Therefore, (4.4) gives a limiting condition on the detection of any low-RCS maneuvering target.

4.3 SEARCH WAVEFORM SELECTION

During search, a primary factor determining the range resolution τ is the issue of signal-processing throughput requirements. Perhaps more importantly, signal-to-noise ratio (SNR) increases linearly with τ in the absence of pulse compression and, thus, a large compressed pulse width may be required to boost the SNR (Barton, 1964). This is especially true if the transmitter cannot support pulse compression or if pulse compression is limited to small compression ratios.

The selection of the dwell time T is determined by the issues of angle coverage and required scan time. The radar must complete the entire scan coverage in the time allotted. Typical ground-based TWS systems require 360° of azimuth scan coverage. Airborne and missile-based tactical radars require from $\pm 20°$ to $\pm 60°$ of forward azimuth scan coverage, although some airborne search radars require up to 360° coverage. The required elevation coverage is generally coverage to detect overhead diving threats. TWS scan times vary anywhere from 1 sec for short-range military applications to 1 min for tracking of quiescent targets; the important criteria is that the scan rate must support the required track accuracy under the assumed target maneuver conditions. With an ESA, since tracking is decoupled from search, the important criterion is that an undetected target cannot penetrate a significant fraction of the detection range between scans. Significantly longer scan times may be encountered in this case.

To formalize the preceding statements, consider a baseline system scanning 360° in azimuth. Let the azimuth beamwidth be 2° such that 180 beam positions result assuming no overlapping is employed. Taking the total scan time to be 5 sec, then the resulting dwell time is 28 msec. In three-dimensional applications there may be up to seven or more elevation beam positions per azimuth bar, limiting the dwell time to 4 msec under this condition.

Overlapping of the antenna beamwidth may be required for several reasons. Since the antenna gain falls off symmetrically from its maximum value, some beam overlapping may be required to prevent beamshape loss (Skolnik, 1980, p. 58). In the case of nonmonopulse radars at least two sequential detections are required for angular measurement, hence the overlapping must be at least 2:1 for this reason alone. Additional beam overlapping may be employed to permit the PRF and/or RF to be switched more frequently. In airborne radars another reason for overlapping is to allow the AGC more time to react to the presence of a strong target or clutter backscatter. For these reasons it is not uncommon to employ 2:1 to 10:1 overlapping, further reducing the time available per dwell.

4.4 PRF SELECTION

In search, and to a lesser extent track, the primary consideration influencing PRF selection is the need to separate the target from clutter and other targets. The PRF does *not* influence the signal-to-noise ratio for a fixed transmitter duty factor and loss budget. The PRF does, however, shift the target and clutter returns around on the range/Doppler grid, so the PRF needs to be selected to minimize clutter on those portions of the range/Doppler grid utilized for detection purposes.

One consideration influencing clutter distribution is the phenomenon of multiple-time-around-echo (MTAE) range ambiguities. This phenomenon, as

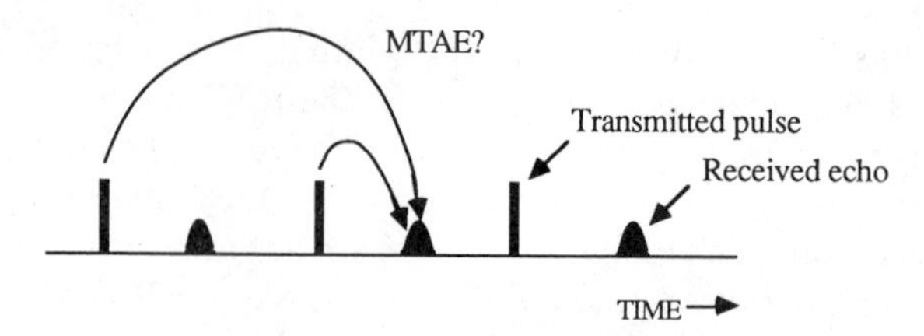

Figure 4.4 Multiple-time-around echos (MTAEs) are a problem in any radar system. Low PRF waveforms resolve this ambiguity by transmitting at such a low PRF that all MTAEs are negligible over the range swath of interest.

illustrated by Figure 4.4, concerns the reception of echoes from targets at longer ranges folding over into the next pulse repetition interval, and hence competing with clutter possessing a different $1/R^4$ free-space propagation factor.

The radar's unambiguous range R_u is, by definition, that range at which a target's echo will arrive back just as the next pulse is being transmitted. Any return from beyond R_u is folded over into the next 1/PRF pulse transmission period and, hence, is range ambiguous. R_u is mathematically expressed as

$$\begin{aligned} R_u &= c/(2\ \text{PRF}) \\ &\cong 150\ \text{km}/(\text{PRF in kHz}) \end{aligned} \tag{4.5}$$

for c the velocity of propagation. The factor 2 appears because of the two-way travel time to the target and back.

The other major constraint concerns the Nyquist rate. From Nyquist sampling theory the radar's sampling frequency (PRF) must be twice the signal's highest frequency component in order to avoid aliasing. Since for radar systems the frequency component is determined by the target's Doppler frequency shift, this implies that

$$f_T < \text{PRF}/2 \tag{4.6a}$$

or, from (4.1),

$$\dot{R} < \lambda\ \text{PRF}/4 \tag{4.6b}$$

Example of PRF Selection. Next consider an application where surface targets are to be detected, such as cars, tanks, ships, or terrain following helicopters. Our purpose here is to illustrate how the problem of MTAEs dictates the choice of waveform for a few baseline applications. In this kind of application all the targets lie on the surface of the earth (a sphere). Thus, selection of R_u is bounded by spherical geometry. Specifically, the radar's horizon limits the maximum detection range at low altitudes. At high altitudes, as illustrated in Figure 4.5, the geometry is bounded due to the finite intersection

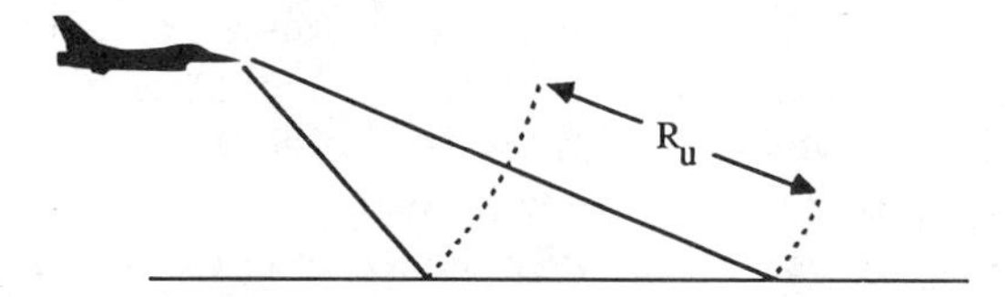

Figure 4.5 Geometric factors may limit the range swath of interest.

of the beamwidth with the earth's surface. If the elevation null-to-null beamwidth is small, then the value of R_u within the confines of the antenna footprint is also small.

Considering the previous geometrical factors, let us take $R_u = 75\,\text{km}$ such that PRF $= 2\,\text{kHz}$ from (4.5). Letting $\lambda = 1.5\,\text{cm}$ (C-band), then the highest ambiguous range rate is $\dot{R} = 108$ km/hr. Thus, the Nyquist rate is indeed satisfied for the typical target velocities of interest, and so unambiguous range/Doppler measurements are available to the tracking filter.

At the higher RFs (X-band and above), the Doppler frequency shift f_T may no longer be unambiguous with respect to another target's doppler f_T. However, the condition *may* still be satisfied that the target's Doppler is unambiguous with respect to stationary clutter at zero frequency, and so f_T is always *clear* of clutter. Although, strictly speaking, this is a weaker condition than the Nyquist rate, for most practical purposes it is equivalent since the target's unambiguous Doppler can always be resolved by switching the PRF between two clear detections.

In detecting aircraft and missile targets, on the other hand, a PRF can rarely be found to satisfy both the Nyquist and the MTAE criteria. This is due to the larger target velocities as well as the larger range swath of interest. Therefore, it is almost always necessary to make some compromises in the choice of PRF. The nature of this compromise determines the system operating characteristics.

4.5 LOW PRF OPERATION

The simplest PRF solution is to transmit at a sufficiently low PRF such that all desired returns are range unambiguous. This solution, referred to as low PRF (LPRF) operation, considerably simplifies the system design since range can then be directly and unambiguously related to the measured pulse delay (called pulse-delay ranging). Clutter rejection is also relaxed since the target is competing only with clutter contained in the same range ring.

In LPRF, the PRF is chosen to disambiguate range for all returns of interest. This includes the required target return out to the longest possible detection range (to ensure a high cumulative probability of detection) plus any addition to R_u needed to disambiguate clutter returns that might interfere with target detection. For example, if a mountain in the mainlobe produces a large specular return, then the PRF may be chosen to disambiguate the mountain return as well as the target return in order to relax the clutter rejection requirements. Other MTAE clutter phenomena that may lower the choice of PRF include rain, chaff, and any large building returns (called *discretes*).

LPRF operation is the best solution for many applications. However, its use does entail some compromises. LPRF undersampling causes *aliasing* as illustrated in Figure 4.6. Aliasing is best visualized as a repetition of the target's Doppler spectrum at multiples of the radar's sampling frequency (PRF). Because of aliasing, the target's measured Doppler frequency bears no simple relationship to its true value. More importantly, aliasing interacts with the clutter rejection procedure to produce speeds for which the radar is *blind*. *Clutter rejection* refers to the process of eliminating all returns observed to have a Doppler shift identical with regions of high clutter concentrations. Since a target can be aliased to have an observed Doppler shift ambiguous with clutter, the net effect is to potentially cause the target to be eliminated along with clutter.

The spectral width of clutter, σ_C, determines the seriousness of this problem. As Figure 4.6 illustrates, the larger the spectral width of clutter, the more region must be "cut out" to reject clutter, and the less clear space there is to detect targets. For ground-based radar systems the spectral width of clutter can be quite narrow. On a clear day the dominant contributor to σ_C is the mechanical motion of the radar antenna itself (Nathanson, 1969). Toomay has derived a spectral width $\sigma_C = 1/(t\sqrt{10})$ Hz for t the time on target to scan over one beam position (Toomay, 1982, p. 114). Since $t \gg T$, the spectral width of clutter σ_C is seen to be a small fraction of a doppler filter width $1/T$, and so the limiting condition is determined by the Doppler resolution of the radar and its ability to reject clutter.

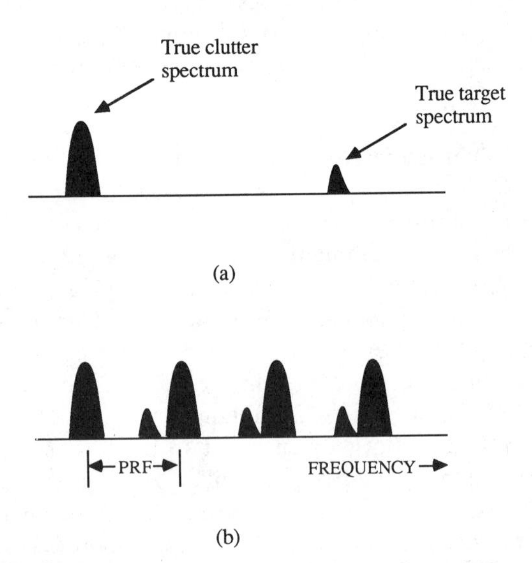

Figure 4.6 (a) True Doppler spectrum. (b) Folded Doppler spectrum due to aliasing.

Wind-blown rain can have a velocity width as wide as ± 50 km/hr (Skolnik, 1980, p. 129) and a mean wind velocity of up to 100 km/hr. Birds and other "sky noise" can have velocities of ± 15 km/hr with respect to the wind. Chaff in the jet stream can have a velocity in excess of 400 km/hr. Finally, for the lower beam positions there is the problem of surface targets that can have velocities continuously spread throughout the region ± 100 km/hr.

Because of these various considerations, ground-based LPRF search radars designed to detect aircraft are generally found at the lower RFs (in the L-, S- or C-band regions) where the blind speeds at λ PRF/2 are larger and the effect of rain backscatter is less (according to the law of Rayleigh scattering). The drawback, of course, is the concomitant poorer angular and velocity resolution for a given size antenna dimensions and dwell time.

4.5.1 PRF Switching

PRF switching involves transmitting two or more PRFs (see Figure 4.7). The PRFs are switched between dwells such that targets blind in the first PRF are clear in the second PRF. This process is also referred to as PRF switching, PRF jittering, PRF block staggering, PRF agility, or PRF diversity (Skolnik, 1980, p. 127; Cartledge and O'Donnell, 1979; Skolnik, 1970, Chapter 17).

Another advantage of PRF switching is that unwanted surface targets and bird detections can be eliminated. These low-speed targets are typically not Doppler ambiguous, and so do not translate into adjacent Doppler bins after switching. Thus, they can be eliminated by the simple procedure of dropping all detections that appear twice in the same range/Doppler bin after PRF switching. In military applications another advantage of PRF switching is that different

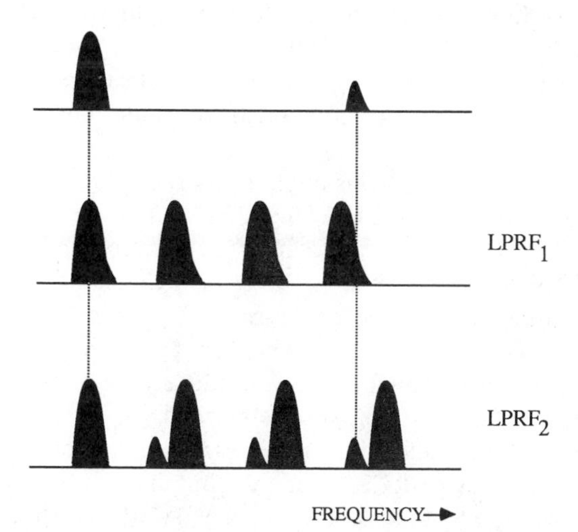

Figure 4.7 To minimize blind zone losses, two or more LPRFs may be transmitted. The second in the pair ensures that any target blind in the first LPRF is seen in the second.

RFs can be commanded between coherent dwells to combat hostile ECM. Finally, PRF switching is also used to screen out MTAEs from targets at long range that would otherwise be falsely identified as a near-by target.

The drawback of PRF switching is that additional time must be spent in transmitting the dwells. Thus, a typical procedure is to automatically activate PRF switching only in those beam positions that contain clutter.

The PRF separation is selected such that any obscured (blind) target return on the first PRF is translated into a clear Doppler filter on the second PRF. However, if the PRF separation is too large, the composite blind speed may pose a problem (e.g., $LPRF_1 = 400$ Hz, $LPRF_2 = 500$ Hz, implies a composite blind frequency of 2000 Hz). Thus, an appropriate PRF separation always involves some compromise, which in some cases may dictate more than two PRFs. Each additional PRF, however, degrades the Doppler resolution and increases the complexity of mechanization. The number of PRFs actually used, therefore, is a compromise between these costs and the cost of having to contend with target speeds for which the radar is blind.

4.5.2 Example of PRF Selection

Since the choice of PRF is so critical in LPRF, the transmitted PRF is normally matched to the expected target and clutter environment. The following example is extracted from Schleher's discussion on a 3-D LPRF TWS ground-based radar.

First, at the higher elevation angles a smaller value for R_u can be selected assuming a fixed target altitude ceiling. At the lower elevation beam positions, where the target altitude ceiling does not limit the maximum detection range, R_u is subdivided into several subregions. A different PRF is then transmitted to cover each subregion. A long-range waveform might cover the range interval from 100 to 300 km using PRF = 500 Hz. Because of the longer transit time, only one or two of these pulses may be afforded per beam position. Fortunately, the processing does not have to contend with ground clutter that is over the horizon at 100 km.

Multiple short-range waveforms might cover the range interval up to 100 km using a nominal PRF of 1500 Hz. The PRF is higher to reject clutter, the pulse width is shorter to maintain a fixed duty factor and to detect close-in targets, and PRF switching is used to reduce the blind zones. In military applications another short-range waveform may be added to cover the region from 0 to 20 km. The purpose of this short-range waveform is to detect high-priority targets such as terrain following aircraft, sea-skimming missiles, or low-altitude helicopters that pop up over the horizon.

Other factors to be considered include the range extent of clutter contained in the beamwidth, any large-RCS discretes contained in the beamwidth, the required duty factor, and the amount of range swath that must be rejected to prevent code sidelobe degradation (Stoker, 1987; Billam, 1985; Bucciarelli et al., 1982; Schleher, 1986, p. 273).

4.5.3 Airborne LPRF Operation

In the case of airborne-based, missile-based, ship-based, or satellite-based radars, the antenna itself possesses a large velocity V_A. As shown in Figure 4.8, the spectral width of clutter increases in proportion to V_A. This increase results because the projected velocity to a single ground patch is roughly given by $V_C = V_A \cos(\theta)$, for V_A the antenna's velocity and θ the angle between the radar's velocity vector and the line-of-sight vector to the patch. Since the angle θ is not the same for each ground patch contained inside the antenna beamwidth θ_{HP}, the backscattered return is spread over a spectrum of frequencies. By differentiating with respect to θ, the resulting clutter spread in velocity is found to be

$$\Delta V_C = V_A \theta_{HP} \sin(\theta) \tag{4.7}$$

assuming $\theta > 0.5\theta_{HP}$, and where θ_{HP} is the 3 dB antenna beamwidth. Taking $V_A = 1000$ km/hr, $\theta = 45°$, and $\theta_{HP} = 2°$, then ΔV_C is 23 km/hr. The corresponding spread in frequency is $\sigma_C = 2\Delta V_C/\lambda$ or $\sigma_C = 425$ Hz at X-band operation ($\lambda \cong 3$ cm).

Of course, if the antenna's mainlobe does not strike the ground, then detection is not limited by clutter considerations. This is the approach that was taken by almost all early airborne radars, including some inexpensive ones still operational today. Satisfactory performance necessarily entails that the target be at a higher altitude than the radar. The second generation of LPRF airborne radars had both a "look-down" as well as "look-up" capability, but the azimuth look angle θ was limited to some small value. This approach provided some look-down capability in the forward sector. However, such a severe look-angle restriction was undesirable.

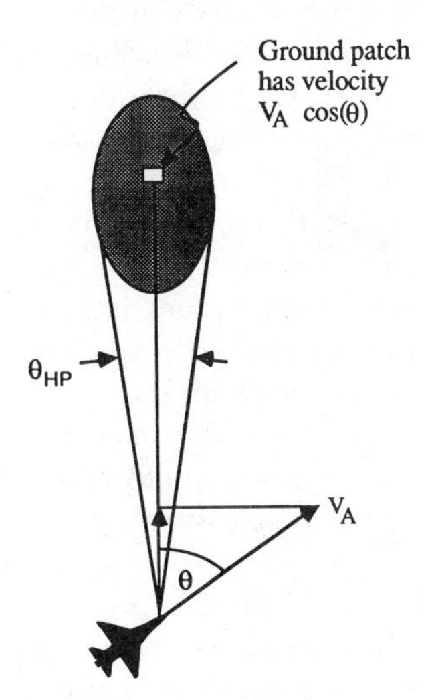

Figure 4.8 If the radar is moving with velocity V_A, the projected ground clutter has mean velocity $V_A\cos(\theta)$ and velocity spread $V_A\sin(\theta)\theta_{HP}$.

Today, LPRF is primarily utilized by large slow-speed early-warning (EW) airborne radars operating at the lower RFs (X-band and lower). LPRF is also used by X-band and Ku-band tactical airborne radars for the express purpose of tracking surface targets, helicopter targets, and aircraft targets close in radial velocity to mainlobe clutter (called crossing targets). In such systems the PRF is typically matched to the aircraft's instantaneous altitude and look angle.

Modern LPRF airborne radars typically divide the signal processing into two categories. Conventional signal processing is used to detect *exo-clutter* targets, that is, those targets characterized by a doppler frequency shift that falls outside the ground clutter spectral region. *Endo-clutter* processing is used to detect targets that compete with the clutter background. This division is convenient since the competing interference is different in the two cases (i.e., thermal noise in the first case and clutter in the second case).

All endo-clutter techniques exploit the fact that the ground clutter is truly stationary while the target is truly moving. A stationary object will exhibit a known functional relationship between the sum signal **S** and difference signal **D** after Doppler filter formation. Endo-clutter processing consists of recognizing this functional relationship and declaring a target when the expected clutter signature is not observed.

Multiple null cancellation (MNC) processing is the simplest endo-clutter technique to describe as well as to implement. MNC is best illustrated by means of Figure 4.8. Since the clutter at angle θ has Doppler frequency shift given by $f_T = 2V_A \cos(\theta)/\lambda$, there exists a known relationship between the observed Doppler frequency shift f_T and the monopulse angle measurement θ. The difference between the Doppler filter's measurement of θ and the monopulse measurement of θ can therefore be used as a detection statistic. An equivalent procedure is to form a weighted sum of the signal from the leading half of the antenna, **A**, and the trailing half of the antenna, **B**, with the weights adaptively selected as a function of Doppler filter index to null out stationary clutter.

Displaced Phase Center Antenna (DPCA) processing is a related technique that is best illustrated by Figure 2.11. Observe that the signal received in the trailing half of the antenna, $\mathbf{B} = \mathbf{Q}_2 + \mathbf{Q}_4$, is delayed in time relative to that received in the leading half of the antenna, $\mathbf{A} = \mathbf{Q}_1 + \mathbf{Q}_3$. Therefore, if the signal in **A** is delayed in time and subtracted from the signal in **B**, the return from clutter will be nulled out [Skolnik, 1980, p. 144]. The amount of delay is matched to the velocity of the platform to make clutter appear stationary. As with any delay-line canceler, the target's signal power is also reduced, but for moving targets this signal loss is hopefully compensated for by the increased clutter cancelation. DPCA processing also compromises the angular information contained in the difference channel, therefore, the target's azimuth angle must be measured by other techniques such as adding an extra subaperture (Stone and Ince, 1980] or by employing sequential lobing.

DPCA performs better than MNC since it does not rely on the Doppler filter's imperfect measurement of angle. Conversely, MNC is easier to implement since a variable PRF is not required, and it does not compromise the angular imformation contained in the monopulse measurement. Because of these conflicting advantages, the approach most often taken is to employ some combination of DPCA and MNC processing techniques.

4.6 HIGH PRF OPERATION

When detecting high-speed targets at the higher RFs against a clutter background, LPRF operation leads to unacceptable performance. This dilemma has led to the development of high PRF operation. In high PRF (HPRF) operation, a PRF is selected such that the target's velocity is unambiguous under all conceivable conditions. At X-band and above, where HPRF operation is most typically employed, PRF values range anywhere between 250 and 400 kHz.

HPRF operation leads to good clutter rejection without potential target signal loss near the blind speeds. In addition, the output of the Doppler filter is a direct indication of target radial velocity, which aids the tracking function. Furthermore, at these high values of PRF, the duty factor can approach 50 percent or more without inducing transmitter burnout. Thus, high average powers can be achieved without having to resort to unreliable high-power transmitters or pulse compression.

Nevertheless, HPRF operation has several drawbacks that make it much more difficult to implement relative to LPRF. Primary among these include (1) the required dynamic range, (2) range eclipsing, (3) range resolving, (4) the required spectral purity, and (5) the inherent need for more signal processing to separate the target from clutter. Because of these various problems, HPRF is primarily used only in those applications where LPRF is deemed impractical.

4.6.1 Dynamic Range Requirement

Dynamic range is the Achilles' heel of HPRF operation. In airborne HPRF radars it is not unusual to have clutter-to-signal ratios at the input to the radar on the order of 90 dB or more after MTAE folding. Even higher ratios may be found in ground-based HPRF systems. Theoretically the clutter return can always be isolated from the target return on the basis of sensed Doppler frequency. As a practical matter, however, it is difficult to build hardware with sufficient dynamic range to isolate returns on the basis of sensed Doppler frequency alone.

The simplest approach to this dilemma is to employ an *analog* clutter filter, such as a bulk quartz-crystal filter.* Dynamic range is then no longer a problem provided that the front-end amplifiers and analog components do not saturate. This is the approach taken by most contemporary HPRF radars.

Equivalent digital filters are, of course, far more reliable and flexible. Clutter rejection can then be accomplished using precisely controlled IIR filters of the Butterworth, Chebycheff, Bessel, or elliptic type. Nevertheless, the added A/D dynamic range typically makes the prior approach preferrable.

*Referred to as a *clear region filter* (in search) or a *speed gate* (in track).

4.6.2 Eclipsing

Another problem with HPRF concerns potential target loss due to *eclipsing*. Eclipsing refers to the phenomenon whereby the target's echo arrives back at the radar's front-end while the receiver is shut down for transmission of the next pulse.

Nevertheless, eclipsing is not as severe a problem as it might seem at first glance. The target is rarely totally eclipsed, as at least a portion of the return usually gets through. Even so, at long range, eclipsing can reduce the signal-to-noise ratio sufficiently to leave periodic "holes" of appreciable size in the radar's range coverage. For this reason, it is often the *average* eclipsing loss that determines the target detection probability.

An approximate expression for the average eclipsing loss may be calculated as follows. First, the instantaneous power loss is proportional to the square of the blanked pulse width (t) to the total pulse width (τ) ratio, or $(t/\tau)^2$. The average loss over an entire PRF cycle is thus

$$L_{ec} = \int_0^{\tau} (t/\tau)^2\,dt + \int_{\tau}^{\tau_V} dt + \int_{\tau_V}^{\tau_V+\tau} (\tau_V + \tau - t)^2/\tau^2\,dt \qquad (4.8)$$

where τ_V is the video integration time and where $\tau_V \geqslant \tau$ has been assumed. After normalization by the PRF, therefore,

$$L_{ec} = (d_R - d_T/3) \qquad (4.9)$$

where $d_T = \tau$ PRF and $d_R = \tau_V$ PRF. Assuming a transmit duty factor $d_T = 0.4$ and a receive duty factor $d_R = 0.4$ (i.e., a TR recovery time of 0.2/PRF) leads to an average eclipsing loss of approximately -5.74 dB. [Note: Eq. (4.9) is only approximate; a more rigorous analysis would incorporate the eclipsing factor inside the probability integral, e.g., as in Figure 3.14 or see Blackman (1986, p. 222).]

In those applications where a high single-scan probability of detection is required, rather than average probability, then detection must be enhanced by switching the PRF among several different values. A pair of HPRFs are transmitted as depicted in Figure 4.9. The second waveform in the pair ensures that any target eclipsed in the first waveform is seen in the second at the range interval of interest.

4.6.3 HPRF Ranging

Another disadvantage of HPRF operation is that added complexity is needed to range the target. One common solution to this problem is to simply not measure range at all. This procedure is called *velocity search* (VS) or *pulse-Doppler search* (PDS).

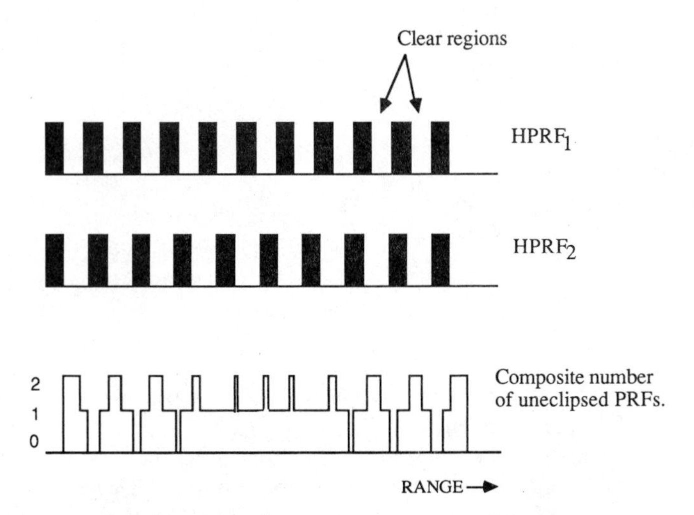

Figure 4.9 Transmitting multiple HPRFs has the effect of increasing the uneclipsed region.

If range is required, on the other hand, more complex techniques must be introduced. One of these techniques is frequency modulation (FM) ranging. In its simplest form, FM ranging is as follows. The transmitted frequency is increased at a constant linear rate as shown in Figure 4.10. This linear modulation is continued for a long time period. Due to the linear nature of the modulation, the instantaneous frequency difference between the received echo and the transmitter, Δf, serves as a linear indication of target range. In particular,

$$t_d = \frac{f_T - \Delta f}{k} \tag{4.10}$$

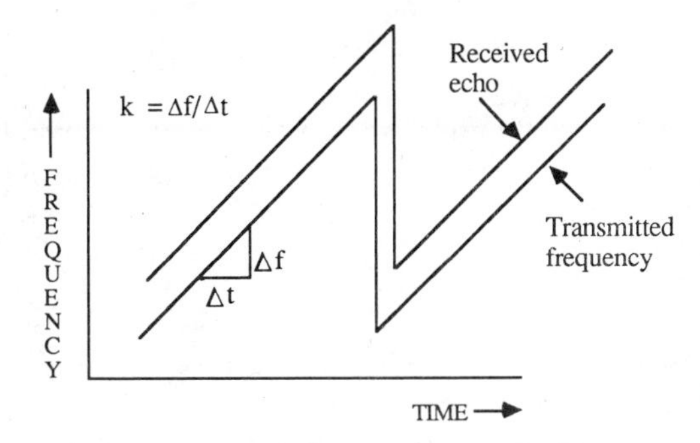

Figure 4.10 FM ranging involves ramping the transmitted frequency at a constant rate k.

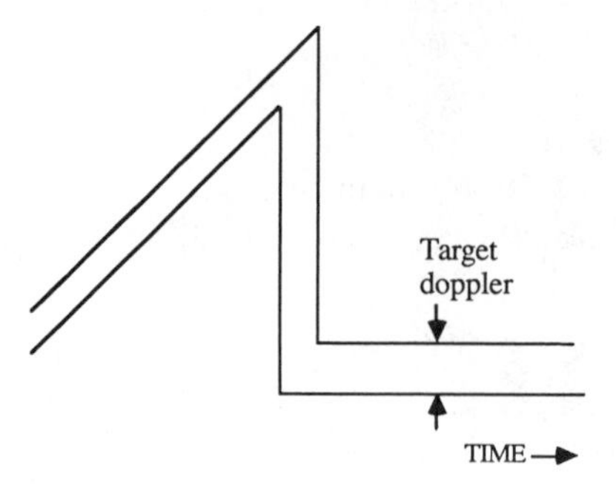

Figure 4.11 The target's Doppler frequency may be measured by adding a constant frequency segment.

for t_d the estimated transit time, k the known rate of frequency change, and f_T the target's true Doppler frequency.

Filter transients can be eliminated by discarding the leading and trailing pulses of each FM segment or dwell. The amount of time discarded is referred to as the *discard time*. The discard time is at least equal to the round-trip transit time to the maximum range of clutter plus target.

The target's Doppler frequency shift f_T can be found by terminating the frequency modulation at the end of each dwell and including a constant-frequency waveform. This "off–on" scheme is portrayed in Figure 4.11. Alternately, the frequency may be ramped in the opposite direction (*polar* ramping). Polar ramping is the most efficient FM method, in the sense that it provides the best range resolution for a fixed bound on the bandwidth kT. Nevertheless, other factors often dictate the FM design, such as clutter masking.

One possible problem with FM ranging concerns *ghosts* (Stimson, 1983). If two targets are detected simultaneously, it may be impossible to correctly pair the measurements. This can be readily understood when it is recalled that the measurement of Δf on the first dwell must be correctly paired to the measurement of f_T on the second dwell to remove the effect of the target's Doppler. Incorrect pairings result in erroneous targets called ghosts. To prevent the problem of ghosts, sometimes a third dwell is added as shown in Figure 4.12. This third dwell resolves any ambiguities that may occur whenever several targets are detected simultaneously, as Figure 4.13 illustrates in the case of two targets.

The range resolution obtained after FM ranging depends on the total frequency excursion kT, where k is the rate at which the transmitter frequency is

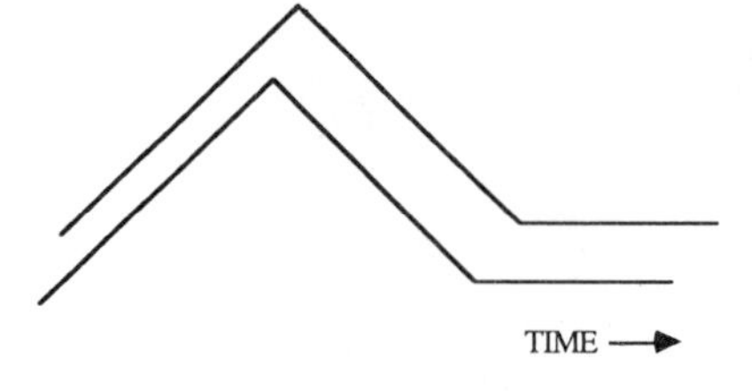

Figure 4.12 The problem of ambiguities can be solved by adding a third segment.

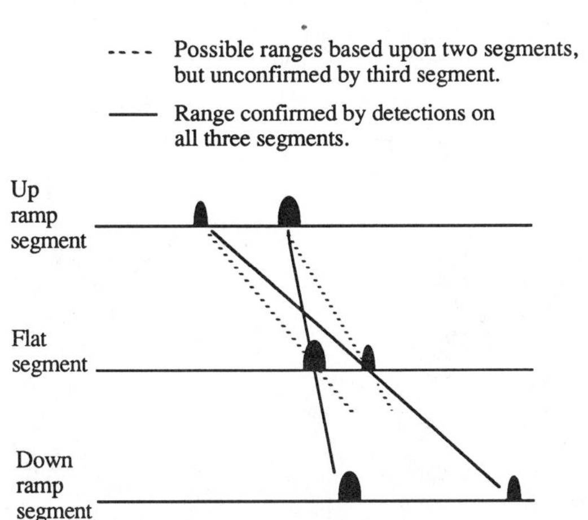

Figure 4.13 Illustration of the ghost problem and its solution.

changed and T is the total dwell time. The exact value is given by $c/(2kT)$. Unfortunately, the value assigned to k is ultimately limited by clutter. As k is increased, clutter returns are smeared over an increasingly broader band of frequencies. If k is too large, a point is quickly reached where clutter blankets the target.

This problem can be simply explained. If the range resolution after FM ranging is X kilometers, then X kilometers must correspond to the resolution with which frequency differences can be resolved through Doppler filtering, that is, equal to $1/T$. Thus, clutter with an extent of Y kilometers is smeared over approximately Y/X Doppler filters during the FM ranging process. Due to this smearing phenomenon, FM ranging is limited to range resolutions on the order of a few percent of the range extent of clutter. In conclusion, FM ranging is inherently less accurate than pulse-delay ranging, as well as being more complex and more time consuming.

4.6.4 Airborne HPRF Operation

In airborne HPRF operation sidelobe clutter poses some special problems. As depicted in Figure 4.14, although backscatter from the antenna's mainlobe is relatively narrow in frequency, backscatter from the antenna's sidelobes can exist anywhere from $2V_A/\lambda$ to $-2V_A/\lambda$ (assuming significant backward-looking sidelobes). For airborne HPRF operation, thus, separating the target return from clutter adds some complications.

Nevertheless, HPRF does play a powerful role in modern airborne radars. If the target happens to be moving toward the radar, its relative Doppler shift will still place it above the competing sidelobe clutter returns, and so the target return can be distinguished from clutter. This type of operation permits long-

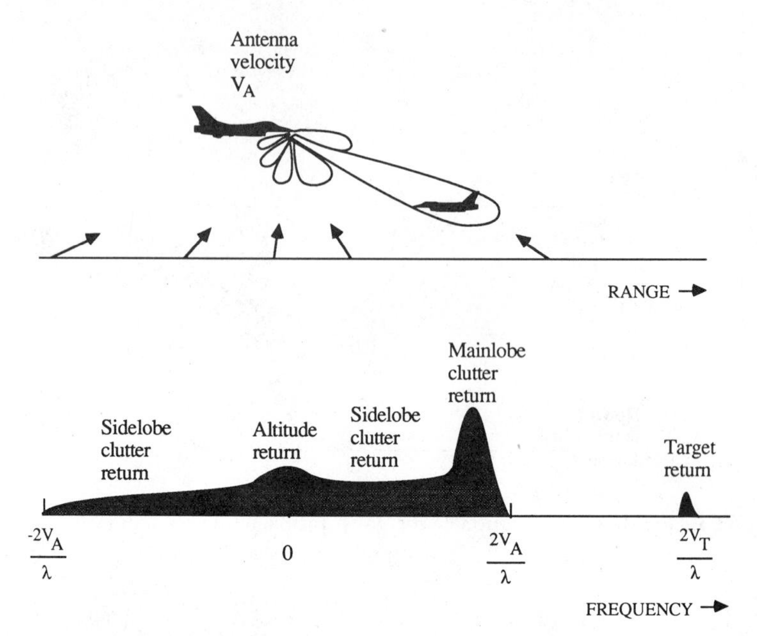

Figure 4.14 Spread in clutter frequencies due to backscatter from clutter located in front of, below, and behind the moving airborne platform.

range detection of nose-aspect (approaching) targets penetrating at low altitudes that cannot be detected by any other means. High-velocity tail-aspect (receding) targets can also be clear of sidelobe clutter if moving with sufficient velocity to be placed below the sidelobe return.

In principle, tail-aspect targets inside the sidelobe clutter region can also be detected. However, this is very difficult to accomplish in practice. Probably the most critical item is a high ownship altitude, h_A, since the sidelobe clutter power scales as $1/h_A^2$ (Hovanessian, 1973, pp. 194–200). Other factors include (1) a relatively short target range R, (2) high range/Doppler resolution, (3) low two-way antenna sidelobes, (4) more FM slopes, and (5) guard processing. Primarily, though, conventional HPRF airborne radars make no attempt to detect targets residing inside sidelobe clutter. Instead, the whole sidelobe clutter region is blanked via an analog clutter filter.

HPRF has no capability to detect targets inside regions of mainlobe clutter in a look-down geometry. Thus, it is incapable of detecting crossing targets that move orthogonal to its line-of-sight vector. It also has no capability to detect co-speed targets whose return falls within the strong "altitude" return reflected from directly underneath the aircraft.

Again, FM ranging is used to range the target. Since the geometry is such that the radar is looking down on the target from above, mainlobe clutter is typically at a longer range than the target. Therefore, clutter masking is minimized by employing FM ranging with all positive FM segments. A typical scheme is to

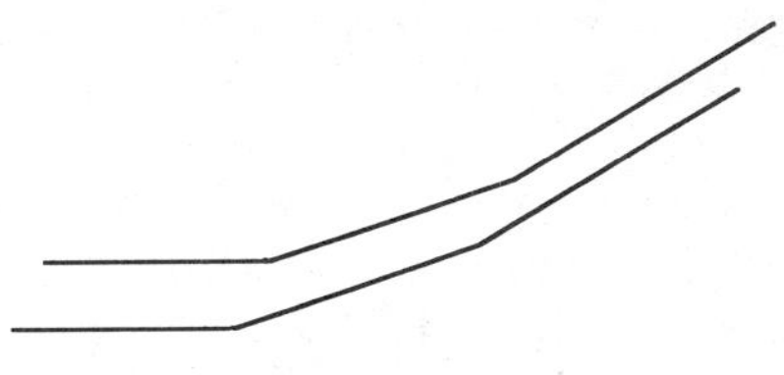

Figure 4.15 FM ramping for airborne radars looking down on clutter.

employ the "off–one–two" scheme shown in Figure 4.15. Each positive FM segment moves the target and clutter to the left, but, assuming the target is at a closer range than the background clutter, clutter masking does not occur since mainlobe clutter is shifted more to the left than the target.

4.6.5 Range-Gated HPRF Operation (RGHPRF)

In conventional HPRF operation, the duty factor is so large that range gating is hardly deemed worthwhile ($35\% < d_T < 50\%$). However, if the duty factor is lowered, as depicted in Figure 4.16, then formation of multiple-range gates becomes feasible. This type of operation is referred to as range-gated HPRF (RGHPRF) operation.

Range gating allows resolution of closely spaced targets and enhances range measurement accuracy (Postema, 1985). Range gating also reduces the level of competing sidelobe clutter. A fourth advantage of range gating is that it allows a more closely matched filter to be applied on receive. This last issue is particularly important if the transmitter is in a performance regime where it can operate at a lower duty factor yet maintain a high average transmitted power. In this case, reception is best implemented by means of a matched filter ($\tau = \tau_V$).

The number of range gates used depends on the PRF, RF, look angle, and sidelobe level. The lowest HPRF must be greater than the sum of:

1. The Doppler frequency of the most rapidly closing target $= 2V_T/\lambda$.
2. The radar's Doppler frequency in the target's line-of-sight $= 2V_A\cos(\theta)/\lambda$.
3. The minimum sidelobe clutter frequency $= -2V_A/\lambda$ (assuming the radar has significant backward-looking sidelobes).

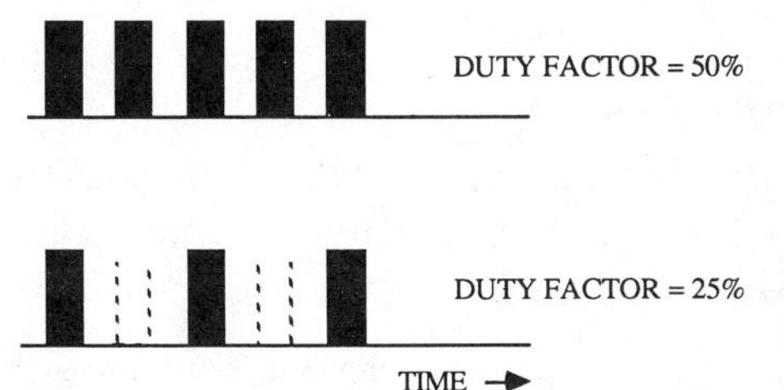

Figure 4.16 Range-gated HPRF essentially amounts to lowering the duty factor and employing multiple gates.

Therefore, the required PRF is $[2 + 2\cos(\theta)]V_A/\lambda + 2V_T/\lambda$. If the radar is moving at 750 m/sec and the target is moving toward the radar at 750 m/sec, for instance, then the required PRF at X-band is greater than 150 kHz. If the range bin size is assumed to be 25 m, and X-band operation is assumed, then the maximum number of range gates is 40. In practice, fewer number of range gates may be selected, particularly if pulse compression is unavailable.

The criterion for ground-based HPRF radars is much less severe. The required PRF is $2V_T/\lambda$, which equals 50 kHz at X-band. This assumes the Doppler unambiguous criterion can be relaxed to a Doppler clear criterion since operating at 50 kHz implies an added ambiguity between opening and closing velocities. The corresponding number of range gates is 120 at X-band. Also, if a lower RF is used, even higher numbers of range gates are feasible.

RGHPRF Ranging Method 1. FM ranging can be used in RGHPRF the same as in conventional HPRF. To illustrate, assume a PRF of 50 kHz, implying from (4.5) that the first range ambiguity occurs at 3 km. If conventional FM ranging can provide a coarse estimate of range to within 3 km, which is not difficult to accomplish, then pulse-delay ranging can unambiguously range the target to within this window.

RGHPRF Ranging Method 2. *Coincidence ranging* is the more popular approach to RGHPRF ranging. Coincidence ranging essentially amounts to switching the PRF and performing range resolving. As illustrated in Figure 4.17, by periodically changing the PRF and measuring the changes in the observed ranges, the true range can be resolved. Coincidence ranging is also referred to as PRF sequencing, PRF switching, or M-out-of-N detection.

In one approach to coincidence ranging, the PRFs are selected to be multiples of the basic range bin unit. That is, if $c\tau/2$ is selected to be the common range bin size, then two R_u's are chosen:

$$\begin{aligned} R_{u1} &= m_1 c\tau/2 \\ R_{u2} &= m_2 c\tau/2 \end{aligned} \tag{4.11}$$

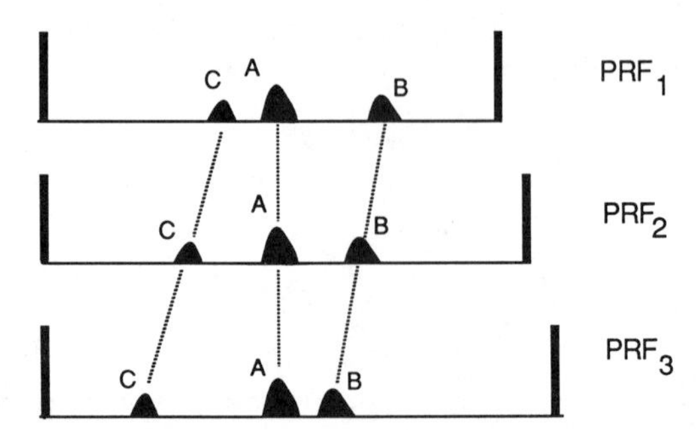

Figure 4.17 The underlying principle of coincidence ranging is that MTAEs will appear in different range bins after changing the PRF. Here, the returns B and C are MTAEs.

where m_1 and m_2 are determined by a fixed countdown from a master clock. By properly choosing the (m_1, m_2) values, up to $m_1 m_2$ unique combinations are possible, hence, the maximum unambiguous range is $m_1 m_2 c\tau/2$ (Skolnik, 1970, pp. 19-15). Assuming, for example, $m_1 = 39$, $m_2 = 40$, and $c\tau/2 = 25\,\text{m}$, then the maximum unambiguous range is 39 km.

The true range, expressed in terms of range bin units, is given by

$$\begin{aligned} R &= (1_1 m_1 + \lambda_1)c\tau/2 \\ &= (1_2 m_2 + \lambda_2)c\tau/2 \end{aligned} \tag{4.12}$$

where (λ_1, λ_2) are the observed target ranges in $c\tau/2$ units. Range resolving is implemented by successively adding m_1 and m_2 to (λ_1, λ_2) until the ranges align within a predetermined window.

The advantage of this method is that range resolving can be easily implemented on a digital computer, and both m_1 and m_2 can be arbitrarily chosen (within limits). The disadvantage is that the duty factor becomes a complicated function of the PRF. A common solution to this dilemma is to change the dwell time T after PRF switching to provide for uniform detection sensitivity.

RGHPRF Ranging Method 3. In a second coincidence ranging method, two R_u's are chosen:

$$\begin{aligned} R_{u1} &= mc\tau_1/2 \\ R_{u2} &= mc\tau_2/2 \end{aligned} \tag{4.13}$$

where $\tau_1 = n_1\,\Delta\tau$, $\tau_2 = n_2\,\Delta\tau$, and (m, n_1, n_2) are determined by a fixed countdown from a master clock. The true range, expressed in terms of R_u, is then given by

$$\begin{aligned} R &= l_1 R_{u1} + \lambda_1 \\ &= l_2 R_{u2} + \lambda_2 \end{aligned}$$

where (λ_1, λ_2) are the observed target ranges. Range resolving is implemented by successively adding R_{u1} and R_{u2} to (λ_1, λ_2) until the ranges align within a window determined by the measurement noise variances.

This method has the advantage that the duty factor can be kept constant from dwell to dwell without timeline changes. The disadvantage is the nonuniform range bin size and associated difficulty in implementing range resolving by any digital method.

RGHPRF Ranging Method 4. In STT a simpler technique can be employed. First, it is convenient to write the range equation in R_u units:

$$R = (l + \lambda)R_u$$

Taking the derivative, the result is obtained

$$\dot{R} = (l + \lambda)\dot{R}_u + \dot{\lambda}R_u$$

The quantity $\dot{R}$ is unambiguously measured and hence is known. Both R_u and $\dot{R}_u$ represent known quantities; they are normally controlled by the data processor to place the target near mid-PRF. The quantity λ is measured and $\dot{\lambda}$ can be derived from the data filter. Therefore, the only unknown in this expression is the range index number l, and so the correct range index can be eventually arrived at by comparison.

4.7 MEDIUM PRF OPERATION

In medium PRF operation (MPRF), both the observed target range *and* Doppler are ambiguous (Stimson, 1983; Aronoff and Greenblatt, 1974; Ringel et al., 1983; Long and Ivanov, 1974; Ethington, 1977). Historically, medium PRF values lie between 8 and 20 kHz, although with modern signal-processing methods it is possible to operate at PRFs of 50 kHz or even higher.

Coincidence detection is used in MPRF to resolve both the target's range and radial velocity. Velocity resolution is desired since it allows surface targets and birds to be rejected. Also, an unambiguous velocity measurement helps shorten the track confirmation process.

The primary motivation for the use of MPRF is better range isolation from competing sidelobe clutter. In fact, it can be easily seen that the intensity of sidelobe clutter in MPRF is midway between LPRF and RGHPRF values. This decrease in sidelobe clutter results not only because of the fewer ambiguous range rings that compete with the target, but also because of better range isolation from the first range ring containing clutter (i.e., the so-called altitude return).*

The drawback of MPRF is that it must contend with blind zones in range as well as Doppler. For a fixed duty factor the percentage of range blind zones is approximately the same for both RGHPRF and MPRF. Hence, MPRF essentially trades added Doppler blind zones for decreased sidelobe clutter. Which waveform performs better, RGHPRF or MPRF, therefore depends on the width of the Doppler blind zones and the aircraft's altitude h_A.

A typical MPRF waveform set is called M/N and cycles through N PRFs, where detection is required on M out of the N for a valid target declaration. The exact value of N is a function of the width of the range/Doppler blind zones, although 3/8 seems to be popular for X-band tactical airborne radar applications (Nevin, 1988). However, there is nothing magical about 3/8, and systems exist that operate with other values of M and N.

* *Altitude return* comprises the sidelobe return from ground located directly underneath the aircraft. The specular component of the altitude return is very large and is localized in range to within 50 m.

Normally all N PRFs are transmitted during a single target illumination time [at least in mechanically scanned antenna (MSA) applications]. All N PRFs are distributed semirandomly over the range of PRFs to be processed. The actual selection of the medium PRFs is normally done through computer simulation techniques. A common MPRF design approach is to start at the highest PRF that can be processed, and then iteratively work downward to find N PRFs that optimize detection and deghosting performance on at least M PRFs.

4.7.1 Medium PRF Tradeoffs

In comparing MPRF vis-á-vis RGHPRF, RGHPRF offers a better stand-alone waveform, especially in those situations where the width of the Doppler blind zone is large. Conversely, MPRF reduces the required dynamic range and signal-processing throughput and achieves better detection of receding targets inside sidelobe clutter.

Due to the lower PRF, MPRF operation does not enjoy the same high duty factor as HPRF or RGHPRF unless some manner of pulse compression is employed. However, there is a limit as to how much pulse compression can be successfully implemented, since the range ambiguities can potentially lead to a target at far range occupying a neighboring range bin to a target at close range thereby masking the weaker target through the code's sidelobes.

Another problem concerns the phenomenon of clutter transients. Because the PRF is switched from dwell to dwell, any clutter that spills into the next dwell will produce undesirable transients. Therefore, a large unused "discard time" must be inserted between dwells to allow for the clutter echo to fade out. The length of this wasted time is proportional to the transit time to the longest range clutter or target. In the lower elevation beam positions, it is not uncommon to have to discard up to 4 msec of data to provide for filter settling, although the exact number is dependent on the geometry and application.

Discard time is a problem for all the waveform categories and, indeed, is the primary motivation for using longer dwell times. However, its effect on MPRF is especially pronounced because of the multiple PRFs. If $N = 8$ is assumed and the discard time is assumed to be 2 msec per dwell, then the total wasted time is 16 msec per PRF cycle. This simple example illustrates the critical nature of the timeline in MPRF.

4.7.2 MPRF Ranging*

MPRF ranging is slightly different from RGHPRF ranging in that both range and velocity must be resolved simultaneously. To illustrate, let us chose the PRFs to facilitate range resolving, specifically, let the PRFs be multiples of a

*This section was co-authored with Tom Zahm.

basic range bin unit:

$$R_{u1} = m_1 c\tau/2$$

$$R_{u2} = m_2 c\tau/2$$

If it is desired to maintain a constant detection sensitivity, then the number of pulses coherently integrated must remain constant. Thus, the Doppler bin *size* changes as the PRF is changed. In addition, the number of Doppler bins formed may remain constant, for convenience of FFT processing, thus implying that the Doppler bin *spacing* may also change as the PRF is changed. In any event it is apparent that the Doppler resolving becomes much more difficult to implement. Hence there is a basic MPRF dilemma—chosing the waveform to facilitate range resolving versus chosing the waveform to facilitate Doppler resolving.

MPRF range and Doppler resolving is a hypothesis testing exercise that attempts to answer the question: *Is it likely that a certain target is in a given ambiguity zone?* This involves making a judgment as to whether the target's observed range and Doppler align within a predetermined tolerance after adding multiples of R_u and D_u on the different PRFs. Uniform bin size and bin spacing considerably simplifies the digital implementation of this hypothesis testing procedure, since the allotted tolerance can be kept constant and integer arithmetic can be applied during the alignment process. Thus, uniform range bin size is desired to facilitate range resolving, whereas uniform Doppler bin size is desired to facilitate Doppler resolving.

For maximum probability of successful resolving, uniform detection sensitivity is desired. This is especially true for current MPRF values since the PRFs can vary by an octave or more. Given this as a constraint, a choice must be made between nonuniform range bin size or nonuniform Doppler bin size. One or the other is kept uniformly quantized in order to reduce the overall complexity of the two tests when performed in series. For example, a common design is to chose one bin size to be uniform over the PRF set, either range or Doppler, and the remaining bin size is chosen to satisfy the desired uniform detection sensitivity. Upon reception, the target is first resolved in that parameter where the bin size is uniform. Since a majority of the false alarms are screened out by this preliminary test, the second (more difficult) test need consider only a reduced rate of detections.

Resolving the target in the presence of nonuniform bin size is typically done by requantizing to a uniform bin size upon reception (so-called pseudobins). Requantization is an attractive solution since, after requantization, the test can proceed in the same manner as for uniform bin size. The drawback, of course, is the introduction of additional quantization noise and hence a larger tolerance on the hypothesis test. For this reason it is typical to select the finest possible requantization level subject to processing limitations. A second disadvantage is the need to requantize the parameter for every hypothesis test made and, consequently, more processing involved for every target return and false alarm considered.

There are various alternative solutions to this dilemma, of which the following is interesting because it solves this particular problem as well as maximizes the duty factor. One possible solution, if it can be mechanized, is to maintain a uniform detection sensitivity by varying the pulse compression ratio, rather than the bin size. For instance, the pulse compression ratio may be chosen:

$$n_i = pm_i$$

where $\{n_i\}$ are the pulse compression ratios and $\{m_i\}$ are the number of range bins per PRF. This solution has the net effect of restoring a fixed duty factor since the ratio of the uncompressed pulse width $(pm_i)c\tau/2$ to range ambiguity zone $(R_u = m_i c\tau/2)$ is always a fixed fraction (p). Thus, the duty factor (p) is always constant, detection sensitivity can be kept uniform while transmitting a fixed coherent dwell time T, and so uniform range bin size and uniform Doppler bin size can be achieved simultaneously. A third advantage of a variable pulse compression ratio is that it reduces the amount of discard time by providing some additional protection against dwell spillover.

The disadvantage of this approach is that the PRF set that results from the available pulse compression ratios $\{n_i\}$ is not optimum with regard to blind zones and ghosts. For this reason a better approach may be to employ a hybrid method that exploits the advantages of both, and reduces the disadvantages of each. For instance, the PRF set may be initially designed assuming both uniform range bin size and uniform Doppler bin size. Then, the detection sensitivity is kept relatively constant within the confines of the available pulse compression ratios.

Whether this proposed solution is feasible or not is open to argument. In any case, the problem of ambiguity resolving in the presence of nonuniform bin size over the PRF set has traditionally been viewed as a difficult problem area.

4.7.3 Airborne MPRF Operation

MPRF was originally developed for use in airborne radars (Aronoff and Greenblatt, 1974). In fact, most modern tactical airborne radars employ HPRF to detect nose-aspect targets and 3/8 MPRF operation for detecting tail-aspect targets inside sidelobe clutter. Sometimes, these different waveforms are interleaved on a bar-by-bar basis or frame-by-frame basis. In other situations different aircraft within a given flight group (or squadron) employ different types of waveforms.

In this application the primary function of the MPRF waveform is to detect tail-aspect targets. Since the closing rate is small for tail-aspect targets, the target often falls within the first Doppler ambiguity zone, and so a high probability of detection exists. Nose-aspect targets can also be detected, but blind zones appear at certain range-velocities.

4.8 PULSE-BURST OPERATION

Pulse-burst operation is a mixture of LPRF and HPRF operation. The basic pulse-burst waveform is illustrated in Figure 4.18. In its simplest form pulse-burst operation can be described as follows. A long pulse of duration τ_B is transmitted called a *burst*. The burst is repeated at a low PRF, referred to as the burst repetition frequency (BRF).

Upon reception, the A/D converter samples the data at a rate sufficiently high to resolve all Doppler ambiguities (HPRF range sampling). The received samples are first sorted into blocks of data $c\tau_B/2$ long. This block sorting enables the target's range to be coarsely measured to within the length of the transmitted burst $c\tau_B/2$ (called burst-delay ranging). Next the samples are sorted according to Doppler by taking the FFT over the dwell of samples contained in each $c\tau_B/2$ block. This allows the target's velocity to be coarsely measured to within $1/\tau_B$ and also implies that a target will be separated from clutter if its Doppler separation from clutter is greater than $1/\tau_B$.

Finally, each range/Doppler bin is integrated over all the burst transmissions. This last step essentially amounts to forming "fine" Doppler bins. Potentially, it also allows some targets to be detected if $f_T < 1/\tau_B$, but blind zones appear just as in LPRF.

The coarse range bins, $c\tau_B/2$, imply that the competing clutter comes from only a single range bin. Thus, the shorter the burst length, the lower the level of competing clutter power. Hence, with pulse burst, the trade-off is one of short bursts for reduced clutter versus long bursts for improved SNR and reduced Doppler bin size.

Because the clutter is separated from the target on the basis of range as well as Doppler, clutter discrimination is better than in HPRF. In particular, pulse-burst operation does not suffer the problem of close-in sidelobe clutter competing with long-range target returns. Another advantage is that pulse-burst operation reduces the required A/D dynamic range by better isolation from nonlocal clutter.

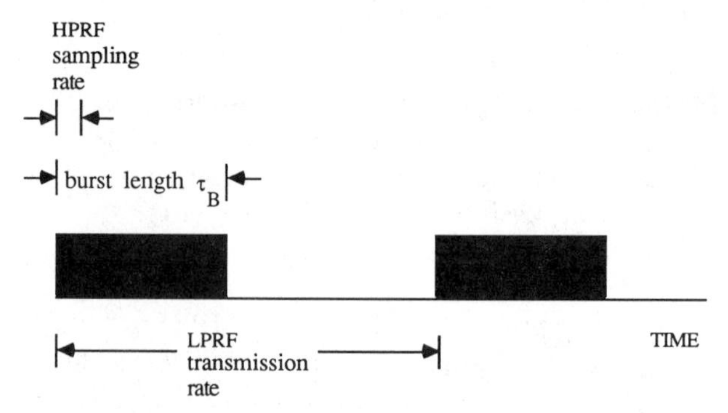

Figure 4.18 The simplest pulse-burst waveform consists of a single long uncoded pulse at the LPRF rate.

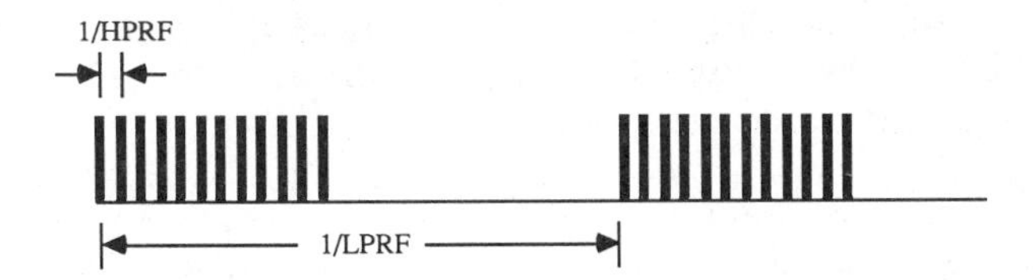

Figure 4.19 Interrupted pulse-burst operation consists of a combination of high and low PRFs.

4.8.1 Pulse-Burst Ranging

The essential low PRF nature of the pulse-burst waveform enables the target's range to be coarsely measured to within the length of the transmitted burst $c\tau_B/2$. However, finer range resolution is usually required—particularly if the burst length τ_B is long.

One possible approach is to employ FM ranging, which is simple to implement and can theoretically attain accuracies on the order of a few percent of the burst length. The disadvantage is the increased clutter smearing that results after ramping.

Coincidence ranging can also be employed. Coincidence ranging requires that a train of pulses be transmitted at the HPRF rate as depicted in Figure 4.19. By varying the intraburst PRF, the target's true range can be resolved. Likewise, the target's Doppler may be resolved by simultaneously varying the interburst PRF (BRF). Forming a train of pulses has the side benefit in that for short-range targets ($< c\tau_B/2$) eclipsing reverts to HPRF values. Hence, some target energy still leaks through, and so the problem of short-range target detection is partially solved.

Pulse compression may be applied to enhance the range resolution. The code length may be selected to match the intraburst HPRF as shown in Figure 4.20. This allows continuous transmission over the burst and avoids transmitter power interruptions. The penalty is the added eclipsing loss at short range.

4.8.2 Pulse-Burst Trade-offs

Pulse-burst waveforms offer a balanced combination of HPRF and LPRF features. The pulse-burst waveform, however, has some unique deficiencies of its own. One disadvantage is that the signal processing and storage requirements

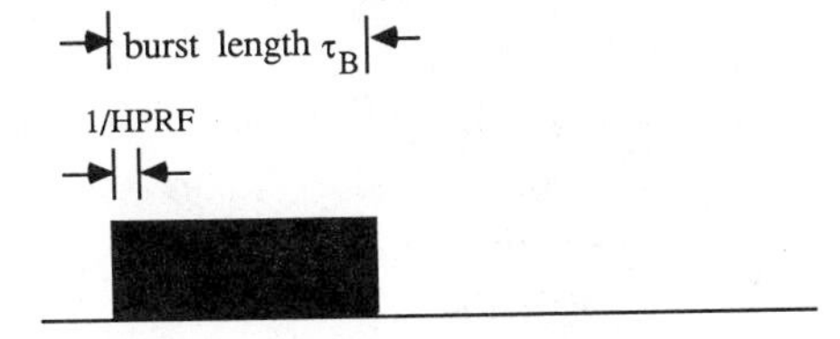

Figure 4.20 If pulse compression is employed, the code length may be selected to match the intraburst PRF to avoid power interruptions.

can become excessive. This objection has limited the prior use of pulse-burst operation, but it is largely overcome with modern methods of signal processing.

Another problem is the need to reject clutter transients. As shown in Figure 4.21, both the target and ground return are modulated in amplitude according to the shape of the transmitted burst. In addition, both may be shifted in time relative to the center of the coarse range gate. This situation can potentially cause clutter transients to occur after Fourier transformation.

Fortunately, the geometry is often such that this objection does not pose a serious problem. If the burst is long, and the target-to-clutter separation is small, then only a single narrow range swath is of interest. Therefore, the clutter transients can be conveniently rejected by discarding the leading and trailing pulses of the received pulse burst. This geometry is duplicated in many long-range applications including space-based radar.

In general, however, distributed clutter may be present. Therefore, it is usually not possible to discard enough pulses to make this approach feasible. It is technically feasible to amplitude weight (or attenuate) the pulse burst on transmit to reduce the transients, as shown in Figure 4.22, thereby matching the transmitted processing to the receive processing. Weighting on transmit to -20 to $-40\,\text{dB}$ sidelobes, corresponding to clutter rejection by -20 to $-40\,\text{dB}$, would provide adequate clutter rejection for most detection purposes.

If weighting on transmit is applied, the specific weights can be standard weights such Dolph–Chebyshev. However, for pulse-burst operation, the pertinent figure of merit is the *signal loss*, rather than amplitude weighting loss, since receiver noise is unaffected by weighting on transmit. Dolph–Chebyshev

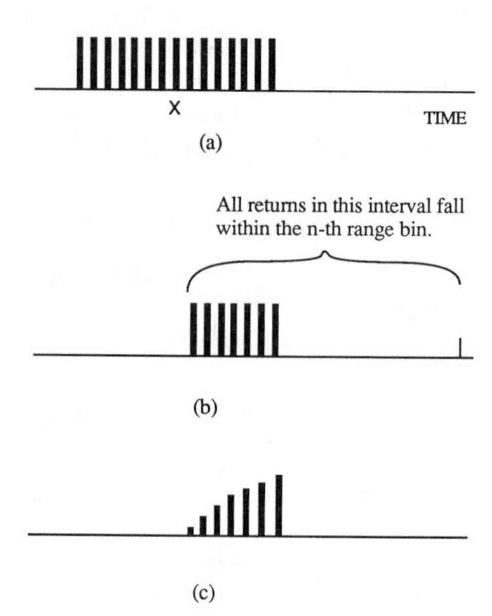

Figure 4.21 Potential for transient behavior after Doppler filter formation: (a) Return from clutter at range X. (b) Portion residing in nth range bin. (c) Return after weighting on receive.

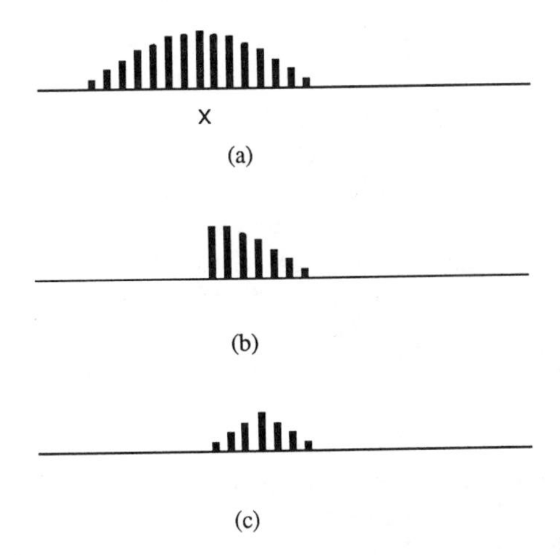

Figure 4.22 Weighting on transmit reduces transient behavior: (a) Return after weighting on transmit is applied. (b) Portion of return residing in nth range bin. (c) Return after weighting on receive.

weights with −40-dB sidelobes yield an average signal loss of 4.8 dB (see page 40). The signal loss associated with Dolph–Chebyshev weights is large because of the gradual Dolph–Chebyshev tapering. In other words the tapering must be very gradual to produce equiripple sidelobes. The amplitude weighting loss AWL is relatively unaffected since the tapering reduces both the signal and receiver noise simultaneously.

These comments seem to imply that the standard weights may not be suitable for pulse-burst operation. Other weights may be investigated with a less gradual taper. A less gradual taper results in less weighting loss, but also results in larger close-in sidelobes. Finally, equivalent weighting should be applied on receive to provide a symmetrical response to clutter.

The interested reader is referred to the literature (Spafford, 1968; Zeoli, 1971b) for a more rigorous approach to this topic. These references present computer search routines to derive the optimum weighting function for a generalized pulse-burst waveform.

4.9 NONCOHERENT PULSE OPERATION

If clutter is not a serious problem and velocity measurements are not required, the waveform design is freed from Doppler considerations altogether. Therefore, a noncoherent pulse (NCP) waveform can be employed. NCP may consist of a single pulse or a dwell of pulses. The essential nature of NCP, however, is that phase coherence need not be maintained and target Doppler is not sensed.

All early radar systems employed NCP operation. Such waveforms can be easily generated with a low-cost magnetron transmitter. However, since Doppler is not sensed, the usual problem occurred of clutter masking the target return.

Today, NCP is largely confined to those applications involving relatively benign clutter-to-signal ratios. An example is tracking aircraft targets beyond the earth's horizon or in a look-up situation (i.e., the antenna beam is looking up and, thus, supposedly free from clutter). Furthermore, some form of NCP must be used when operating against stationary targets or targets moving too slowly to be separated from clutter on the basis of Doppler processing alone. For example, airborne sea surveillance radars routinely use NCP when scanning off the velocity vector, and missile-based radars use NCP to detect stationary tanks.

Clutter discrimination can be enhanced by commanding different RFs between pulses. RF agility acts to statistically decorrelate the clutter returns from pulse to pulse and thus ensures that the PDI process helps the signal-to-clutter statistics in the same manner that it helps the signal-to-noise statistics. Circular polarization may also be applied to reduce the level of rain clutter backscatter (Schleher, 1986, p. 506; Cartledge and O'Donnell, 1977, p. 395). Finally, perhaps the most critical item in NCP is intelligent use of range processing (Emerson, 1954).

4.9.1 Military NCP

In military applications another well-known advantage of NCP operation is its relative immunity to spot jamming and deception jamming. LPRF and HPRF waveforms require both time and phase coherence. Therefore, a responsive spot jammer can measure the RF of the first transmitted pulse and center its jamming to spot jam the following pulses (Loppnov, 1983).

NCP has the advantage here in that RF agility can be used between consecutive pulses to deny the jammer RF information. Also, pseudorandom pulse staggering (PRS) can be commanded to deny the jammer PRF information. Finally, NCP can passively receive (sniff) before each transmission to test for the presence of jamming, and discard or spoof known jammed RFs in a pulse-by-pulse manner (Johnson and Stoner, 1976; Strappaveccia, 1987).

The goal of waveform selection is to choose a waveform that maximizes the signal-to-junk ratio, where "junk" refers to all noise contributors that compete with the signal. This suggests that a combination of waveforms may be appropriate. If detection is limited by clutter or noise, then the regular waveform

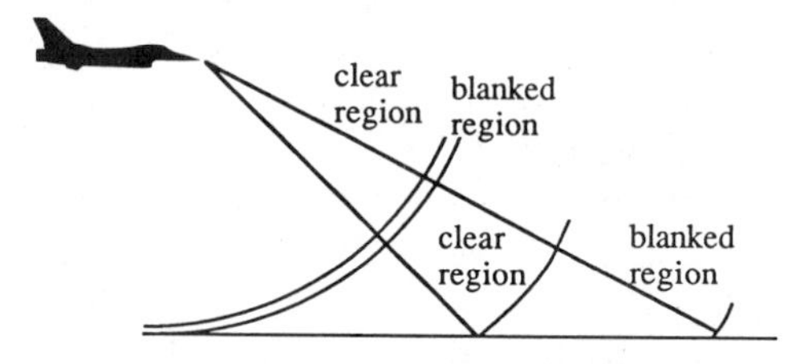

Figure 4.23 Airborne NCP has a limited look-down capability.

may be employed. If detection is limited by hostile interference, then NCP may be switched in. If both clutter and jamming are encountered, then RF agility may be employed between short coherent dwells.

Figure 4.23 illustrates an example of using NCP in an airborne radar looking down. As shown, the airborne radar has a limited ability to utilize NCP provided the target is at a high enough altitude.

4.10 COHERENT RF AGILITY

Coherent RF agility (CRFA) is an interleaved mixture of LPRF and NCP operation (Petrocchi et al., 1978). It is useful in those applications requiring coherent Doppler processing for clutter reasons, and RF agility for ECCM, PDI, MTAE suppression, or range resolution reasons. The drawback is the requirement for state-of-the-art exciter technology.

To set up a concrete example, consider the waveform shown in Figure 4.24. A single dwell is transmitted composed of a series of pulses. The RF is switched (hopped) before transmitting each pulse. The transmitted PRF is random for ECCM purposes but is roughly matched to the range swath of interest. This process is continued over N pulses and N frequencies. At the end of the Nth pulse the radar switches back to its original RF, transmits and receives another pulse on that RF, and the entire sequence repeats.

Two kinds of PRFs exist in this type of operation. The inner PRF is matched to the range swath of interest. The outer PRF (or the rate at which the RF is revisited) may be selected to suppress MTAE ambiguities or to ensure that enough RFs are visited to perform the function required.

Upon reception, all returns from the same RF are first coherently integrated together in the signal processor. Coherent integration allows coherent clutter rejection just as in LPRF. The only difference is that now there exist N returns for each range/Doppler output.

Noncoherent integration (PDI) over the N returns can be used to enhance detection of targets competing with mainlobe clutter. Coherent integration, on

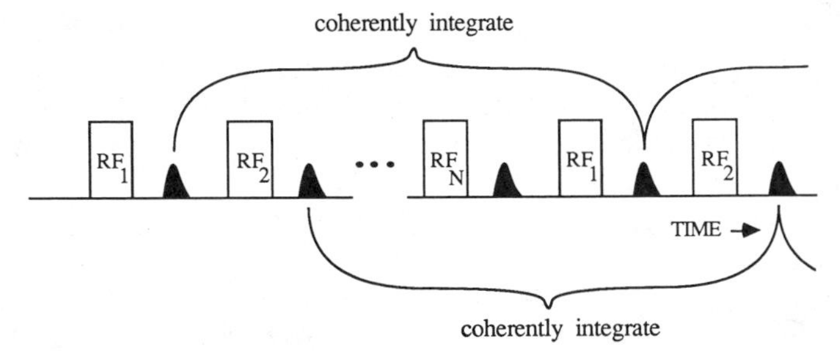

Figure 4.24 Coherent frequency hopping involves changing the RF frequency from pulse to pulse, and coherently integrating on each RF and between RFs.

the other hand, may be used to enhance the range resolution. Coherent integration begins by arranging the N returns into a linear sequence wherein the transmitted RF of each return differs from its neighbors by at most $1/(2\tau)$. An apparent pulse-to-pulse frequency shift is produced that is a linear function of the target's relative position within the pulsewidth (Currie and Brown, 1987, p. 631). FFT processing over the ensemble of N returns then provides fine range resolution.

Phase stability over the RF bandwidth is maintained by replacing the exciter with a frequency synthesizer. By synthesizing all the transmitted RFs from a common base frequency, the frequency synthesizer is able to maintain coherent phase relationships after frequency hopping (Manassewitsch, 1987).

4.11 SUMMARY

During target tracking the waveform parameters are typically matched to the target characteristics. In any search application the waveform parameters are typically dictated by the need to search a given amount of space in a given amount of time and the processing implications of that goal.

The generic PRF categories were discussed including low PRF (LPRF), medium PRF (MPRF), and high PRF (HPRF). LPRF is well suited for most ground-based applications. Some airborne systems use LPRF to detect aircraft targets, especially at the lower RFs, as well as to detect surface targets below X-band.

HPRF is better suited for most clutter-limited search applications, such as faced by an airborne tactical X-band radar. It is also better suited for some ground-based applications, particularly those requiring the detection of low-RCS aircraft targets near the horizon, in weather, or against competing sky noise. Nevertheless, due to the required dynamic range, HPRF will never achieve the long-range detection performance of the other waveform categories.

MPRF provides better detection in sidelobe clutter and relaxes hardware requirements relative to HPRF. MPRF is used on most modern airborne radars as an adjunct to HPRF. It is also used on a few ground-based systems, particularly at the higher RFs where it avoids some of the problems with sky noise.

Pulse-burst operation is theoretically possible at the current stage of radar technology. It is perhaps most appropriate for those applications where LPRF would normally be employed, but some additional Doppler capability is desired from a single pulse.

Noncoherent pulse (NCP) is primarily used in clutter-free conditions or to detect targets that stand out against the clutter background. It may also be used in a cued fashion for ECCM reasons. Circular polarization and RF agility are sometimes used to enhance clutter discrimination. However, perhaps the most critical item in NCP is intelligent use of range processing.

Coherent RF agility is an interleaved mixture of LPRF and NCP waveforms on a pulse-by-pulse basis. It is useful in those applications requiring coherent Doppler processing for clutter reasons, and a large RF bandwidth for ECCM, PDI, MTAE, or high range resolution reasons.

PROBLEMS

4.1 Calculate the blind speeds v_b and unambiguous range R_u given a PRF of 1 kHz. Assume a scan time of 5 sec, a beamwidth of 2°, and a transmitter frequency of 3 GHz.

4.2 Let the single-dwell R_{50} be equal to 75 percent of R_u. Calculate the single-dwell P_d at R_u by appropriate scaling. Is this P_d low enough to neglect in evaluating the potential for MTAEs?

4.3 Consider a radar operating at 5 GHz, a PRF of 500 Hz, and a Doppler filter width of 100 Hz. Calculate the blind speeds up to 325 km/hr. If a second PRF is employed equal to 600 Hz, will the target at 325 km/hr be detected?

4.4 Consider a radar scanning $\pm 45°$ in azimuth and 10° in elevation. Let the azimuth beamwidth be 2° and the total scan time be 4 sec. Calculate the resulting dwell time assuming 2:1 beam overlapping and 0.1 seconds turnaround time at the end of each bar.

4.5 Show that a target acceleration a_T perpendicular to the velocity vector V_T produces a turn rate $\omega_T = a_T/v_T$.

4.6 Consider a radar operating at 5 GHz, a PRF of 150 kHz, and a duty factor of 0.25. Calculate the range zones between 20 and 30 km that are totally uneclipsed. Assume it is desired to detect a target at 30 km under all conditions, and pick a second PRF to achieve this goal.

4.7 An X-band (10-GHz) STT radar is tracking an approaching target with velocity 300 km/hr. The nominal PRF is 2 kHz. Toggle this PRF value to place the target Doppler frequency at mid-PRF.

4.8 Consider a scanning ground-based LPRF radar with beamwidth of 2° and scanning rate of 50°/sec. Calculate the frequency bandwidth of clutter.

4.9 A pulsed police radar has a requirement to detect cars with maximum velocity 200 km/hr and range 10 km. Assume it is required that R_u be twice the maximum detection range. Select the highest RF such that the target's range and velocity are unambiguously measured.

4.10 Show that the width of mainlobe clutter σ_C in an airborne radar is invariant to RF given a constant antenna aperture diameter D.

4.11 An X-band (10-GHz) airborne radar transmits with a PRF of 150 kHz, duty factor of 0.25, and a Doppler filter bandwidth of 200 Hz. Assume a target radial velocity of 200 m/sec, ownship velocity of 300 m/sec, and an azimuth scan angle of 30°. Calculate (a) the center frequency of mainlobe clutter, (b) the boundaries of sidelobe clutter, (c) the frequency of the return from directly underneath the radar (so-called altitude return), and (d) the target frequency.

4.12 If you were a enemy aircraft illuminated by the radar in Problem 4.11, what trajectory would you fly, relative to the radar, to avoid detection?

4.13 In FM ranging there is a direct correspondence between range resolution and Doppler resolution. For a range resolution of 5 km and a Doppler resolution of 200 Hz, calculate the corresponding value of modulation slope. For a maximum detection range of 100 km, calculate the number of Doppler filters that must be discarded due to clutter spreading after FM ranging.

4.14 Certain short range warning radars are designed to sense unambiguous target range and Doppler. An example includes short-range missile warning radars (MWR) that are located in the tail section of military aircraft. Assuming $R_u = 10$ km, $V_T = 1500$ m/s, and negligible ownship velocity, calculate the highest RF and associated PRF for which the target's range and radial velocity are simultaneously unambiguous.

CHAPTER 5

PULSE COMPRESSION

Pulse compression is employed to achieve a short range resolution while simultaneously achieve a large duty factor. All pulse compression methods are matched filter operations where a long pulse is coded to have a wide bandwidth, and the received pulse is compressed in time by passing it through a matched filter. This matched filter may be located either in the signal processor (digital pulse compression) or before the A/D converter (analog pulse compression).

In principle, the coding can be achieved using either amplitude modulation (AM) or phase modulation (PM). In practice, however, phase modulation is far more compatible with conventional exciter/transmitter technology. Therefore, the term *pulse compression* is almost always taken to refer to some manner of phase modulation.

Applications that require large amounts of pulse compression are those characterized by:

1. A small range resolution requirement.
2. A low PRF.
3. Transmitters that are peak power limited such that large average amounts of power can be sustained but the instantaneous peak power is limited.
4. A detection requirement that cannot be met by matching the transmitted pulsewidth to the desired range resolution.

Thus, to demonstrate, early-warning LPRF radars may require large amounts of pulse compression, perhaps a ratio of 1000 or more. Even higher values may be found in space surveillance LPRF applications. Only moderate amounts of pulse compression is employed in present-day medium-range air traffic control (ATC) radars, which generally rely instead on a low-cost magnetron capable of high instantaneous peak power. Pulse compression is not used in conventional HPRF, and only moderate amounts are used in MPRF or RGHPRF.

The use of a traveling-wave tube, klystron, or solid state transmitter often implies the necessity for pulse compression. Pulse compression enables these types of transmitters to achieve performance comparable to a magnetron transmitter yet operate at a lower peak power. Also, the implementation of pulse compression is simplified since the phase modulation can be applied before the power amplification.

5.1 MAXIMUM DUTY FACTOR

The duty factor after pulse compression cannot be made arbitrarily large and indeed is ultimately limited by the range blind zones. Range blind zones occur because the receiver is shut down while the transmitter is active, and also because some range bins must be discarded to prevent excessive degradation to the sidelobes after pulse compression. For example, if one assumes for the sake of argument that 100 percent of the code must be received prior to pulse compression to prevent code sidelobe degradation, and one assumes a transmitter duty factor of 10 percent, i.e., $d_T = 0.1$, then 10 percent of both the beginning and ending range swath must be discarded to prevent sidelobe degradation. The net effect is to produce range blind zones comprising 20 percent of the range swath R_u.

Furthermore, the finite TR switching time can produce additional range blind zones which may be significant at the higher PRFs such as MPRF and RGHPRF. Finally, additional time may be deliberately thrown away to prevent targets that lie within a range blind zone from spilling over into the clear region after pulse compression. The overall effect is to produce range blind zones that must be eventually eliminated by additional PRF switching.

5.2 TIME–BANDWIDTH PRODUCT

Given a total transmitted pulse of length τ_u and a compressed pulse width τ, then from conservation of energy the total power gain due to pulse compression is given by

$$N_{pc} = \tau_u/\tau$$

N_{pc} is called the *pulse compression gain.* The compressed pulsewidth, τ, is approximately equal to the reciprocal of the radar's operational bandwidth B, that is, $\tau \cong 1/B$. Therefore, the gain becomes

$$N_{pc} \cong \tau_u B \tag{5.1}$$

The factor $\tau_u B$ is called the *time–bandwidth product.*

The pulse compression gain is a function only of the time–bandwidth product of the radar, hence, it is invariant to the actual code selected. The duty factor d_T is only a function of the uncompressed pulselength $d_T = \tau_u$ PRF. The range resolution is only a function of the radar's operational bandwidth B. In selecting a particular pulse compression code over another, therefore, the primary issues to address are:

1. The code's sidelobe performance and any associated loss due to mismatched filtering.
2. The ECM performance of the code.
3. Complexity of implementation.
4. Variety of codes available.

No code performs best in all three performance classes. This trade-off gives rise to the study of pulse compression.

Several commonly known ways to accomplish pulse compression are discussed next, including chirp, digital chirp, Barker codes, compound Barker codes, pseudo-noise codes, ripple suppressed codes, Frank codes, and complementary codes. Figure 5.1 surveys some of the tradeoffs between these codes.

- Chirp
  most common type of code
  easy analog implementation
  doppler tolerant
  suitable to high pulse compression ratios
  easy to suppress sidelobes, requires weighting loss
  Frank, P3 and P4 are digital versions of chirp.

- Binary phase codes (Barker, compound Barker, PN)
  second most common type of code
  easy digital implementation, apply separately to I/Q channels
  doppler sensitive
  Barker codes have good sidelobes for low comp. ratios.
  further sidelobe suppression possible, but difficult
  variety of codes available.

- Low sidelobe codes (ripple suppressed, complementary, Frank)
  requires more processing and costs
  doppler sensitive
  complementary codes reduce the amount of doppler clear region
  Frank codes require more complexity in the phase encoder.

Figure 5.1 Survey of pulse compression trade-offs.

5.3 CHIRP

Chirp is the oldest and best known type of pulse compression code. With chirp, as shown in Figure 5.2, the radio frequency is linearly increased throughout the transmitted pulse (linear in frequency, quadratic in phase). The term *chirp* arises because of its parallel to the linear chirping of a bird.

Consider the following matched filter equation (Klauder et al. 1960)

$$\mathbf{S}_0(t) = \int \mathbf{S}_i(r)\mathbf{h}^*(t - r)\, dr \tag{5.2}$$

$\mathbf{S}_0$ is the compressed pulse, $\mathbf{S}_i$ is the received signal, and $\mathbf{h}(t)$ is the matched filter's impulse response. In the case of chirp the signals are swept in frequency, so the transmitted signal is of the form

$$e^{j(\omega_{\mathrm{RF}}t + \mu t^2/2)}p(t) \tag{5.3}$$

for μ the FM slope constant ($\mu = 2\pi B/\tau_u$, B = total frequency excursion, and τ_u = uncompressed pulsewidth); $p(t)$ is the pulsewidth function that models the transmitted signal as being zero outside the pulse width τ_u.

The matched filter response at IF, $\mathbf{h}^*(t)$, is the time reverse of the transmitted signal

$$\mathbf{h}(t) = e^{j(-\omega_{\mathrm{IF}}t + \mu t^2/2)}p(-t) \tag{5.4}$$

The received IF signal will be the same as the transmitted signal except shifted in frequency proportional to the Doppler shift, $\omega_T = 2\pi f_T$,

$$\mathbf{S}_i(t) = e^{j[(\omega_{\mathrm{IF}} - \omega_T)t\, +\, \mu t^2/2]}p(t) \tag{5.5}$$

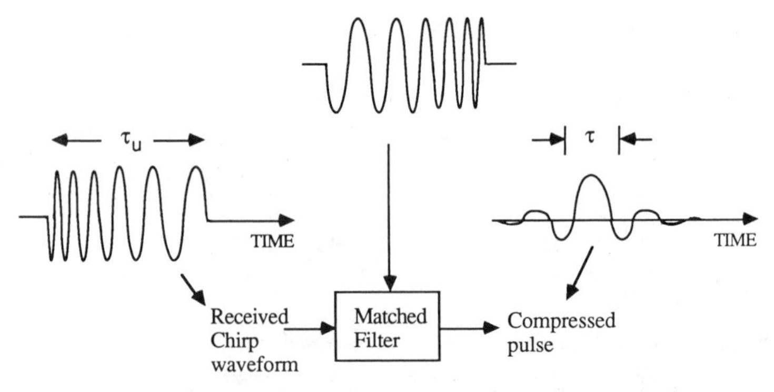

Figure 5.2 Chirp comprises a sweep frequency waveform. The return is passed through a matched filter, and the subsequent output is the compressed pulse.

The output of the convolution integral is

$$S_0(t) = \int_{-\tau_u/2+t}^{\tau_u/2} e^{j[(\omega_{IF}-\omega_T)r+\mu r^2/2]} e^{j[\omega_{IF}(t-r)-\mu(t-r)^2/2]} dr$$

for positive t. By eliminating common terms, and inserting $\mu\tau_u = 2\pi B$, this is straightforwardly reduced to

$$S_0(t) = (\tau_u - t)e^{j(\omega_{IF}-\omega_T-\mu t/2)t} \operatorname{sinc}[(f_T + Bt/\tau_u)(\tau_u - t)]$$

where $\operatorname{sinc}(x) \triangleq \sin(\pi x)/(\pi x)$. Thus, the unweighted response is essentially governed by the sinc( ) function and has (approximate) compressed pulsewidth $1/B$.

5.3.1 Doppler Sensitivity

The peak value after matched filtering occurs at

$$t = -\tau_u f_T/B \tag{5.6}$$

where f_T is the Doppler frequency shift and B is the bandwidth. Thus, the Doppler shift does not change the basic nature of the received signal—it only introduces a slight time delay proportional to the ratio $\tau_u f_T/B$. If the Doppler shift f_T is a fraction of the bandwidth B (as needed to remain inside the receiver's passband), then the delay is a fraction of the pulse width τ_u.

For longer pulse widths τ_u or smaller bandwidths B, it may be necessary to model this time–Doppler relationship in the Kalman filter (Fitzgerald, 1974a). Another procedure is to alternate up-ramps with down-ramps and so resolve the ambiguity in a manner similar to FM ranging. But, in general, Doppler only begins to have a significant impact whenever the Doppler becomes large enough to push the received signal outside of the receiver's passband B.

5.3.2 Time Sidelobes and Weighting

The compressed pulse of the unweighted chirp has a characteristic sinc( ) shape, with resultant time sidelobes having a peak of -13.2 dB. These large sidelobes are often objectionable since a large target can mask a neighboring small target.

Time sidelobes can be reduced by amplitude weighting on receive (Klauder et al., 1960; Cook and Bernfeld, 1967, Chapter 7). The applied weighting is of the same general type as described in Table 3.1, the only difference is the need to replace Dolph–Chebyshev weights with Taylor weights. Taylor weights are an approximation to the Dolph–Chebyshev weights that is realizable for continuous signals. The cost of amplitude weighting on receive is the degraded resolution and increased SNR loss.

The time sidelobes can also be reduced by phase weighting on both transmit and receive (Key et al., 1959a; Johnston and Fairhead, 1986; Skolnik, 1970, p. 20-15). This approach, called *nonlinear FM*, avoids the mismatching SNR loss associated with the previous method.

5.3.3 Conclusion to Chirp

Chirp is the most common type of pulse compression code as well as being the best for many applications. It is fairly simple to generate and compress using IF analog techniques, for example, surface acoustic wave (SAW) devices (Brookner, 1977, Chapter 12). Another advantage of chirp is that large pulse compression ratios can be achieved. Typical time–bandwidth products are 50–300, although values as large as 10^5 have been attained (Brookner, 1977, p. 175). In addition, chirp is relatively insensitive to uncompensated Doppler shifts and can be easily weighted for sidelobe reduction.

The analog nature of chirp sometimes limits its flexibility. Analog technology cannot as easily be programmed to support multiple range resolutions, pulse compression ratios, or codes using common hardware. In terms of today's technology, a separate matched filter (mounted on a common substrate) is usually required for each code and range resolution required. The length of the code can sometimes be changed (and hence the pulse compression ratio) by electronically selecting from a set of transducers spaced along the SAW device (Matthews, 1977, pp. 460–463). General purpose SAW convolvers are also being developed that may in the future provide more flexibility (Brookner, 1977, p. 186). Nevertheless, any analog device is inherently not as competitive as digital technology when programmability is considered.

The very predictability of chirp makes it a poor choice for ECCM purposes. Chirp represents a well-known code that changes its phase history in a regular, continuous, and predictable fashion. In addition, it is easily detected by ESM receivers since—although its total frequency excursion B may be large—its instantaneous bandwidth is small.

5.4 BINARY PHASE CODES

Binary coding is the second most common type of pulse compression code, and the most common type of digital code. With binary coding the transmitted pulse length is divided into N_{pc} segments each of duration τ. Thus, the actual transmitted pulse length is of duration $\tau_u = N_{pc}\tau$. The phase of each segment is switched between two binary values, 0 or π radians, in accordance with a predetermined code sequence. This is graphically portrayed in Figure 5.3.

Upon reception, the signals are compressed in time by passing it through a matched filter. If there are N_{pc} segments per transmitted pulse, all of width τ, then ideally the total system response will be τ seconds long, with an amplitude N_{pc} times greater than the amplitude of the received signal. The actual output consists of a central spike plus spurious sidelobes on either side of the peak.

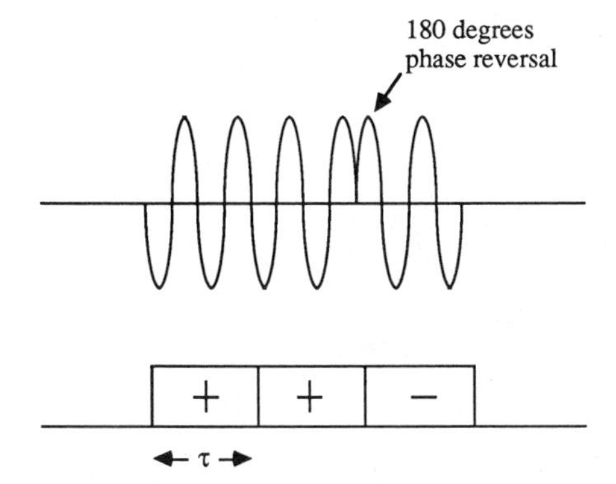

Figure 5.3 Example of binary phase coding. Transmitted pulse is divided into N_{pc} discrete segments (here, $N_{pc} = 3$) and certain segments have their phase reversed by 180°.

Binary phase codes are easy to implement with digital techniques. The matched filter can be conveniently broken up into separate I/Q parts, and the encoder accuracy is reduced relative to digital chirp. Binary codes have also been implemented by analog means (e.g., SAW-tapped delay lines). The advantage of the analog approach is that it is sometimes more economic, particularly if the required number of codes is small.

5.5 BARKER CODES

Barker codes are the most frequently used binary code, primarily because they achieve sidelobes that are small relative to other binary codes. The known list of Barker codes is given in Table 5.1, where (+) and (−) is shorthand for 0 and π radians phase, respectively. Note that plus and minus signs can be reversed; for example, + + − can be changed to − − +, and the ordering of the digits can be reversed, for example, + + − changed to − + +.

All the Barker codes have the desirable property of producing constant sidelobes, with each sidelobe having an amplitude no greater than the long pulse. This is shown in Figure 5.4 for the Barker code of length 3. For the Barker code of length 13, which is the longest Barker code, this leads to a sidelobe level −22.3 dB [$=20\log(\frac{1}{13})$] below the peak. The associated pulse compression gain is 11.1 dB [$=10\log(13)$].

TABLE 5.1 List of Barker codes.

N_{pc}	CODE ELEMENTS	PEAK SIDELOBE, DB
2	+ -, + +	-6.0
3	+ + -	-9.5
4	+ + - +, + + + -	-12.0
5	+ + + - +	-14.0
7	+ + + - - + -	-17.0
11	+ + + - - - + - - + -	-20.8
13	+ + + + + - - + + - + - +	-22.3

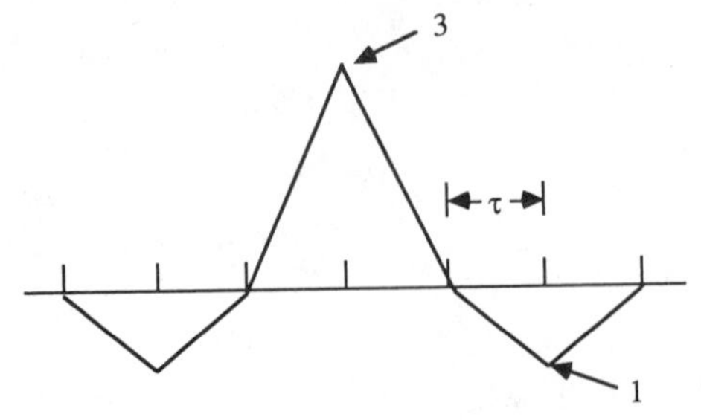

Figure 5.4 Matched filter output for Barker code of length 3.

As already stated, the longest Barker code is of length 13. When more gain is required, alternative codes must be investigated. One alternative, as illustrated in Figure 5.5, is to form a compound Barker code by concatenating two or more Barker codes. Compound Barker codes can be constructed to have a very large pulse compression gain, but unfortunately the peak sidelobe levels are not proportionally decreased (as shown in Figure 5.6).

The 169 compound Barker code is frequently used (a 13 Barker code inside a 13 Barker code) because it offers the longest code from a single concatenation. The 169 compound Barker code has a peak sidelobe ratio of −22.3 dB and a pulse compression gain of 22.3 dB.

The primary reason for the widespread use of compound Barker codes is their ease of digital implementation. In particular, the processing for the entire code can be conveniently decomposed into independent processing of the inner code followed by processing the outer code. This implies that the number of operations performed increases as the sum of the constituent codes, as opposed to the product (Eaves and Reedy, 1987, p. 482).

Finally, compound Barker codes exhibit the desirable property that only a few of the sidelobe peaks are large. This property makes it easier to perform sidelobe blanking or ripple suppression after pulse compression.

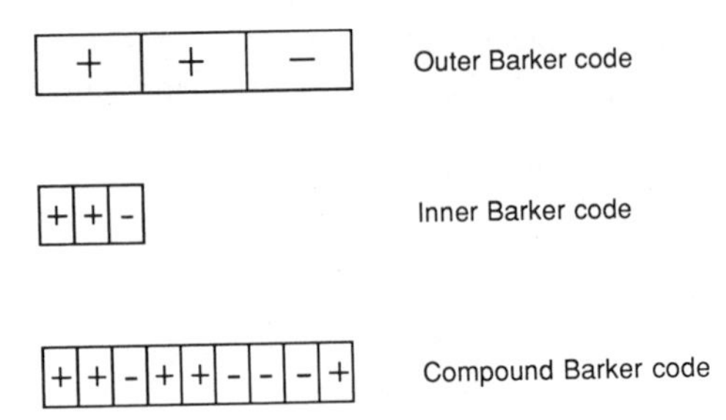

Figure 5.5 Longer codes may be formed by concatenating one Barker code within another (here, we show an inner Barker code of length 3 and an outer Barker code of length 3).

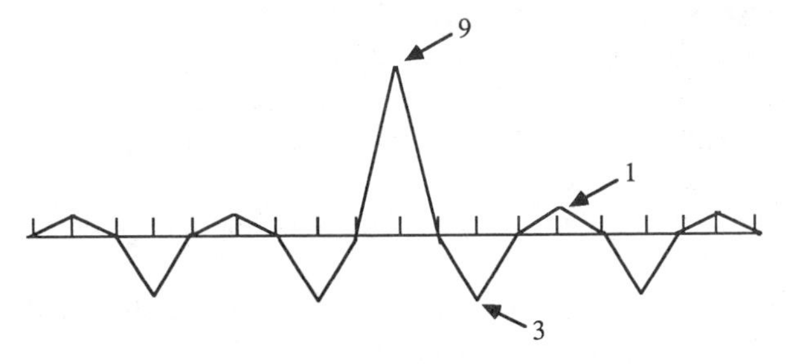

Figure 5.6 Matched filter output of the 3/3 compound Barker code.

5.5.1 Conclusion to Barker Codes

The principal limitation of the Barker codes is their short length. Barker codes longer than 13:1 have not been found. Compound Barker codes can provide more gain, but the sidelobe levels are not proportionally reduced. This limits the peak sidelobe level to −22.3 dB, which may not be satisfactory in a dense target environment.

Another limitation is their sensitivity to uncontrolled target Doppler frequency shifts and exciter phase drift (Cook and Bernfield, 1967). Any uncontrolled phase drift will produce spurious sidelobes of average magnitude below the peak:

$$\text{Average sidelobes} \cong 10 \log[\Delta\psi^2/(12N_{pc})] \tag{5.7}$$

for $\Delta\psi = 2\pi f_T \tau_u$. Assuming −22.3-dB average sidelobes is the important criterion and $N_{pc} = 13$, then (5.7) limits $f_T \tau_u$ to less than 0.15.

At the higher RFs, such as C-band and above, an attempt is usually made to correct as much of the Doppler shift as possible before passing it through the matched filter. In STT, it is typical to apply a Doppler correction near the second IF by commanding a frequency offset equal to the target's Doppler estimate. The required Doppler estimate is obtained from the Doppler filter (in HPRF) or the range-rate filter (in LPRF and NCP).

Search sometimes uses a process of *Doppler tuning* (Cook and Bernfeld, 1967, p. 298) which amounts to using a bank of matched filters. A typical procedure is to divide the compressing into two or three subregions and then tune the matched filter in each subregion to a different Doppler frequency. In the limit of a large number of subdivisions, this procedure can be shown to be equivalent to performing the pulse compressing after the Doppler filtering.

In summary, for low pulse compression ratios (such as 13:1 or 169:1) where the full Doppler merits of the chirp code may not be warranted, Barker codes provide the merits of better suitability to digital processing over a large range swath. Digital coding also implies that a variety of codes can be generated for ECCM purposes or to provide protection against MTAEs. However, at the higher RFs, the Doppler sensitivity of the Barker codes may pose a problem.

5.6 PSEUDO-NOISE SEQUENCES

Pseudo-noise (PN) codes are binary-valued sequences similar to Barker codes. The name *pseudo-noise* stems from the fact that they resemble a noiselike sequence of coin tossings for which $+1$ represents a head and -1 represents a tail. Hence, they are ideal for those applications where randomness is a desired property.

Another advantage of PN codes is that they can be easily generated using feedback shift registers. Indeed, one type of PN code, called *maximum-length codes*, possess the longest code length for this manner of generation. Figure 5.7 and Table 5.2 illustrates the generation of a maximum-length PN code of length $N_{pc} = 7$, which is accomplished with the use of a shift register of length 3. After each shift of the contents of the shift register to the right, the contents of the second and third stages are used to produce an input to the first stage through an exclusive-or (XOR) operation (i.e., binary add without carry). In general, by using an N-stage shift register with proper feedback connections, maximum-length PN sequences of length $N_{pc} = 2^N - 1$ may be obtained.

The specific case $N = 3$ is interesting since the resulting PN sequence can be identified as the Barker code of length 7. Therefore, the sidelobes are the same as the Barker code, namely -16.9 dB. For larger values of N, however, the sidelobes are not proportionally decreased. Indeed, for large N the sidelobes approximately converge to $1/N_{pc}$ (relative to the peak) as compared to $1/N_{pc}^2$ for Barker codes.

It can be shown that maximum-length PN codes exist for every shift register of length N. The feedback connections that result in maximum-length sequences have been tabulated in the literature (Braasch and Erteza, 1966; Skolnik, 1970, p. 20-20; Dixon, 1976, p. 81) and are reproduced in Table 5.3. The use of this table is illustrated by means of the following example. Consider the case $N = 3$, for which the maximum length is 7. Table 5.3 lists [3, 1], which corresponds to the connection Register #3 XOR Register #1. In addition, all mirror images of this connection produce maximum-length sequences. The mirror image of $[n_1, n_2, n_3, \ldots]$ is given by $[n_1, n_1 - n_2, n_1 - n_3, \ldots]$. That is, the mirror image of [3, 1] would be [3, 2]. Furthermore, different initial conditions produce different PN sequences after truncation (Cook and Bernfeld, 1967, p. 253).

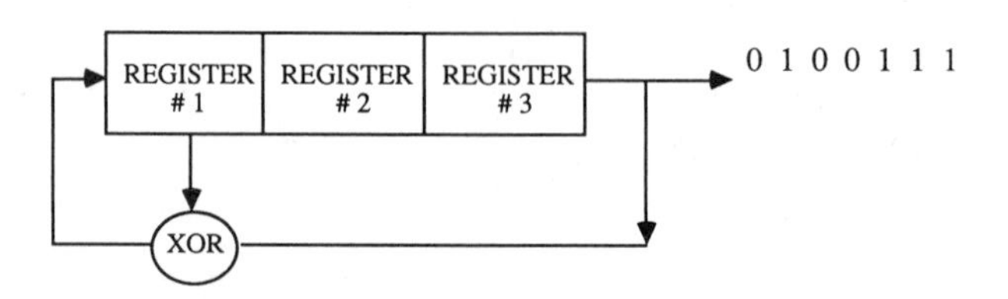

Figure 5.7 Generation of 7-bit PN sequence.

TABLE 5.2 Example of shift register contents: (a) XOR truth table. (b) Generation of 7-bit PN sequence.

INPUT A	INPUT B	XOR OUTPUT
1	1	0
1	0	1
0	1	1
0	0	0

(a)

REGISTER # 1	REGISTER # 2	REGISTER # 3
0	1	0
0	0	1
1	0	0
1	1	0
1	1	1
0	1	1
1	0	1
0	1	0

(b)

TABLE 5.3 Feedback connections for maximum length PN sequences.

N	Feedback Connections
2	[2,1]
3	[3,1]
4	[4,1]
5	[5,2] [5,4,3,2] [5,4,2,1]
6	[6,1] [6,5,2,1] [6,5,3,2]
7	[7,1] [7,3] [7,3,2,1] [7,4,3,2]
8	[7,6,4,2] [7,6,3,1] [7,6,5,2] [7,6,5,4,2,1] [7,5,4,3,2,1]
9	[9,4] [9,6,4,3] [9,8,5,4] [9,8,4,1] [9,5,3,2] [9,8,6,5] [9,8,7,2] [9,6,5,4,2,1] [9,7,6,4,3,1] [9,8,7,6,5,3]
10	[10,3] [10,8,3,2] [10,4,3,1] [10,8,5,1] [10,8,5,4] [10,9,4,1] [10,8,4,3] [10,5,3,2] [10,5,2,1] [10,9,4,2]
11	[11,1] [11,8,5,2] [11,7,3,2] [11,5,3,2] [11,10,3,2] [11,6,5,1] [11,5,3,1] [11,9,4,1] [11,8,6,2] [11,9,8,3].

5.6.1 Conclusion to PN Codes

PN codes possess the desirable property that longer codes can be generated and the sidelobes eventually reduce. Conversely, the primary drawback of PN codes is that the pulse compression ratio N_{pc} is typically not large enough to achieve the required peak sidelobe ratio and Doppler insensitivity (Eaves and Reedy, 1987, p. 486).

PN codes are less susceptible to sidelobe degradation in the presence of small Doppler frequency shifts. This can be seen from (5.7) since the average degradation due to Doppler shifts approximately reduce as $1/N_{pc}$ for a fixed pulsewidth τ_u. In fact, Doppler shifts only begin to have a significant impact whenever they begin to reduce the peak signal at the match point (which occurs whenever $f_T \tau_u > 0.15$).

Another advantage of PN codes is that cyclic codes can be easily generated by letting the shift register run on. Indeed PN codes form near perfect codes for periodic sequences. This property may be of interest in future applications where periodic pulse-burst codes are desired.

In military applications a desirable feature of the PN codes is that they appear noiselike and so blend into the background. However, as with all binary codes, PN codes are susceptible to a type of ESM technique called *frequency doubling*. Frequency doubling amounts to squaring the signal in the ESM receiver before integration, thereby stripping off the code from the pulse. Use of the frequency doubler strongly favors some manner of polyphase coding for military applications (Johnson, 1977; Carlson, 1988). An example is using a PN sequence to drive the I/Q settings of a polyphase encoder. This would solve the frequency doubler problem at the expense of more processing costs and, as with all PN codes, sidelobes that approximately converge to $1/N_{pc}$.

5.7 RIPPLE-SUPPRESSED CODES

When very low sidelobes are required, one approach is to employ conventional digital coding on transmit, but augment the receive processing with some manner of ripple suppression. *Ripple suppression* has been variously called *inverse filtering, equalization,* or *signal restoration.* Ripple suppressed codes are described next since they follow easily from our prior discussion on binary codes.

The theory behind ripple-suppressed codes is best described in the frequency domain. Let us denote the received spectrum of the backscattered signal as

$$\mathbf{S}_0(\omega) = \mathbf{S}_i(\omega)\mathbf{H}(\omega) \tag{5.8}$$

where $\mathbf{S}_i(\omega)$ is the Fourier transform of the received voltage return and $H(\omega)$ is the Fourier transform of the receiver's matched filter. If a Barker code is transmitted, for example, then $\mathbf{S}_i(\omega)$ would represent the Barker spectrum.

Assuming no Doppler degradations in transit, then $\mathbf{S}_i(\omega) = \mathbf{H}^*(\omega)$, and so

$$\mathbf{S}_0(\omega) = |\mathbf{H}(\omega)|^2 \tag{5.9}$$

Due to the discrete nature of digital processing, $|\mathbf{H}(\omega)|^2$ will possess the following z-transform format:

$$|\mathbf{H}(\omega)|^2 = 1 + ae^{j\omega\tau} + \cdots$$

In the time domain the corresponding output would be $1 + a\delta(t - \tau) + \cdots$. Thus the "ripples" in the frequency domain are equivalent to "sidelobes" in the time domain. This equivalence is graphically portrayed in Figure 5.8.

If the received waveform is passed through a second filter, whose frequency characteristic is $1/|\mathbf{H}(\omega)|^2$, then the output would be the desired uniform impulse response spectrum (Nathanson, 1969, p. 489). In other words all the sidelobes would be suppressed to zero. This mathematically describes the concept of ripple suppression.

Ripple suppression has also been described in terms of a Fourier series expansion (Rihaczek and Golden, 1971), a transversal filter implementation (Lathi, 1983, p. 205; Key et al., 1959b), and a minimum peak sidelobe implementation (Zoraster, 1980). At any rate the first 30 terms are shown in Table 5.4 for the case of the 13-bit Barker code (after Rihaczek and Golden, 1971).

5.7.1 Conclusion to Ripple-Suppressed Codes

Ripple suppression represents a convenient way to reduce the receive sidelobe level without affecting the transmitted processing. Ultimately, however, the sidelobe performance after ripple suppression is still (roughly) governed by $10\log(\Delta\psi^2/12N_{pc})$ due to the presence of Doppler-type phase errors. Therefore, the merits of reducing the "theoretical" sidelobe level far below this Doppler-controlled sidelobe floor is questionable.

Another perceived objection to ripple suppression is that the mismatch filter's impulse response, $\mathbf{H}(\omega)/|\mathbf{H}(\omega)|^2$, can in theory be very long in the time domain. Since $|\mathbf{H}(\omega)|^2$ is of the form $|\mathbf{H}(\omega)|^2 = 1 + ae^{j\omega\tau}$, its inverse $1/|\mathbf{H}(\omega)|^2$ is essentially an infinite series expansion. Thus, in the time domain, the receiver's impulse response is theoretically infinite. This restricts the minimum target range at which ripple suppression can be successfully applied.

To circumvent this problem, a truncated version of $\mathbf{H}(\omega)/|\mathbf{H}(\omega)|^2$ may be formulated to meet the required minimum detection range. The truncated

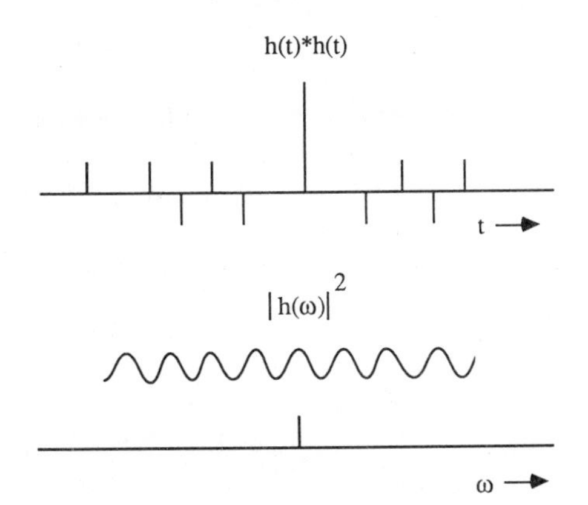

Figure 5.8 Overall transfer function after pulse compression, illustrating undesirable "ripple" effect.

Table 5.4 Coefficients for ripple suppression of 13-bit Barker code. The odd terms are zero.

Index	Value
0	1.0000
2	-0.0383
4	-0.0448
6	-0.0506
8	-0.0556
10	-0.0598
12	-0.0630
14	0.0207
16	0.0178
18	0.0142
20	0.0100
22	0.0054
24	0.00035
26	-0.0050
28	-0.0033
30	-0.0018

response is typically two or three times the response of the original code. One way to visualize this is to consider keeping only the lower-order terms in a Taylor series expansion.

Another well-known problem with ripple suppression relates to its basic incompatibility with the automatic gain control (AGC). All matched filter techniques assume time invariance. However, if a time-varying AGC is inserted prior to the A/D, this assumption is not valid. Thus, all digital pulse compression techniques are basically incompatible with the use of time-varying AGC. This incompatibility is compounded if small sidelobes are sought since small sidelobes are more sensitive to the invariance assumption. If the total AGC amplitude variation across the pulse is 5 percent, for example, then the lowest sidelobe attainable is equal to 43 dB.

Theoretically speaking it is possible to "undo" the adverse effects of the AGC by removing the time variance immediately after the A/D. However, in practice, this procedure encounters some difficulties. First, the AGC is not a single operation; rather, it consists of a time-varying attenuation, video filtering, and then A/D quantization. Therefore, to undo the effects of the AGC would amount to undoing the effects of the A/D, undoing the effects of the video filtering, and finally undoing the effects of the AGC. The middle step in particular is very difficult to accomplish in practice.

5.8 FRANK CODES

Frank codes are a family of polyphase codes that are closely related to the chirp and Barker codes. In fact, Frank codes may be generally described by stating that the shorter Frank codes somewhat resemble Barker codes, and the longer Frank codes somewhat resemble the chirp code.

Frank codes are implemented as follows (Frank, 1963). The number of phase increments to be used, N, is first established. The range resolution, τ, is found by subdividing the total pulse length τ_u by the square of the number of phase increments N (i.e., $\tau_u/N^2 = \tau$). Within each segment of length $N\tau$, the frequency is held constant. However, from segment to segment, the frequency is increased in a piecewise linear fashion. This produces N frequencies or "tones."

Mathematically this is written

$$f_j = (j - 1)/(N\tau) \qquad 1 \leqslant j \leqslant N \tag{5.10}$$

such that the total frequency excursion over the N segments is equal to $(N - 1)/(N\tau) \cong 1/\tau \cong B$. The phase of the ith element in the jth segment is thus

$$\begin{aligned} \beta_{ij} &= 2\pi \int_0^{(i-1)\tau} f_j \, dt \\ &= (2\pi/N)(i - 1)(j - 1) \qquad 1 \leqslant i \leqslant N \qquad 1 \leqslant j \leqslant N \\ &= \delta(i - 1)(j - 1) \end{aligned} \tag{5.11}$$

where $\delta = 2\pi/N$ is the fundamental phase increment.

In the case of the four-phase Frank code, for example, $N = 4$ and $\delta = 360°/4 = 90°$, making the phases 0°, 90°, 180°, and 270°. The code consists of 4 segments of 4 elements—a total of 16 elements. The transmitted signal looks like Figure 5.9.

The analogy between the Frank codes and the chirp and Barker codes may be seen by comparing phase histories. Figure 5.9 compares the 16:1 Frank code to an equivalent chirp code. Figure 5.10 compares the 16:1 Frank code to an equivalent Barker code. A Barker code of length 16 does not exist, but a Barker code of length 13 suffices to show the general trend.

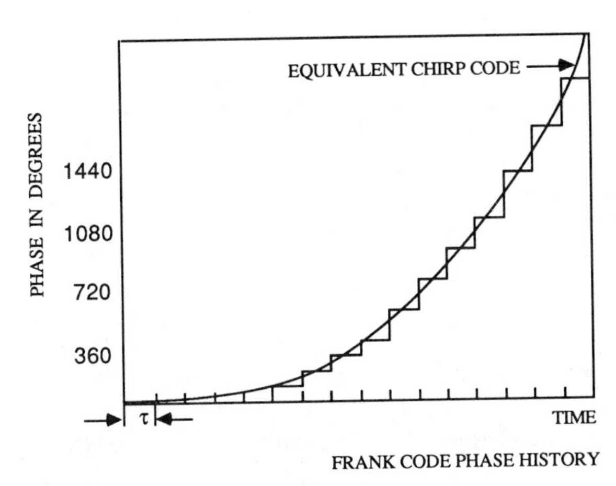

Figure 5.9 Phase history of 16:1 Frank polyphase code and chirp code.

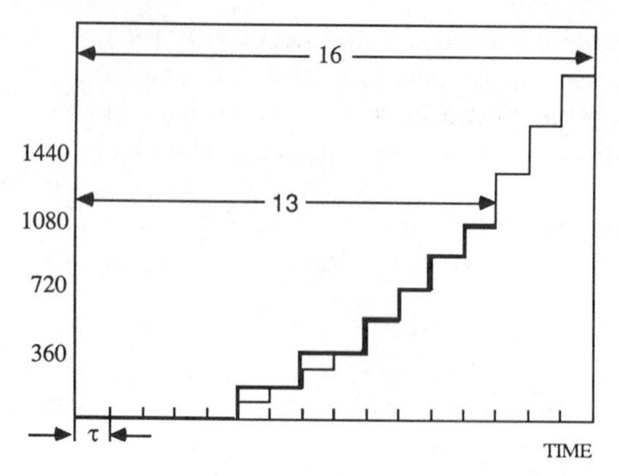

Figure 5.10 Phase history of 16 : 1 Frank polyphase code and 13 : 1 Barker code.

Given the strong similarity between the Frank and chirp codes, it is reasonable to expect that the sidelobe performance of the Frank code is similar to the sidelobe performance of the chirp code. In fact, however, Frank codes achieve superior sidelobe performance (assuming no Doppler frequency shifts). From (Cook and Bernfeld, 1967, p. 261), the sidelobe performance is (approximately) given by

$$\text{Peak sidelobe ratio} = 10 \log(\pi^2 N_{pc}) \tag{5.12}$$

for large N, where $N_{pc} = N^2$. This sidelobe performance may be compared to the value $10 \log(N_{pc})$ for PN codes and $10 \log(N_{pc}^2)$ for Barker codes. The actual peak sidelobe ratios for selected Frank codes are as follows:

N	Sidelobe Ratio
3	19.08
4	21.16
5	23.88
6	25.11
7	26.76
8	27.82

5.8.1 Conclusions to Frank Codes

Frank codes are basically a digital version of chirp that are (relatively) easier to implement using near-term technology. An additional benefit is that the time sidelobes are smaller relative to the unweighted chirp. However, Frank codes do not achieve the Doppler tolerance of the chirp code, although they do perform slightly better than the Barker codes. To be specific, the Frank codes are relatively insensitive to Doppler frequency shifts provided $f_T \tau_u < 0.3$ (Cook and Bernfeld, 1967, p. 262; Kretschmer et al., 1983, p. 527; Di Vito et al., 1985, p. 189, Figure 4). For $f_T \tau_u > 0.3$, secondary peaks occur, the main peak cycles through 4 dB for every 0.5 increase in $f_T \tau_u$, and, for large Doppler shifts, a large image lobe appears such that the output breaks up into two match points.

The primary objection to Frank codes relates to their polyphase implementation (and cost). In particular, as N increases, the size of the phase increment decreases, imposing a greater restriction on the phase modulator accuracy. The stability must also be greater since the difficulty of real-time calibration increases with the number of encoder bits.

Corresponding implementation difficulties are found on reception. Each of the matched filter weights possesses both real and imaginary values, hence the matched filter cannot be separated into independent I/Q operations as with the binary codes. Besides doubling the number of multiplications, this also implies that the I and Q components must be mixed at an earlier stage of processing.

Also, since the instantaneous phase shifts between successive elements (δ) is smaller than π, the instantaneous A/D bandwidth (δ/τ) is less than an equivalent binary code. A small instantaneous bandwidth (δ/τ) implies poorer spread spectrum performance since conventional ESM receivers respond to instantaneous bandwidth.

5.9 DIGITAL CHIRP

Conventional chirp, as mentioned previously, is relatively insensitive to Doppler frequency shifts provided the signal is within the passband of the receiver, that is, $f_T \tau \ll 1$. Frank codes are a digital version of chirp that is relatively insensitive to Doppler frequency shifts provided that $f_T \tau_u < 0.3$. The P3 and P4 codes are digital versions of chirp that are more Doppler tolerant than the Frank codes but less Doppler tolerant than the analog chirp code. This property is attained at the expense of more implementation difficulties.

Similar to the chirp code, the instantaneous angular frequency of the P3 code is given by $2\pi f = \mu t$ for μ the FM slope constant ($\mu \tau_u = 2\pi B$). The phase of the ith element is therefore

$$\begin{aligned}\beta_i &= \int_0^{(i-1)\tau} 2\pi f(t)\, dt \\ &= \int_0^{(i-1)\tau} \mu t\, dt \\ &= (\mu/2)(i-1)^2\tau^2 \qquad 1 \leqslant i \leqslant N_{pc} \end{aligned} \tag{5.13}$$

which is approximately the same as the chirp phase history (after accounting for modulo 2π phase wraparound). Substituting $\mu = 2\pi B/\tau_u$, and recognizing that

$B \cong 1/\tau$ and $N_{pc} = \tau_u/\tau$, the digital phase history becomes

$$\beta_i = \delta(i-1)^2 \qquad 1 \leqslant i \leqslant N_{pc} \tag{5.14}$$

for $\delta = \pi/N_{pc}$ the basic phase quantization level.

The P4 code is similar to the P3 code except that the phase history is symmetrical about $i = N_{pc}/2$. Acknowledging that a shift in time does not change the salient features of the code, we can introduce a fixed time offset

$$\begin{aligned}\beta_i &= \delta(i-1-N_{pc}/2)^2 \\ &= \delta[(i-1)^2 - N_{pc}(i-1) + N_{pc}^2/4]\end{aligned}$$

After dropping a constant phase term, this becomes

$$\beta_i = \delta(i-1)^2 - \pi(i-1) \qquad 1 \leqslant i \leqslant N_{pc} \tag{5.15}$$

Except for a π shift in phase every even code element, this sequence is the same as the P3 code (Lewis and Kretschmer, 1982, p. 638).

Digital matched filtering may be implemented either by direct convolution or by taking the FFT over the range swath (Wehner, 1987, pp. 138–142; Brookner, 1977, p. 161; Lewis and Kretschmer, 1982). Direct convolution is preferred for smaller code lengths N_{pc}, whereas FFT implementation is more efficient for large code lengths. The technical difficulties associated with implementing an FFT over a large range swath is an area of current interest.

5.9.1 Conclusions to Digital Chirp

Digital chirp offers some of the Doppler tolerance of the chirp code, as well as the flexibility and accuracy of digital processing. This is attained at the cost of more implementation difficulties.

The implementation of digital chirp can be divided into three classes:

1. Digital on transmission, analog on reception.
2. Analog on transmission, digital on reception.
3. Digital on both transmission and reception.

The primary difficulty with transmitting a digital chirp code is the finer quantization required. For Frank codes the basic quantization level is equal to $\delta = 2\pi/N$, for $N = \sqrt{N_{pc}}$. For P3 and P4 codes the basic phase quantization level is $\delta = \pi/N_{pc}$. Thus, to implement a code with $N_{pc} = 16$, 32 quantization levels are required versus 4 for the Frank code. The corresponding number of encoder values are 5 versus 2, respectively. For this reason, large values of N_{pc} are difficult to attain using P3 and P4 on transmission.

Digital chirp codes with coarser phase quantization may of course, be generated. The analysis in Di Vito et al. (1985) assumed a nonlinear digital chirp code with $N_{pc} = 100$ and 16 quantization levels (4 encoder values). The analysis in Iglehart (1978) assumed an approximate P4 code with 16 quantization levels.

Upon reception the matched filtering may be implemented digitally or by means of a SAW analog device. The problems associated with digital polyphase matched filtering have been discussed in the prior section on Frank codes. The problems associated with analog SAW devices have been discussed on the prior section on chirp.

Systems that employ P3 and P4 codes with high compression ratios are usually those that employ a SAW device to generate the code, and use digital processing only for the receive matched filtering (matched to the equivalent P3 or P4 version of the code). This type of processing is perhaps more closely related to the analog chirp code.

5.10 COMPLEMENTARY CODES

Complementary codes consist of a pair of codes with complementary sidelobes, that is, their sidelobes are equal and opposite (Golay, 1961). The simplest complementary code is the Barker code of length 2. The two Barker codes of length 2 are $(1, 1)$ and $(1, -1)$, and their corresponding matched filter output is $(1, 2, 1)$ and $(-1, 2, -1)$.

If these two complementary codes are transmitted alternately—first one code and then the other—the sidelobes will be shifted by PRF/2 during the coherent integration process. Hence, as shown in Figure 5.11, the sidelobes can be conveniently removed by bandpass filtering.

By chaining several complementary codes together, codes of almost any length can be built. For example, if a single complementary pair is available each of length 2, denoted A_2 and B_2, then a new pair of length 4 may be generated by simply forming

$$\begin{aligned} A_4 &= A_2 B_2 \\ B_4 &= A_2 \bar{B}_2 \end{aligned} \tag{5.16}$$

where $\bar{B}_2$ refers to the inverse of B_2. For example, in the case of the previous Barker code, this would be

$$\begin{aligned} A_2 &= +,\ + \\ B_2 &= +,\ - \\ A_4 &= +,\ +,\ +,\ - \\ B_4 &= +,\ +,\ -,\ + \end{aligned}$$

which can be recognized as the four-digit Barker codes. Hence the four-digit Barker codes also have the complementary feature.

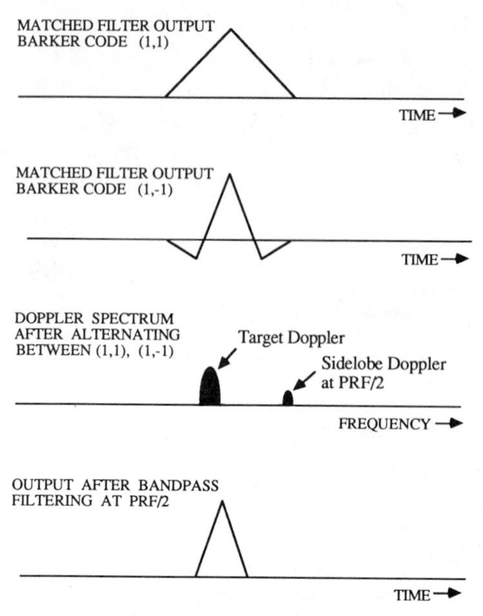

Figure 5.11 Example of complementary codes. Codes with complementary sidelobes are transmitted on alternate 1/PRF intervals (i.e., code A, then code B, then code A, etc.). When the received signals are integrated in the Doppler filter bank, the time sidelobes are shifted relative to the target by PRF/2 and so are removed.

This procedure can be rigorously proven quite simply. First, let us denote $x * y$ as the convolution of x with y. Therefore, what we wish to show is that $A_4 * A_4$ is the complement of $B_4 * B_4$, where A_4 and B_4 are given by (5.16). First, note that the convolution $A_4 * A_4$ consists of four parts, the convolutions $A_2 * B_2$, $B_2 * A_2$, $A_2 * A_2$, and $B_2 * B_2$. The convolutions $A_2 * A_2$ and $B_2 * B_2$ contribute equal and opposite components and, hence, contribute nothing except to the mainlobe. The $A_2 * B_2$ and $B_2 * A_2$ convolutions are solely responsible for the sidelobes. Likewise, only the $A_2 * \bar{B}_2$ and $\bar{B}_2 * A_2$ convolutions contribute to the B_4 sidelobes. But, the $A_2 * B_2$ and $A_2 * \bar{B}_2$ convolutions have only one of the members inverted (B_2 to $\bar{B}_2$). In conclusion, the results must also be inverted (or complementary).

5.10.1 Conclusions to Complementary Codes

In theory, complementary codes produce zero sidelobes (or, at least, sidelobes that are limited only by hardware nonlinearities). In theory, Doppler shifts have no impact on the sidelobe level. However, this theoretical performance is attained only if the coherent integration time is long enough to filter out the PRF/2 shifted sidelobe targets. For shorter integration times, it becomes more difficult to remove the PRF/2 shifted sidelobes, and Doppler frequency shifts begin to have a significant impact on the code's time sidelobes. For this reason complementary codes are not commonly considered suitable for any search application.

Nevertheless, complementary codes do have a place in search if the dwell time T can be made long enough or the PRF high enough. This is especially true if the added blind zones at PRF/2 do not pose a serious drawback to detection. A good example of this is detecting surface targets at the lower RFs (where Doppler blind zones are not a major concern). Complementary codes are useful in RGHPRF for the same reason. Finally, since tracking is not so limited with regards to blind zones, complementary codes offer a good track code at almost all RFs.

5.11 SUMMARY

Chirp is the most commonly employed code as it enables a large compression ratio to be achieved, is fairly simple to implement, is Doppler tolerant, and can be weighted for sidelobe reduction. Chirp possesses the property that a mean Doppler frequency causes a delay in the matched filter response, but otherwise leaves the shape of the response relatively unchanged. The various digital codes offer the advantage of flexibility and digital processing at the expense of more Doppler sensitivity and/or processing cost.

Binary phase codes are the most extensively used type of digital code. Binary codes are easy to implement using binary phase shifters and the hardware is relatively easy to calibrate. Binary codes can achieve (unaided) a sidelobe level that is acceptable for shorter code lengths, and ripple suppression can permit further sidelobe reduction.

The P3 and P4 codes are digital versions of chirp that are more compatible with digital processing (on both transmit and receive). They are becoming increasingly popular as longer Doppler-tolerant digital codes are being sought. The Frank code is another digital version of chirp that permits some hardware simplifications at the expense of Doppler sensitivity.

Complementary codes offer the advantage of zero sidelobes and Doppler tolerance. However, their use is limited to those applications where the added blind zones at PRF/2 offset do not pose a serious drawback.

PROBLEMS

5.1 A chirp code occupies the bandwidth $B = 1\,\text{MHz}$ and is of duration $\tau_u = 100\,\mu\text{sec}$. The target's Doppler frequency shift corresponds to 300 m/sec at 10 GHz. What is the bias in the range measurement?

5.2 Find the matched filter and autocorrelation function of the 5-bit Barker code and 5-bit reverse Barker code.

5.3 Consider the PN generator shown in Figure 5.7. Let the XOR connection be [3, 2] and the initial state of the registers be 100. Find the output sequence.

5.4 Repeat 5.3 for the case [4, 1] and initial state 1000.

5.5 Multipath can be treated as a linear filter of the form

$$\mathbf{H}(\omega) = 1 + ae^{j\omega\tau} \qquad a \ll 1$$

Using the concept of ripple suppression, design an inverse filter that removes the first-order degrading effects of the multipath condition. Give your answer in both the time and frequency domains.

5.6 Similar to Figure 5.10, compare the phase histories of the Barker code of length 7, the Frank code of length 9, and the digital chirp code of length 7. Also compare the sidelobe level in each case.

5.7 Find a pair of complementary codes of length 8 by concatenating two Barker codes of length 4. Sketch the autocorrelation functions of the complementary pair.

5.8 List at least one advantage and disadvantage of the following codes: chirp, digital chirp, Barker codes, Frank codes, PN codes, and complementary codes.

CHAPTER 6

MEASUREMENT THEORY

Radar tracking is traditionally divided into three steps: measurement of the target parameters, correlation of the measurements to tracks, and then filtering the measurements. This three-part division can be shown to be nearly optimum provided the signal plus noise can be modeled as Gaussian.

In this chapter we present some measurement formation algorithms and derive associated accuracy expressions. Also, use of the measurement theory is illustrated by means of examples drawn from monopulse angle, range, Doppler, and clutter measurements. The tracking steps are covered in subsequent chapters.

6.1 MEASUREMENT MODEL

Consider a series of signals denoted as (**A**), (**B**), ..., (**Z**). In general these signals may be construed as voltages taken from consecutive cells separated in time, angle, range, or frequency. Measurement theory involves inverting this set of signals [(**A**), (**B**), ..., (**Z**)] to obtain a measurement of the target's true position, x.

Only two signals are necessary to measure x provided that a single target is present. This is equivalent to saying that only two equations are necessary to determine two unknowns (target position and target strength). Therefore, a logical first step is to collapse the data into just two signals. Strictly speaking, this step is not necessary but simplifies the processing and leads to negligible performance loss provided the two signals are properly chosen.

For convenience it is assumed that these two signals are in (**S**, **D**) format. Again, this assumption is not necessary but avoids having to treat all notations separately. In the specific case of amplitude comparison monopulse signals, the relationship is

$$\mathbf{S} \triangleq \mathbf{A} + \mathbf{B}$$
$$\triangleq \mathbf{V}(x) \tag{6.1a}$$

where **A** is the signal in the top (left) beam, **B** is the signal in the bottom (right) beam, and **V** is the signal in the sum channel. A difference signal is also formed:

$$\mathbf{D} \triangleq \mathbf{A} - \mathbf{B}$$
$$\triangleq F(x)\mathbf{V}(x) \tag{6.1b}$$

For $F(x)$ the *error function* defined by

$$F(x) \triangleq \frac{A(x) - B(x)}{A(x) + B(x)} \tag{6.2}$$

The functions $A(x)$ and $B(x)$ are known deterministic functions that describe the response of the (**A**, **B**) signals to a target at position x. After collapsing into (**S**, **D**) format, the resulting sum and difference responses are illustrated in Figure 6.1.

As a final assumption, the receiver is assumed to supply coherent signals. That is, the signals (**S**, **D**) can be written using complex notation as follows:

$$\mathbf{S} = S_I + jS_Q \qquad \mathbf{D} = D_I + jD_Q$$

where (I, Q) are the in-phase (real) and quadrature phase (imaginary) components, respectively, $j = \sqrt{-1}$ is the imaginary number, and (in this chapter) boldface denotes complex valued.

6.1.1 Additive Noise

The observations are corrupted by various disturbances. Primary among these is the presence of additive noise:

$$\mathbf{S} = \mathbf{V} + \mathbf{N} \tag{6.3a}$$

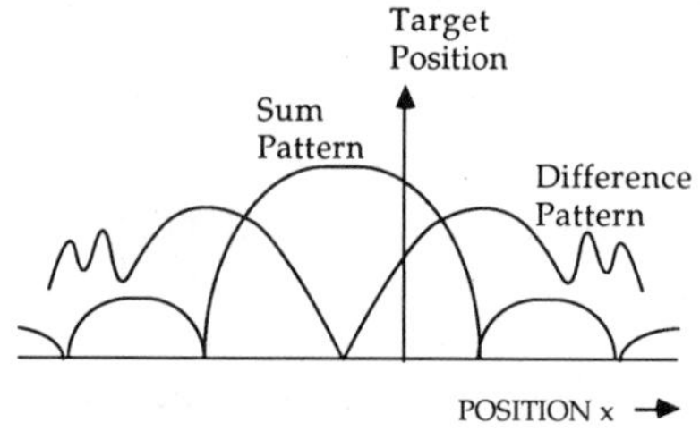

Figure 6.1 Typical sum and difference patterns.

for **N** denoting zero-mean additive Gaussian noise. Similarly,

$$\mathbf{D} = F(x)\mathbf{V} + \mathbf{M} \tag{6.3b}$$

for **M** noise in the difference channel.

6.1.2 Target Glint

If the target backscatter is distributed over a finite extent, such as illustrated in Figure 6.2, then another source of disturbance occurs called target *glint*. To model glint, it is necessary to recognize the fact that the target energy is distributed over a finite position:

$$\mathbf{S} = \int \mathbf{V}(\gamma)\, d\gamma$$

$$\mathbf{D} = \int F(\gamma)\mathbf{V}(\gamma)\, d\gamma \tag{6.4}$$

where $\mathbf{V}(\gamma)$ is the radar response function describing the response of the radar to a point target at γ. The position variable, γ, is modeled as having a mean component, x, and a distribution about that mean as given by $\Delta\gamma$:

$$\gamma = x + \Delta\gamma$$

Consider next the truncated Taylor series expansion around x of the error function:

$$F(\gamma) \cong F(x) + F'(x)(\gamma - x)$$

for $F'(x)$ the derivative of $F(x)$ with respect to x. Substituting this series into (6.4), the result is obtained

$$\mathbf{S} = \mathbf{V} \tag{6.5a}$$

$$\mathbf{D} = F(x)\mathbf{V} + F'(x)\sigma_W\mathbf{T} \tag{6.5b}$$

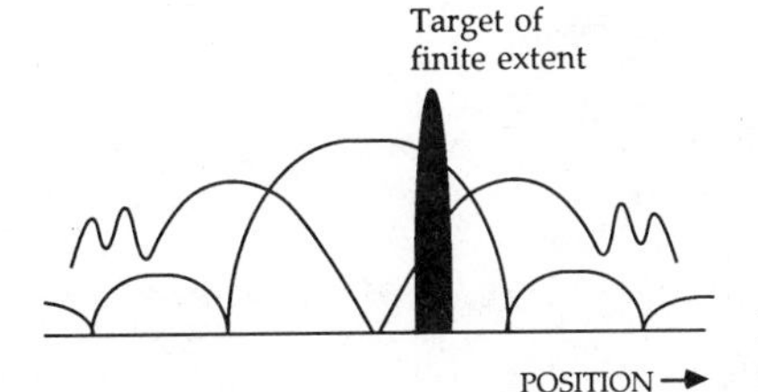

Figure 6.2 Sum and difference patterns with target of finite extent.

where the parameter **T** is (by definition)

$$\mathbf{T} \triangleq \int (\gamma - x)\mathbf{V}(\gamma)\, d\gamma/\sigma_W \tag{6.5c}$$

T is called the *glint noise*. By definition of σ_W, **T** has a variance equal to **V**, that is, $E(\mathbf{T} * \mathbf{T}) = E(\mathbf{V} * \mathbf{V})$. Furthermore, **T** is independent of **V** and is also Gaussian (given our assumption that **V** is Gaussian). The parameter σ_W is a statistical quantity describing the average level of glint. If the scattering is dominated by two independent point scatterers separated by W, for instance, then σ_W would equal $W/2$. If a multitude of scatterers are present, uniformly spread over W, then σ_W would equal $W/\sqrt{12}$.

The statistic σ_W can be large, particularly if multiple targets, multipath, or deception jamming is present. Also, all targets possess a finite extent that induces glint, and potentially **T** can "scintillate up" to quite large values (Barton, 1964). The Kalman filter can handle this situation, however, provided the variances fed to the measurement model are correct. Therefore, it is important that this error source be correctly included in the modeling process.

6.1.3 Channel Imbalances

Imbalances between the **S** and **D** channels also act to disturb the measurement process. Consider letting **D** have a relative (unknown) imbalance of α,

$$\mathbf{S} = \mathbf{V} \tag{6.6a}$$

$$\mathbf{D} = F(x)e^{\alpha}\mathbf{V} \tag{6.6b}$$

where

$$\alpha \triangleq \beta + j\theta$$

for β the log of the amplitude imbalance and θ the phase imbalance. For small values α can be approximately treated as a random variable. Substituting in a truncated Taylor series expansion,

$$e^{\alpha} \cong 1 + \alpha$$

into (6.6), the result is obtained

$$\mathbf{S} = \mathbf{V} \tag{6.7a}$$

$$\mathbf{D} = F(x)\mathbf{V} + F(x)\sigma_{\alpha}\mathbf{T}' \tag{6.7b}$$

where the parameter $\mathbf{T}'$ is given by

$$\mathbf{T}' \triangleq \alpha\mathbf{V}/\sigma_{\alpha} \tag{6.7c}$$

By definition of σ_α, $\mathbf{T}'$ has a variance equal to $\mathbf{V}$, that is, $E(\mathbf{T}' * \mathbf{T}') = E(\mathbf{V} * \mathbf{V})$; $\mathbf{T}'$ is modeled as independent of $\mathbf{V}$ (valid for small α).

6.1.4 Target Scintillation

Sometimes it is possible for the $(\mathbf{A}, \mathbf{B})$ signals to become slightly decorrelated prior to $(\mathbf{S}, \mathbf{D})$ formation, usually because the $\mathbf{A}$ and $\mathbf{B}$ signals are sequentially separated in time. This decorrelation induces an additional noise term. To incorporate this into our model, let us introduce ρ to be the correlation coefficient describing signal decorrelation between the $(\mathbf{A}, \mathbf{B})$ channels:

$$\rho = \frac{E(\mathbf{A} * \mathbf{B})}{E(\mathbf{A} * \mathbf{A})}$$

Thus, when the signals are eventually combined for $(\mathbf{S}, \mathbf{D})$ processing,

$$\begin{aligned} \mathbf{S} &\triangleq \mathbf{A} + \mathbf{B} \\ &\triangleq \mathbf{V} \end{aligned} \tag{6.8a}$$

and

$$\begin{aligned} \mathbf{D} &\triangleq \mathbf{A} - \mathbf{B} \\ &= F(x)\mathbf{V} + G(x, \rho)\mathbf{T}'' \end{aligned} \tag{6.8b}$$

By definition of $G(x, \rho)$, $\mathbf{T}''$ has a variance equal to $\mathbf{V}$, that is, $E(\mathbf{T}'' * \mathbf{T}'') = E(\mathbf{V} * \mathbf{V})$, and is independent of $\mathbf{V}$. It can be shown that

$$G(x. \rho) \cong \frac{(1 - \rho)^{1/2}}{(1 + \rho)^{1/2}} \qquad \text{when } F(x) = 0 \tag{6.9}$$

6.1.5 Conclusions

To briefly review, the signals $(\mathbf{S}, \mathbf{D})$ can be related to the target's position by

$$\begin{aligned} \mathbf{S} &= \mathbf{V} + \mathbf{N} \\ \mathbf{D} &= F(x)\mathbf{V} + F'\sigma_W\mathbf{T} + F\sigma_\alpha\mathbf{T}' + G(x, \rho)\mathbf{T}'' + \mathbf{m} \end{aligned} \tag{6.10}$$

for $\mathbf{V}$ the signal in the sum channel, $F(x)$ the error function, $\mathbf{N}$ and $\mathbf{M}$ zero-mean Gaussian random noise, σ_W the statistic describing target glint, σ_α the statistic describing channel imbalances, and $G(x, \rho)$ the statistic describing signal decorrelation prior to $(\mathbf{S}, \mathbf{D})$ formation. The random noise variables $\mathbf{T}$, $\mathbf{T}'$, $\mathbf{T}''$, and $\mathbf{V}$ all have equal variances and are jointly independent.

6.2 MEASUREMENT STRATEGY

The **S** and **D** signals can be put in the general form of a Cholesky decomposition (Bierman, 1977) as shown below:

$$\begin{vmatrix} \mathbf{S} \\ \mathbf{D} \end{vmatrix} = \begin{vmatrix} 1 & 0 \\ \chi & \mu \end{vmatrix} \times \begin{vmatrix} \mathbf{S} \\ \mathbf{U} \end{vmatrix} \tag{6.11}$$

where both variables **S** and **U** are random, statistically independent, and of equal variance. By definition, $\chi\mathbf{S}$ is that component of **D** that is correlated with **S**, and $\mu\mathbf{U}$ is that component of **D** that is uncorrelated with **S**. Both parameters χ and μ are deterministic for a given target scenario.

In the approach taken here, the Cholesky parameter χ is the primary target parameter of interest, denoted as χ_m. The Cholesky measurement, χ_m, is mapped into a position measurement, x_m, by means of a table look-up function that compensates for the finite SNR and nonlinear radar response. This procedure is done to simplify the derivation and processing, and to decouple the SNR estimate from the position measurement. The invariance property of maximum-likelihood estimators assures us that no loss in optimality accrues as a result of this simplification (Van Trees, 1968, p. 70).

Solving the above decomposition for the joint maximum-likelihood estimate (see Problem 6.5), the result is easily obtained:

$$\chi_m = \frac{\mathrm{Re}(\mathbf{D} * \mathbf{S})}{\mathbf{S} * \mathbf{S}} \tag{6.12a}$$

$$\mathrm{SNR}_m = \frac{\mathbf{S} * \mathbf{S} - \sigma^2}{\sigma^2} \tag{6.12b}$$

$$\mu_m^2 = \frac{\mathbf{D} * \mathbf{D}}{\mathbf{S} * \mathbf{S}} - \chi_m^2 \tag{6.12c}$$

assuming $F(x)$ is real after calibration.

The problem remains of how to relate the Cholesky parameters to the desired target parameters. The appendix to this chapter goes through the necessary transformations and shows that

$$F(x) = \chi \frac{\mathrm{SNR} + 1}{\mathrm{SNR}} \tag{6.13a}$$

$$\mu^2 = \frac{[F'^2\sigma_W^2 + F^2\sigma_\alpha^2 + G^2(x, \rho)]\mathrm{SNR} + 1 + \chi^2(1 + \mathrm{SNR})/\mathrm{SNR}}{\mathrm{SNR} + 1} \tag{6.13b}$$

The transformation (6.13a) is used to relate the measurements (χ_m, SNR_m) to the desired quantity x. The transformation (6.13b) is used to generate the variance μ^2 assuming the quantities (σ_W^2, σ_α^2, ρ) are known. Hypothetically, (6.13b) can also be used to determine ($\sigma_W^2, \sigma_\alpha^2, \rho$) provided a measurement of μ^2 is formed, μ_m^2, but an ambiguity exists between the various statistical quantities that must be sorted out.

6.3 ERROR ANALYSIS

The Kalman filter requires values for the measurement statistics. First substitute

$$\mathbf{D} = \chi \mathbf{S} + \mu \mathbf{U}$$

into (6.12a), thereby obtaining

$$\chi_m = \frac{\mathrm{Re}(\mathbf{D} * \mathbf{S})}{\mathbf{S} * \mathbf{S}} = \chi + \mu \frac{\mathrm{Re}(\mathbf{U} * \mathbf{S})}{\mathbf{S} * \mathbf{S}} \tag{6.14}$$

Recognizing that **U** and **S** are uncorrelated, then the expected value of χ_m is equal to χ. Thus, χ_m is an unbiased estimator of χ. Next the variance is calculated. The variance is determined by the term $\mathrm{Re}(\mathbf{U} * \mathbf{S})/\mathbf{S} * \mathbf{S}$. Since **U** and **S** are uncorrelated and identically distributed, it follows that

$$\mathrm{Re}(\mathbf{U} * \mathbf{S}) = |\mathbf{U}|\,|\mathbf{S}| \cos(\Psi)$$

where the random variable Ψ is uniformly distributed within $(0, 2\pi)$. Also, note that $\cos(\Psi)$ has an average variance of $(\frac{1}{2})$. Therefore, the χ_m measurement has a net variance:

$$\sigma_\chi^2 = \mu^2 E(|\mathbf{U}|^2) E\left(\frac{1}{|\mathbf{S}|^2}\right)\left(\frac{1}{2}\right) \tag{6.15a}$$

The term $E(1/|\mathbf{S}|^2)$ has a theoretical value of infinity since **S** can take on all values (e.g., consider the case $\mathbf{S} = 0$). However, assuming magnitude thresholding is correctly applied, $|\mathbf{S}| > T_h$, then the denominator will be well behaved. Thus, we can approximately state

$$\sigma_\chi^2 \cong \mu^2/2 \tag{6.15b}$$

This last step is not rigorous but avoids the necessity of introducing the exponential integral function $E_i(\,)$ as a function of the threshold T_h.

From (6.13a),

$$F(x) \cong \chi$$

assuming large SNR. Thus, taking the derivative, it follows that

$$\sigma_x \cong \frac{\sigma_\chi}{F'(x)} \tag{6.16}$$

such that, substituting in (6.15b) and (6.13b),

$$\sigma_x^2 \cong \frac{1}{2F'^2(\mathrm{SNR}+1)} + \frac{\sigma_W^2}{2} + \frac{F^2\sigma_\alpha^2}{2F'^2} + \frac{1-\rho}{2F'^2(1+\rho)} \tag{6.17}$$

The Kalman filter may directly use this expression to model the measurement accuracies. The only exception is the need to replace the actual statistical quantities (SNR, $\sigma_W, \sigma_\alpha, \rho$) by their corresponding estimates.

6.4 PHASE COMPARISON MONOPULSE

In the case of phase comparison monopulse, the desired parameter is the direction cosine to the target, ζ. From (2.3c), $F(\zeta)$ is approximately given by

$$F(\zeta) \cong \tan(\psi/2) \tag{6.18}$$

where

$$\frac{\psi}{2} = \frac{\pi d \zeta}{\lambda}$$

Thus, F' is approximately given by

$$F' \cong \pi d/\lambda \qquad \text{(on boresight)} \tag{6.19}$$

where d is the effective phase center separation.

Most phase comparison antenna examined by the author consistently displayed the following "invariance" property:

$$F' \cong k/\theta_{\mathrm{HP}} \tag{6.20}$$

where $\theta_{\mathrm{HP}} \triangleq$ antenna two-sided half-power beamwidth (one way)
$k =$ monopulse slope factor

and where the constant of proportionality, k, is approximately equal to 1.35. This relationship is important as it relates the unknown parameter, F', to a known parameter, θ_{HP}, via a simple constant.

A crude justification for (6.20) can be offered as follows. To begin, consider a rectangular-aperture antenna with uniform weighting. In this specific case d in

(6.19) is given by $d = D/2$, for D the total aperture diameter. Furthermore, the beamwidth θ_{HP} is given by $0.886\lambda/D$, such that

$$d = D/2$$
$$\theta_{HP} = 0.886\lambda/D$$

Substituting in for d and θ_{HP},

$$F' = 1.39/\theta_{HP} \tag{6.21}$$

for the rectangular-aperture antenna.

To provide a second example, consider next a circular-aperature antenna. In this case d, the effective phase center separation, must be determined by integrating x over the semicircular aperture. When this is done, the end result obtained is $d = 4D/3\pi$. Thus,

$$d = 4D/3\pi$$
$$\theta_{HP} = 1.028\lambda/D$$

for the circular-aperature antenna. Substituting into (6..19), we find

$$F' = 1.37/\theta_{HP} \tag{6.22}$$

When realistic antenna weighting is taken into account, the calculated k values are found to lie between $1.30 < k < 1.42$. The author verified this by examining the difference-to-sum ratio of numerous phase comparison antennas that were designed and manufactured at X-band (e.g., as in Figure 6.3). This invariance is also verified by Problem 6.4 in the example set.

The true significance of this invariance property is that a conventional four-port phase comparison antenna possesses only a limited number of degrees of freedom. Unlike amplitude comparison monopulse, a phase comparison mono-

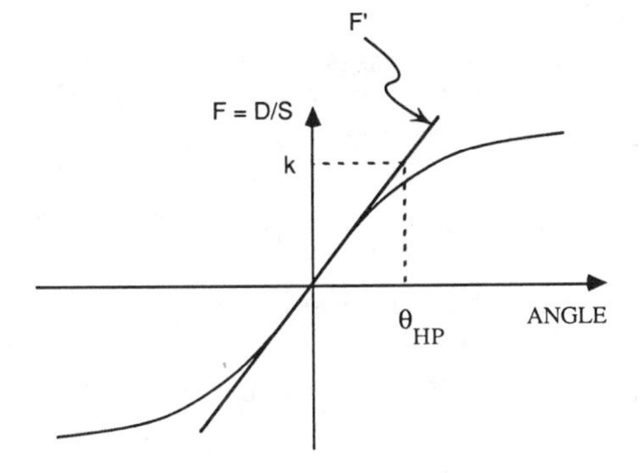

Figure 6.3 The monopulse slope factor k equals the ordinate of the F function at the beamwidth.

pulse antenna cannot easily trade off resolution versus accuracy, nor can it as easily change its feed structure to optimize performance (Hannan, 1961, p. 445). Unlike monopulse antennas with independent feeds, a phase comparison monopulse antenna cannot easily trade off accuracy versus sidelobe level in the difference pattern, since the sum and difference feeds are not independently constructed. Although it is theoretically possible to conceive of a phase comparison monopulse antenna with any value of slope coefficient, when practical sidelobe considerations are taken into account, values near 1.35 are normally attained in practice.

6.4.1 Angular Glint

So far, we have shown that F' can be simply related to the antenna beamwidth. Choosing a reasonable value for angular glint σ_W is more difficult since angular glint is dependent on the target environment. If a single target is thought to be present, then the main contributor to σ_W is due to the target's finite angular extent:

$$\sigma_W = (\text{target extent in radians})/\sqrt{12} \tag{6.23}$$

If multiple targets lie in the same resolution cell, then the dominant contributor to σ_W is intertarget glint:

$$\sigma_W = 0.5\ (\text{target separation in radians})$$

The presence of multiple targets can be inferred from correlation logic or from various μ estimation schemes (Bogler, 1986).

6.4.2 Channel Imbalances

Channel imbalances refer to gain/phase imbalances between the sum and difference channels. Imbalances occur due to the amplifier or because of path-length differences in the radar's frontend.

Interruptive built-in-test (BIT) calibration procedures are one means by which errors of this sort are reduced. In the context of a monopulse system, BIT refers to the process of injecting a reference pulse into the radar's front end and measuring the resultant gain/phase imbalance at the received output. Any sensed imbalances are calibrated out by applying corrections to subsequent difference signal outputs.

Nevertheless, channel imbalances are difficult to eradicate entirely. A value after BIT of $\sigma_\alpha = 0.005$ is considered good for ground-based systems. For smaller radars or radars on moving platforms, a value $\sigma_\alpha = 0.05$ is considered acceptable.

The following example illustrates the use of our expressions. Assume $\sigma_\alpha = 0.05$ and $F = 0.25$. From (6.17) this leads to

$$\begin{aligned}\text{Error} &= F\sigma_\alpha/\sqrt{2}F' \\ &= (0.25) * (0.05) * \theta_{\mathrm{HP}}/(\sqrt{2}1.35) \\ &= 0.013\theta_{\mathrm{HP}}\end{aligned}$$

where the invariance relationship (6.20) has been used.

In a properly designed monopulse radar, the value assigned to σ_α is normally a good indication of the overall accuracy of the radar at its nominal operating range. That is, the benefits of reducing the σ_α error source far below comparable error sources is not economically justifiable given the law of diminishing returns. Therefore, one can ordinarily ascertain the overall fidelity of a particular radar system by learning its σ_α value.

6.4.3 Conclusions

In conclusion, the values required by (6.17) for phase comparison monopulse are

$$\begin{aligned}F' &\cong 1.35/\theta_{\mathrm{HP}} \\ \sigma_W &= (\text{target extent})/\sqrt{12} \\ \rho &= 1 \\ 0 < \sigma_\alpha &< 0.05 \qquad 0 < F < 1\end{aligned} \tag{6.24}$$

6.5 AMPLITUDE COMPARISON MONOPULSE

For amplitude comparison monopulse the monopulse ratio is

$$F(\zeta) \triangleq \frac{A(\zeta) - B(\zeta)}{A(\zeta) + B(\zeta)} \tag{6.25}$$

For covariance calculations it is necessary to develop a simple analytical model for $A(\zeta)$ and $B(\zeta)$. The approach taken here is to use a Gaussian beam shape. It should be noted that this approach is closely similar to that taken by Sherman and probably numerous others.

The Gaussian assumption is written as

$$A(\zeta) = \exp\left[-1.38\,\frac{(\zeta - \Delta)^2}{\partial_{\mathrm{HP}}^2}\right] \tag{6.26a}$$

$$B(\zeta) = \exp\left[-1.38\,\frac{(\zeta + \Delta)^2}{\partial_{\mathrm{HP}}^2}\right] \tag{6.26b}$$

where $\zeta \triangleq$ direction cosine
$\Delta \triangleq$ squint angle
$\partial_{\mathrm{HP}} \triangleq$ beamwidth of the Gaussian beams

The constants ∂_{HP} and Δ, which transform the Gaussian function into the desired shape, are computed in the following manner. The half-power point of the Gaussian function is denoted ∂_{HP}, that is,

$$\exp\left(-1.38\frac{\zeta^2}{\partial_{\mathrm{HP}}^2}\right) = 0.707 \qquad \text{at } \zeta = \partial_{\mathrm{HP}}/2 \tag{6.27}$$

Next, it is assumed that the beams undergo an angular shift Δ but remain otherwise unchanged (Sherman, 1984, p. 133). Subject to these assumptions, F becomes

$$F = \frac{\exp(-1.38(\zeta - \Delta)^2/\partial_{\mathrm{HP}}^2] - \exp[-1.38(\zeta + \Delta)^2/\partial_{\mathrm{HP}}^2]}{\exp[-1.38(\zeta - \Delta)^2/\partial_{\mathrm{HP}}^2] + \exp[-1.38(\zeta + \Delta)^2/\partial_{\mathrm{HP}}^2]} \tag{6.28a}$$

$$= \frac{\exp[-1.38(\zeta^2 + \Delta^2)/\partial_{\mathrm{HP}}^2]\sinh(2.76\zeta\Delta/\partial_{\mathrm{HP}}^2)}{\exp[-1.38(\zeta^2 + \Delta^2)/\partial_{\mathrm{HP}}^2]\cosh(2.76\zeta\Delta/\partial_{\mathrm{HP}}^2)} \tag{6.28b}$$

$$= \tanh\left(\frac{2.76\zeta\Delta}{\partial_{\mathrm{HP}}^2}\right) \tag{6.28c}$$

Thus, the parameter F' is given by

$$F' = 2.76\Delta/\partial_{\mathrm{HP}}^2 \quad \text{(on-boresight)} \tag{6.29}$$

6.5.1 Optimization of Squint Angle

In most amplitude comparison monopulse systems, the value assigned to Δ, or *squint angle*, is approximately equal to one-half the beamwidth (Sherman, 1984). A larger value for squint angle increases F', as shown in (6.29), thereby improving angular accuracy. However, the squint angle cannot be made arbitrarily large since the width of the sum pattern also increases. The proper value of squint angle should therefore balance between these two considerations (Hannan, 1961; Sherman, 1984, p. 132).

Barton, Sherman, and Skolnik examined this trade-off in detail and proposed the following criterion for optimum squint angle selection (Sherman, 1984, p. 140):

$$\max_{\Delta} F' \, \mathrm{SNR}^{1/2} \tag{6.30}$$

From (6.17), this just amounts to minimizing the angular error due to additive noise. Based on this criterion and using

$$F' = 2.76\Delta/\partial_{\mathrm{HP}}^2$$

and a value for SNR proportional to $|A(\zeta) + B(\zeta)|^4$,

$$\text{SNR} \propto \left| \exp\left[-1.38 \frac{(\zeta^2 + \Delta^2)}{\partial_{\text{HP}}^2} \right] \cosh\left(\frac{2.76\zeta\Delta}{\partial_{\text{HP}}^2} \right) \right|^4$$

the optimum parameters are found to be

$$\Delta = 0.43\partial_{\text{HP}} \tag{6.31a}$$

$$\theta_{\text{HP}} = 1.33\partial_{\text{HP}} \tag{6.31b}$$

such that

$$F' = 1.58/\theta_{\text{HP}} \tag{6.32}$$

The corresponding SNR loss is 4.43 dB two way (for this example).

6.5.2 Comparative Figure of Merit

The performance of the four-feed amplitude comparison antenna is slightly inferior to phase comparison monopulse. In particular, the beamwidth gets broader, and the signal-to-noise ratio suffers relative to the phase comparison values. This phenomenon is well known and results due to the two-way effects of transmitting and receiving on the squinted beams.

To combat this, an alternate approach is to use a five-horn feed configuration, consisting of a central feed transmitting the sum pattern and four surrounding feeds generating the elevation and azimuth difference patterns (Hannan, 1961). This configuration avoids the problems previously noted since the sum feed is now independent of the location of the difference feeds. In addition, some of the other antenna characteristics (such as sidelobe levels and spillover radiation) are considerably improved.

6.5.3 Conclusions

In conclusion the values required by (6.17) for the case of amplitude comparison monopulse are

$$\begin{aligned} &F' \cong k/\theta_{\text{HP}} \qquad 1.2 < k < 2.0 \\ &\alpha_W = (\text{target extent})/\sqrt{12} \\ &\rho = 1 \\ &0 < \sigma_\alpha < 0.05 \qquad 0 < F < 1 \end{aligned} \tag{6.33}$$

The bound $1.2 < k < 2.0$ is conservative and is taken from Skolnik (1970, pp. 21–32).

6.6 COMMUTATIVE MONOPULSE

A well-established practice in the STT community is to commute the monopulse channels to reduce the channel imbalances (Kirkpatrick, 1952, p. 84). In this procedure two identical waveforms are transmitted and received, denoted here as $(\mathbf{S}_1, \mathbf{D}_1)$ and $(\mathbf{S}_2, \mathbf{D}_2)$, respectively. Between the two transmissions, an RF switch in the radar's front end is used to commute the channel paths. Due to this channel commutation, the $\mathbf{D}_2$ signal is propagated down the channel previously used by $\mathbf{S}_1$, and the $\mathbf{S}_2$ signal is propagated down the channel previously used by $\mathbf{D}_1$.

Due to the channel commutation, the signals take the form

$$\mathbf{D}_1 = \chi e^{\alpha} \mathbf{S}_1 \tag{6.34a}$$

$$\mathbf{D}_2 = \chi e^{-\alpha} \mathbf{S}_2 \tag{6.34b}$$

where $\alpha \triangleq \beta + j\theta$, for β the amplitude imbalance and θ the phase imbalance. Afterward, a composite estimator is taken:

$$\chi_m = \frac{\mathrm{Re}(\mathbf{D}_1 * \mathbf{S}_1)}{2\mathbf{S}_1 * \mathbf{S}_1} + \frac{\mathrm{Re}(\mathbf{D}_2 * \mathbf{S}_2)}{2\mathbf{S}_2 * \mathbf{S}_2} \tag{6.35}$$

in analogy to (6.12a).

Substituting (6.34) into (6.35), the expected value becomes

$$E(\chi_m) = \frac{\chi e^{\beta} \cos(\theta)}{2} + \frac{\chi e^{-\beta} \cos(-\theta)}{2} \tag{6.36a}$$

$$= \chi \cos(\theta) \cosh(\beta) \tag{6.36b}$$

In conclusion, the net effect of the channel commutation is to reduce the imbalance error β. The price one pays for this error reduction is to double the time to complete the measurement. Hence, this strategy is not normally considered compatible with TWS.

6.7 SEQUENTIAL ANGLE MEASUREMENT

In any sequential measurement scheme, two or more voltages **A** and **B** are observed at separate times. Again, the maximum-likelihood estimator is of the form

$$\chi_m = \frac{\mathrm{Re}(\mathbf{D} * \mathbf{S})}{\mathbf{S} * \mathbf{S}} \tag{6.37}$$

for

$$\mathbf{S} = \mathbf{A} + \mathbf{B}$$

$$\mathbf{D} = \mathbf{A} - \mathbf{B}$$

and where

$$F(\zeta_m) = \chi_m \frac{\text{SNR} + 1}{\text{SNR}}$$

For $F(\ )$ the error function, and ζ_m is the target's measured angular position relative to the average of the two boresight values taken at beam positions **A** and **B**.

In terms of the actual observables, (6.37) becomes

$$\chi_m = \frac{\text{Re}(\mathbf{A} + \mathbf{B}) * (\mathbf{A} - \mathbf{B})}{(\mathbf{A} + \mathbf{B}) * (\mathbf{A} + \mathbf{B})}$$

$$= \frac{|\mathbf{A}|^2 - |\mathbf{B}|^2}{(\mathbf{A} + \mathbf{B}) * (\mathbf{A} + \mathbf{B})} \tag{6.38}$$

The evaluation of the slope constant closely follows that of amplitude comparison monopulse. The only difference is the need to account for the two-way effects of transmitting and receiving on the sequential beams, as opposed to monopulse which transmits on the sum beam and uses the monopulse configuration for receive only. Correspondingly, we replace (6.26) by the two-way patterns

$$A(\zeta) = \exp\left[-2.776 \frac{(\zeta - \Delta)^2}{\theta_{\text{HP}}^2}\right] \tag{6.39a}$$

$$B(\zeta) = \exp\left[-2.776 \frac{(\zeta + \Delta)^2}{\theta_{\text{HP}}^2}\right] \tag{6.39b}$$

In analogy to (6.31), we find

$$F' = 5.55\Delta/\theta_{\text{HP}}^2$$

$$\Delta = 0.43\theta_{\text{HP}}$$

such that the theoretical value is

$$F' = 2.38/\theta_{\text{HP}} \tag{6.40}$$

In practice, a slightly smaller value of Δ may be chosen to offset the SNR loss, hence, a conservative value for F' lies between 1.7 and 2.4.

The prominent source of error is the amplitude scintillation that occurs between the sequential observations. It is convenient to model this decorrelation as exponential (Walters, 1958):

$$\rho = e^{-T/\tau_c} \tag{6.41}$$

for T the time between discrete observations and τ_c the target coherence time as per Figure 4.2. Given $T/\tau_c = \frac{1}{10}$ and F' given by (6.40), then the contribution to σ_ζ is $0.06\theta_{\mathrm{HP}}$.

6.8 RANGE MEASUREMENT

The methods already developed for the measurement of angle are directly applicable to the measurement of range. The only difference is the need to replace the antenna pattern by its mathematical equivalent.

The ideal video filter response depicted in Figure 6.4 can be used to calculate the range processing error. Assuming two samples separated by $c\tau/2$, denoted as the *early* (**A**) and *late* (**B**) samples, respectively, the linear nature of the response allows us to immediately write

$$\mathbf{A} = (1 - F)\mathbf{V}/2 + \mathbf{N}_A \tag{6.42a}$$

$$\mathbf{B} = (1 + F)\mathbf{V}/2 + \mathbf{N}_B \tag{6.42b}$$

where F is linearly related to the range error, Δr, by

$$F(\Delta r) = (4\Delta r)/(c\tau) \qquad -1 < F(\Delta r) < 1.$$

for Δr the range error relative to $\mathbf{A} = \mathbf{B}$.

Given this model it is straightforward to show that

$$F' = 4/(c\tau) \tag{6.43}$$

The resulting variance is

$$\sigma_r^2 = \frac{(c\tau/2)^2}{8(\mathrm{SNR} + 1)} + \frac{\sigma_W^2}{2} \tag{6.44a}$$

where SNR refers to the SNR after collapsing into (**S**, **D**) format.

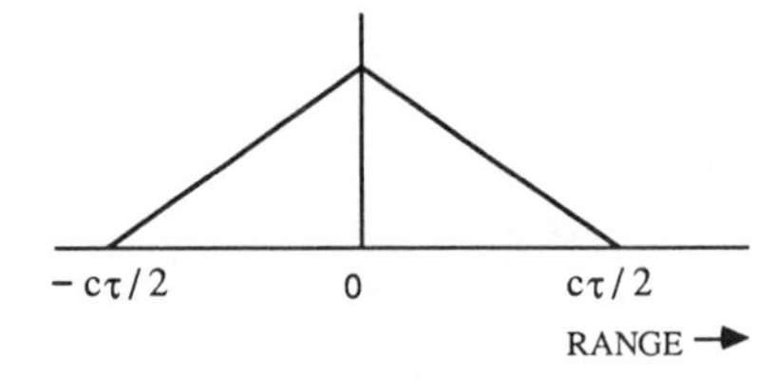

Figure 6.4 Ideal video filter response to point target.

The traditional approach has been to reference the range accuracy relationship to the *peak* SNR (Barton, 1964, p. 41), which is by definition the SNR in the **A** or **B** channel if sampled at the peak. The idea is that peak SNR values are less sensitive to new system design. Recognizing that this results in an additional factor of 2 in the ideal case, the final answer is obtained:

$$\sigma_r^2 = \frac{(c\tau/2)^2}{4(\mathrm{SNR}_p + 2)} + \frac{\sigma_W^2}{2} \tag{6.44b}$$

where SNR_p now refers to the peak detection SNR.

Finally, a value for σ_W must be inserted based on the range extent of the target,

$$\sigma_W = (\text{target extent})/\sqrt{12} \tag{6.45}$$

6.9 FREQUENCY MEASUREMENT

As shown in Figure 6.5, a modified cosine function is proposed for the Doppler filter shape. The target's position relative to the Doppler filter spacing determines its attenuation according to

$$\mathbf{A}(\Delta f) = \cos\left[\frac{\pi(\Delta f - \mathrm{DFS}/2)}{2\,\mathrm{DFW}}\right] \mathbf{V} \tag{6.46a}$$

$$\mathbf{B}(\Delta f) = \cos\left[\frac{\pi(\Delta f + \mathrm{DFS}/2)}{2\,\mathrm{DFW}}\right] \mathbf{V} \tag{6.46b}$$

where $\Delta f \triangleq$ Doppler frequency offset
DFS $\triangleq$ Doppler filter spacing
DFW $\triangleq$ Doppler filter width in Hertz

and where the known phase shift between successive Doppler filters is omitted for notational convenience.

Forming pseudosum $(A + B)$ and pseudodifference $(A - B)$ samples, the F

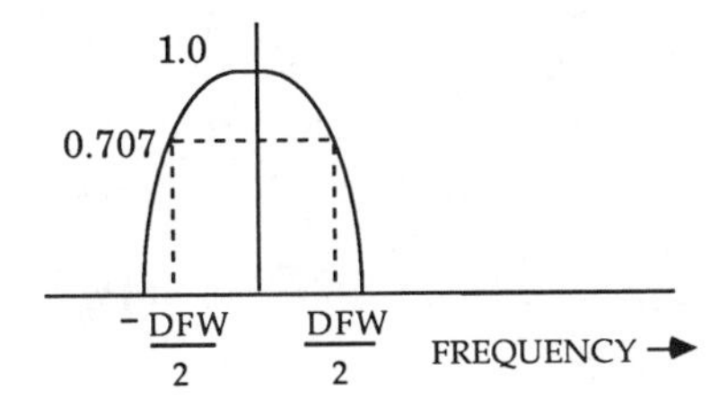

Figure 6.5 Simplified Doppler filter shape.

ratio becomes

$$F = \frac{\cos[\pi(\Delta f - \mathrm{DFS}/2)/2\,\mathrm{DFW}] - \cos[\pi(\Delta f + \mathrm{DFS}/2)/2\,\mathrm{DFW}]}{\cos[\pi(\Delta f - \mathrm{DFS}/2)/2\,\mathrm{DFW}] + \cos[\pi(\Delta f + \mathrm{DFS}/2)/2\,\mathrm{DFW}]}$$

$$= \tan\left(\frac{\pi\,\mathrm{DFS}}{4\,\mathrm{DFW}}\right)\tan\left(\frac{\pi\,\Delta f}{2\,\mathrm{DFW}}\right) \tag{6.47}$$

where the trigonometric identity

$$\cos(A + B) = \cos(A)\cos(B) - \sin(A)\sin(B) \tag{6.48}$$

has been inserted. The specific parameter of interest to error analysis is

$$F' \cong \tan\left(\frac{\pi\,\mathrm{DFS}}{4\,\mathrm{DFW}}\right)\frac{\pi}{2\,\mathrm{DFW}} \tag{6.49}$$

Equation (6.49) shows that there is a functional relationship between DFS, DFW, and accuracy. The filter weighting applied to ensure good sidelobes broadens the Doppler filter width DFW, implying that DFS is usually much less than DFW thus reducing the measurement accuracy. To overcome this effect, three or more Doppler filters may be utilized in succession, or a middle filter may be skipped. In any case no standard algorithms are used for the Doppler measurement, and so no standard accuracies can be quoted.

A further complication is the fact that the noise in adjacent Doppler filters are correlated with value ρ_N. This is just another consequence of the filter overlapping. After forming pseudosum and difference samples, therefore, the noise in the difference sample is reduced by $(1 - \rho_N)/(1 + \rho_N)$ over the noise in the sum sample. Neglecting the noise in the sum sample in evaluating the overall measurement error, the total error is reduced by $(1 - \rho_N)/(1 + \rho_N)$ over the uncorrelated case. This effect is not widely reported in the literature.

6.10 DIVERSITY COMBINERS

An implied assumption of all the previous analysis was that the target's parameters were to be estimated on the basis of a single sample. The analysis is now generalized to include multiple samples. Physically, this might correspond to the situation where multiple samples (or dwells) are integrated into a single measurement. In addition, the samples are assumed to be independent. Thus, the analysis in this section is most applicable if frequency agility is employed to decorrelate the samples, or if thermal noise is the dominant source of noise.

The maximum-likelihood estimator is given by (see Problem 6.5)

$$\chi_m = \frac{\sum_{i=1}^{N} \text{Re}(\mathbf{D}_i * \mathbf{S}_i)}{\sum_{i=1}^{N} \mathbf{S}_i * \mathbf{S}_i} \tag{6.50a}$$

$$\text{SNR}_m = \frac{\sum_{i=1}^{N} \mathbf{S}_i * \mathbf{S}_i - \sigma^2}{\sigma^2} \tag{6.50b}$$

$$\mu_m^2 = \frac{\sum_{i=1}^{N} \mathbf{D}_i * \mathbf{D}_i}{\sum_{i=1}^{N} \mathbf{S}_i * \mathbf{S}_i} - \chi_m^2 \tag{6.50c}$$

for $\{\mathbf{S}_i, \mathbf{D}_i, 1 \leqslant i \leqslant N\}$. This type of estimator is called a *diversity combiner* since independent powers are being combined.

Again, the variance of the χ measurement is found by substituting

$$\mathbf{D}_i = \chi \mathbf{S}_i + \mu \mathbf{U}_i$$

into Eq. (6.50a), obtaining

$$\chi_m = \chi + \mu \frac{\sum_{i=1}^{N} \text{Re}(\mathbf{U}_i * \mathbf{S}_i)}{\sum_{i=1}^{N} \mathbf{S}_i * \mathbf{S}_i} \tag{6.51}$$

The variance is proportional to the second term:

$$\frac{\sum_{i=1}^{N} \text{Re}(\mathbf{U}_i * \mathbf{S}_i)}{\sum_{i=1}^{N} \mathbf{S}_i * \mathbf{S}_i}$$

where $\mathbf{U}_i$ and $\mathbf{S}_i$ are uncorrelated. The term $\Sigma\,\text{Re}(\mathbf{U}_i * \mathbf{S}_i)$ can be interpreted as the real part of the projection of the vector $\mathbf{U}_i$ onto $\mathbf{S}_i$, that is,

$$\sum_{i=1}^{N} \text{Re}(\mathbf{U}_i * \mathbf{S}_i) = \sqrt{\sum_{i=1}^{N} \mathbf{U}_i * \mathbf{U}_i} \sqrt{\sum_{i=1}^{N} \mathbf{S}_i * \mathbf{S}_i} \cos(\Psi) \tag{6.52}$$

for $\cos(\Psi)$ the direction cosine between the vectors $\{\mathbf{U}_i\}$ and $\{\mathbf{S}_i\}$. From this, it follows that

$$E[\cos^2(\Psi)] = 1/(2N)$$

since $\cos(\Psi)$ represents one direction cosine element out of $2N$. Hence, the average magnitude squared of one element, $E[\cos^2(\Psi)]$, must equal $1/(2N)$. Thus, χ_m has variance:

$$\sigma_\chi^2 = \mu^2 E\left(\sum_{i=1}^{N} \mathbf{U}_i * \mathbf{U}_i\right) E\left\{\frac{1}{\sum_{i=1}^{N} \mathbf{S}_i * \mathbf{S}_i}\right\} \frac{1}{2N} \tag{6.53}$$

Both $\mathbf{U}_i$ and $\mathbf{S}_i$ are Gaussian random variables with variance $\sigma^2(\text{SNR} + 1)$. Therefore, the probability density function (pdf) of $\Sigma\,|\mathbf{S}_i|^2$ is distributed as a $2N$ chi-squared pdf, that is,

$$E\left(\frac{1}{\sum_{i=1}^{N} \mathbf{S}_i * \mathbf{S}_i}\right) = \int^{\infty} \frac{1}{\sum_{i=1}^{N} \mathbf{S}_i * \mathbf{S}_i} \{\text{chi-squared pdf}\}$$

$$= \frac{1}{\sigma^2(\text{SNR} + 1)} \frac{1}{\Gamma(N) P_D} \int^{\infty} Z^{N-2} e^{-Z}\, dZ \tag{6.54}$$

where the integration variable Z is given by

$$Z \triangleq \frac{\sum_{i=1}^{N} \mathbf{S}_i * \mathbf{S}_i}{\sigma^2(\text{SNR} + 1)}$$

and where, by definition,

$$\Gamma(N) \triangleq \text{gamma function}$$
$$= (N - 1)!$$

If we can assume that the probability of detection P_D is close to unity, then

$$E\left(\frac{1}{\sum_{i=1}^{N} \mathbf{S}_i * \mathbf{S}_i}\right) \cong \frac{1}{\sigma^2(\text{SNR} + 1)} \frac{1}{\Gamma(N)} \int_0^{\infty} Z^{N-2} e^{-Z}\, dZ$$

$$= \frac{\Gamma(N - 1)}{\sigma^2(\text{SNR} + 1)\Gamma(N)}$$

$$= \frac{1}{\sigma^2(\text{SNR} + 1)(N - 1)} \tag{6.55}$$

In conclusion, (6.53) and (6.55) and the relationship

$$E\left(\sum_{i=1}^{N} \mathbf{U}_i * \mathbf{U}_i\right) = N\sigma^2(\mathrm{SNR} + 1)$$

yield

$$\sigma_\chi = \frac{\mu^2}{2(N-1)} \tag{6.56}$$

Noncoherent integration reduces the error by a factor of $1/(N-1)$. Thus, it may be said that noncoherent integration is more efficient than coherent integration since *both* the SNR-induced and glint-induced errors are uniformly reduced by the same factor $1/(N-1)$. In practice, the $1/(N-1)$ reduction factor applies only to those error terms where statistical independence is a valid assumption, so the actual variance reduction ratio lies somewhere between 1 and $1/(N-1)$. Nevertheless, this is still superior to coherent integration, which does not help to reduce glint error whatsoever.

The drawback of noncoherent integration is that it is only weakly effective at low values of SNR. As the per-sample SNR becomes low, the problem of detection supersedes the problem of measurement, and detection is more effectively implemented with coherent integration.

6.11 CLUTTER MEASUREMENT

The discussion in this section concerns measuring the mean component of the clutter's frequency. The presence of a mean frequency component to clutter, uncompensated for by any prior processing, adversely affects the clutter filter's ability to notch out clutter. Hence, an attempt is normally made to measure mainlobe clutter and translate it down to zero-mean frequency during the second IF stage (or alternatively during I/Q detection).

Assuming the signals are already digitized, a preliminary estimate of mean clutter signal can be simply formed by coherently summing the clutter over M pulses:

$$\mathbf{A} = \sum_{i=0}^{M-1} \mathbf{S}_{ij}''' \tag{6.57}$$

where $\mathbf{S}_{ij}'''$ refers to the input to the clutter tracker (see Figure 2.19).

If the clutter truly has zero mean frequency, then $\mathbf{A}$ will remain constant from one time to the next (on the average). If not, it will change, with a rate of change proportional to the frequency error. Thus, by comparing the $\mathbf{A}$ signal to a delayed version of itself, the mean frequency component of the received clutter can be estimated.

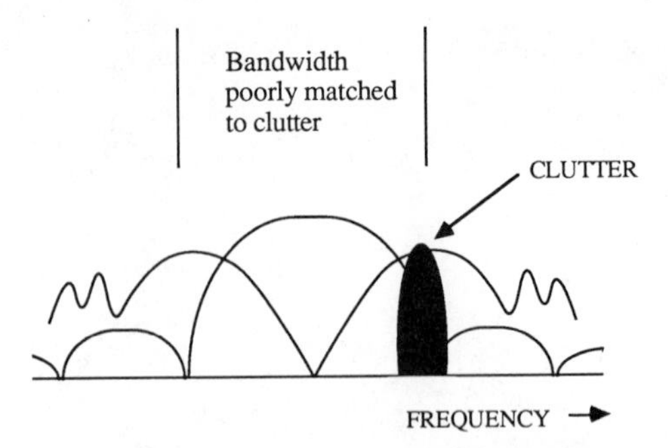

Figure 6.6 During the process of clutter frequency measurement, pseudosum and difference patterns may be formed. Too small a sum gain results in loss of sensitivity and isolation from targets, whereas too large a gain results in aliasing.

Let us form two successive signals denoted by **A** and **B**,

$$\mathbf{A} = \sum_{i=0}^{M-1} \mathbf{S}_{ij}''' \qquad \mathbf{B} = \sum_{i=M}^{2M-1} \mathbf{S}_{ij}'''$$

and pseudosum and pseudodifference signals

$$\mathbf{S} \triangleq \mathbf{A} + \mathbf{B}$$

$$\mathbf{D} \triangleq j(\mathbf{B} - \mathbf{A})$$

If $N * M$ is the total number of samples available to the clutter measurement, then there are $N/2$ samples of (**S**, **D**) data available. From (6.50a), the maximum-likelihood estimator is, therefore, of the form

$$\chi_m = \frac{\sum_{i=1}^{N/2} \mathrm{Re}(\mathbf{D}_i * \mathbf{S}_i)}{\sum_{i=1}^{N/2} \mathbf{S}_i * \mathbf{S}_i} \tag{6.58}$$

As proposed here, clutter measurement involves a combination of coherent integration and noncoherent integration. Coherent integration (6.57) is used to improve the per-sample clutter-to-noise ratio as well as reduce any interferences due to noise, targets, or jamming contained on the PRF interval. The drawback of coherent integration, as illustrated in Figure 6.6, is that it aliases the clutter. Thus, the amount of coherent integration is limited to match the a priori clutter boundaries. Noncoherent integration over the remaining samples is used to increase the measurement accuracy as per (6.58).

6.12 SPLIT-GATE MEASUREMENT

The following type of suboptimum estimator is frequently used (Skolnik, 1970, pp. 21-40):

$$\chi_m = \frac{|\mathbf{A}| - |\mathbf{B}|}{|\mathbf{A}| + |\mathbf{B}|} \tag{6.59}$$

This estimate is called a *magnitude split-gate* measurement. Because it ignores phase, the split-gate measurement is often used in those applications where phase is missing. Examples include range or Doppler tracking systems where magnitude only is stored, as well as some kinds of "stacked-beam" angle tracking systems for which significant corruption occurs in the phase information.

Simulation is required to precisely quantify the performance of this type of estimator. However, it can be easily shown that the performance is near optimal for large SNR (SNR > 10 dB). To show this, multiply both the numerator and denominator by $|\mathbf{A}| + |\mathbf{B}|$:

$$\begin{aligned}\chi_m &= \frac{(|\mathbf{A}| - |\mathbf{B}|)(|\mathbf{A}| + |\mathbf{B}|)}{(|\mathbf{A}| + |\mathbf{B}|)(|\mathbf{A}| + |\mathbf{B}|)} \\ &= \frac{|\mathbf{A}|^2 - |\mathbf{B}|^2}{|\mathbf{A}|^2 + |\mathbf{B}|^2 + 2|\mathbf{A}||\mathbf{B}|}\end{aligned} \tag{6.60}$$

If we make the reasonable assumption that $|\mathbf{A}||\mathbf{B}| \cong \text{Re}[\mathbf{A} * \mathbf{B}]$ for SNR > 10 dB, then

$$\begin{aligned}\chi_m &\cong \frac{|\mathbf{A}|^2 - |\mathbf{B}|^2}{|\mathbf{A}|^2 + |\mathbf{B}|^2 + 2\text{Re}[\mathbf{A} * \mathbf{B}]} \\ &= \frac{\text{Re}[\mathbf{D} * \mathbf{S}]}{\mathbf{S} * \mathbf{S}}\end{aligned} \tag{6.61}$$

which reproduces the maximum-likelihood χ estimator. Thus, under the assumption of large SNR, the split-gate estimator and the maximum-likelihood estimator are identical. Identical performance is also obtained if $|\mathbf{A}| \cong |\mathbf{B}|$.

6.13 POWER-CENTROID MEASUREMENT

The power-centroid measurement is a three-sample version of the previous noncoherent estimator. Denote three filters in succession as **A**, **B**, and **C**. The power-centroid measurement is then conventionally written as

$$\chi_m = \frac{|\mathbf{A}|^2 - |\mathbf{C}|^2}{|\mathbf{A}|^2 + |\mathbf{B}|^2 + |\mathbf{C}|^2} \tag{6.62}$$

This type of estimator is used in some Doppler tracking systems. Its performance is somewhat better than the performance of the two-filter maximum-likelihood estimator, but it is worse than the performance of the three-filter maximum-likelihood estimator.

Another measurement that is sometimes used is the weighted linear estimate:

$$\chi_m = \frac{F_A|\mathbf{A}|^2 + F_B|\mathbf{B}|^2 + F_C|\mathbf{C}|^2}{|\mathbf{A}|^2 + |\mathbf{B}|^2 + |\mathbf{C}|^2} \tag{6.63}$$

where **A**, **B**, and **C** are the voltage outputs of three filters in succession, and F_A, F_B, and F_C are the filter (center) values. Assuming three filters equally spaced by DFS, then

$$F_A = F_B - \mathrm{DFS}$$

$$F_C = F_B + \mathrm{DFS}$$

The following demonstrates that the estimator appearing in (6.63) and the estimator appearing in (6.62) are equivalent except for a scale factor. To begin,

$$\begin{aligned}
\chi_m &= \frac{F_A|\mathbf{A}|^2 + F_B|\mathbf{B}|^2 + F_C|\mathbf{C}|^2}{|\mathbf{A}|^2 + |\mathbf{B}|^2 + |\mathbf{C}|^2} \\
&= F_B + \frac{(F_A - F_B)|\mathbf{A}|^2 + (F_C - F_B)|\mathbf{C}|^2}{|\mathbf{A}|^2 + |\mathbf{B}|^2 + |\mathbf{C}|^2} \\
&= F_B - \mathrm{DFS}\,\frac{|\mathbf{A}|^2 - |\mathbf{C}|^2}{|\mathbf{A}|^2 + |\mathbf{B}|^2 + |\mathbf{C}|^2}
\end{aligned}$$

which is the same form as (6.62).

6.14 LOGARITHMIC ESTIMATOR

This type of estimator is used whenever logarithmic processing (rather than an AGC) is used to provide the necessary receive dynamic range (Leonov and Fomichev, 1986, p. 76; Jacovitti, 1983). Begin by passing the signals (**S** − **D**), (**S** + **D**) through logarithmic amplifiers and subtracting the two. The output is thus

$$\log\left(\frac{\mathbf{S}-\mathbf{D}}{\mathbf{S}+\mathbf{D}}\right)=\log\left(\frac{1-\mathbf{D}/\mathbf{S}}{1+\mathbf{D}/\mathbf{S}}\right)=\frac{2\mathbf{D}}{\mathbf{S}}+\frac{(\mathbf{D}/\mathbf{S})^3}{3}+\cdots$$

For small $\mathbf{D}/\mathbf{S}$ the approximation is made that the output equals $2\mathbf{D}/\mathbf{S}$.

6.15 SUMMARY

Conventional radar measurement theory is concerned with estimating the target's parameters such as to maximize the likelihood (by definition, the likelihood is the conditional probability when viewed as a function of the parameter). Using the invariance property of maximum-likelihood estimators, this problem can be recast into estimating the cross-correlation coefficient between two or more signals, for example, as in (6.11)–(6.13). The signals are corrupted by various disturbances, including additive thermal noise, target glint, channel imbalances, and scintillation. Assuming Gaussian statistics, these various disturbances can be grouped together into a position variance on the estimate as given by (6.17).

Next, use of the measurement theory was illustrated by means of simple examples drawn from angle, range, and Doppler measurement. Phase comparison monopulse angle measurement was treated first—both because of its current importance and because simple examples are lacking in the current literature as needed to demonstrate the concept. Amplitude comparison monopulse was briefly covered next and the work of Hannan summarized. Following this, channel commutation was discussed and the performance after commutation related to STT phase and amplitude comparison monopulse systems. Finally, sequential (non-monopulse) angle measurement was briefly reviewed.

The methods developed previously for accurate angle measurement are directly applicable to the measurement of range and Doppler, with the only difference being the need to account for the different response patterns and amount of overlapping. Both optimal as well as suboptimal measurement strategies were reviewed. Finally, the theory of diversity combining was presented and used to measure the mean clutter Doppler frequency. The tradeoff of coherent integration versus noncoherent integration was presented from the measurement perspective.

APPENDIX

This appendix goes through the necessary transformations to find the parameters needed for the measurement model.

Cholesky Parameter χ. Equating the relationships from (6.11) and (6.10),

$$\mathbf{D} = \chi \mathbf{S} + \mu \mathbf{U} \tag{A6.1a}$$

$$= F(x)\mathbf{V} + F'\sigma_W \mathbf{T} + F\sigma_\alpha \mathbf{T}' + G(x, \rho)\mathbf{T}'' + \mathbf{M} \tag{A6.1b}$$

Multiplying both sides of (A6.1a) by **S** and taking the expected value,

$$\begin{aligned} E(\mathbf{D} * \mathbf{S}) &= E[(\chi \mathbf{S} + \mu \mathbf{U}) * \mathbf{S}] \\ &= \chi E(\mathbf{S} * \mathbf{S}) \\ &= \chi E[(\mathbf{V} + \mathbf{N}) * (\mathbf{V} + \mathbf{N})] \\ &= \chi \{E(\mathbf{V} * \mathbf{V}) + E(\mathbf{N} * \mathbf{N})\} \\ &= \chi \{E(\mathbf{V} * \mathbf{V}) + \sigma^2\} \end{aligned} \tag{A6.2a}$$

And likewise for (A6.1b),

$$\begin{aligned} E(\mathbf{D} * \mathbf{S}) &= F(x)E(\mathbf{V} * \mathbf{S}) \\ &= F(x)E[\mathbf{V} * \mathbf{V} + \mathbf{V} * \mathbf{N})] \\ &= F(x)E(\mathbf{V} * \mathbf{V}) \end{aligned} \tag{A6.2b}$$

Let us define the signal-to-noise ratio to be the total power in the sum signal, $|\mathbf{V}|^2$, divided by the noise power, σ^2. With this definition,

$$\text{SNR} \triangleq \frac{E(\mathbf{V} * \mathbf{V})}{E(\mathbf{N} * \mathbf{N})} = \frac{E(\mathbf{V} * \mathbf{V})}{\sigma^2} \tag{A6.3}$$

In conclusion,

$$F(x) = \chi \frac{\text{SNR} + 1}{\text{SNR}} \tag{A6.4}$$

Cholesky Parameter μ. Combining (A6.4) and (A6.1),

$$\mu \mathbf{U} = \mathbf{F}'\sigma_W \mathbf{T} + F\sigma_\alpha \mathbf{T}' + G(x, \rho)\mathbf{T}'' + \mathbf{M} + \chi[\mathbf{V}/\text{SNR} - \mathbf{N}] \tag{A6.5}$$

Upon squaring (A6.5), taking the ensemble average, and recognizing that

$$E(\mathbf{T} * \mathbf{T}) = E(\mathbf{T}' * \mathbf{T}') = E(\mathbf{T}'' * \mathbf{T}'') = E(\mathbf{V} * \mathbf{V}) = \sigma^2 \text{SNR}$$

$$E(\mathbf{U} * \mathbf{U}) = E(\mathbf{S} * \mathbf{S}) = \sigma^2(\text{SNR} + 1)$$

the expression results

$$\mu^2 = \frac{[F'^2\sigma_W^2 + F^2\sigma_\alpha^2 + G^2(x, \rho)]\text{SNR} + 1 + \chi^2(1 + \text{SNR})/\text{SNR}}{\text{SNR} + 1} \tag{A6.6}$$

In conclusion, the observables (**S**, **D**) can be conveniently written in the form of a Cholesky decomposition. The Cholesky coefficient χ is a measure of target position x [from (A6.4)]. The coefficient μ is a measure of total noise power [from (A6.6)].

PROBLEMS

6.1 Derive the expression $\sigma_W = W/\sqrt{12}$, for W the physical width of the target(s). This expression assumes the scatterers are uniformly distributed over W. Derive an equivalent expression for the case of three equal scatterers placed at $-W/2$, 0, and $W/2$. Physically, this might correspond to a scatterer located on the left wingtip of the target, on the aircraft fuselage, and on the right wingtip, respectively.

6.2 Derive (6.9).

6.3 Derive (6.22).

6.4 Repeat the derivation leading up to (6.21) for the case of a rectangular antenna aperture with interlaced dipole density as shown in Figure P6.4. Assume $b = 4a$ and leave your answer in terms of the unknown a.

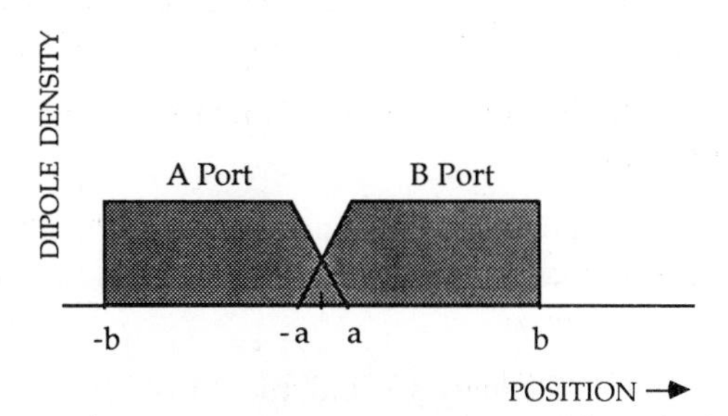

Figure P6.4 Simplified example of dipole density after interlacing (as per Figure 2.12). Interlacing is commonly employed in phase comparison monopulse antennas to reduce the sidelobe level in the difference pattern.

6.5 Assuming Swerling II, the joint probability density function (pdf) of the observables ($\mathbf{S}_i$, $\mathbf{D}_i$; $1 \leqslant i \leqslant N$) conditioned on χ, μ, and $\mathrm{P}_S \triangleq \mathrm{E}\{\mathbf{S}_i * \mathbf{S}_i\}$

can be written as a complex Gaussian pdf:

$$L(\mathbf{S}_i, \mathbf{D}_i | \chi, \mu, \mathbf{P}_S)$$

$$= \frac{1}{(\pi\mu \mathbf{P}_S)^{2N}} \exp\left\{-\frac{\sum_{i=1}^{N} \mathbf{D}_i * \mathbf{D}_i + (\chi^2 + \mu^2)\mathbf{S}_i * \mathbf{S}_i - 2\,\mathrm{Re}[\chi * \mathbf{D}_i * \mathbf{S}_i]}{\mu^2 \mathbf{P}_S}\right\}$$

where Re[] stands for the real part of the quantity between the parentheses. Derive the joint maximum likelihood estimates of χ, μ, and $\mathbf{P}_S$ from the system of likelihood equations given by

$$\frac{d\ \ln(L)}{d\chi} = 0$$

$$\frac{d\ \ln(L)}{d\mu} = 0$$

$$\frac{d\ \ln(L)}{d\mathbf{P}_S} = 0$$

conditioned on $\mathbf{P}_S > 0$ and $\mu > 0$.

6.6 What is the loss in SNR associated with the configuration outlined in (6.40)?

6.7 Assume a gain imbalance exists between the sum and difference channels as given by $\sigma_\alpha = 0.1$. Let $F = 0.5$, slope constant $= 1.5$, and antenna beamwidth $= 2°$. Find the resultant angular error due to gain imbalance. Is it possible to reduce this error by employing $(\mathbf{S} + j\mathbf{D}, \mathbf{S} - j\mathbf{D})$ processing? Assume no source of phase imbalance exists.

6.8 How is the scintillation time τ_c in (6.41) related to the scintillation time discussed in Chapter 4?

6.9 Assume $\mathbf{A} = 3 + j$ and $\mathbf{B} = 1 - j$, and compute the corresponding split gate estimate and maximum likelihood estimate. Are they the same? Explain any discrepancy.

6.10 Sketch a reasonable strategy to find the M.L. estimate of the target's position in Cartesian coordinates, given our prior discussion on M.L. estimation of the target's position in range, angle, and Doppler. (Hint: use the M.L. invariance property.)

CHAPTER 7

KALMAN FILTERING

Central to the idea of data filtering is smoothing a sequence of measurements and predicting future target kinematic behavior. The particular filter to be discussed here is the Kalman filter. The Kalman filter is the general solution to the recursive minimum mean-square error (MMSE) data-filtering problem within the class of linear estimators (Gelb, 1974). In addition to minimizing the mean-squared error, the Kalman filter has a number of other important advantages that make it particularly attractive for radar tracking. These advantages include (Blackman, 1986):

1. The filter's bandwidth is automatically chosen based on the assumed target maneuver and sensor measurement models. Thus, as long as the target and sensor are accurately modeled, the filter's bandwidth optimally changes to account for varying measurement and dynamical statistics.
2. The filter's bandwidth automatically changes to account for variable detection histories. This includes a variable ESA sampling interval as well as the possibility of missed detections.
3. The filter's bandwidth automatically adapts to both transient and steady-state conditions.
4. The Kalman filter provides a convenient measure of estimation accuracy through its covariance matrix. This measure is required in performing the gating, correlation, maneuver detection, and allocation functions to be discussed in later chapters.
5. Other common types of estimators can always be viewed as special cases of the Kalman filter. For example, simple precomputed filter gains (such as those of the $\alpha\beta\gamma$ type) can always be viewed as a special case of the steady-

state Kalman filter in the limit of constant measurement and dynamical noise covariances (Benedict and Bordner, 1962; Farina and Studer, 1985a, p. 174) and assuming uncoupled dimensions. Weighted least-squares filters can always be viewed as a special case of the Kalman filter.

This chapter introduces Kalman filtering in one dimension only. Hence, the pertinent quantities of interest will be one-dimensional parameters such as target position (x), velocity (v_x), and acceleration (a_x). This restriction allows the salient features of Kalman filtering to be explained without having to resort to the full complexities of three-dimensional coordinates.

A range tracker is a good example of a one-dimensional tracker. A range tracker measures the target's range and, in some applications, measures the target's radial velocity (or range rate). The function of the data filter, in this case, is to filter (smooth) the range measurements and to derive associated range derivatives. Knowledge of higher-order derivatives is attained by modeling them as state variables in the data filter.

7.1 INTRODUCTION

Suppose a target exists at unknown range (x). The radar provides a time sequence of range measurements, denoted here as $m(1)$, $m(2)$, $m(3)$, and so on, where each measurement is modeled as being in error by some random amount:

$$m(1) = x + n(1)$$

$$m(2) = x + n(2)$$

$$m(3) = x + n(3)$$

assuming (for now) the x is constant, and where n represents the measurement error. The error is assumed to Gaussian distributed. From our prior discussion in Chapter 6, it is evident that this Gaussian assumption is not valid but is convenient as it obviates the need to propagate the probability density itself.

In general, the kth measurement can be modeled as

$$m(k) = x + n(k)$$

where

$$E[n(k) * n(k)] = \sigma_k^2$$

for σ_k^2 the measurement variance. Also, it is assumed (for now) that $n(k)$ is independent from measurement to measurement.

When the first range measurement is made, our best estimate of range is simply the measurement itself:

$$\begin{aligned}\hat{x}(1) &= m(1)\\ &= x + n(1)\end{aligned}$$

where ^ denotes estimate, and where it is useful to index the estimates so that $\hat{x}(k)$ is the estimate after the kth measurement is received. The index k can be considered to denote increasing time.

Associated with each new estimate is an estimation error that is, by definition, the difference between the estimate and the true value of the parameter. A standard notation for this error is a tilde placed over the variable. Thus,

$$\begin{aligned}\tilde{x}(1) &\triangleq \hat{x}(1) - x\\ &= n(1)\end{aligned}$$

so that the estimation error statistics become

$$P(1) = \sigma_1^2$$

for $P(k)$ the variance of the filter's kth output (by definition).

When the second measurement is received, we have our first opportunity to improve upon the accuracy of the raw measurements. In Kalman filtering this is done recursively, that is, each new estimate is formed by combining the old estimate and the new measurement in a linear recursive fashion. Thus, as each new measurement is received, a series of recursive estimates and their variances are generated. Recursive processing is used to reduce the computer memory required to implement the data filter.

The equation for the new estimate is conventionally written as

$$\hat{x}(2) = \hat{x}(1) + K(2)[m(2) - \hat{x}(1)] \tag{7.1}$$

where $K(\,)$ is a weighting factor referred to as the *filter gain*. This equation is derived as follows. We desire a recursive estimate as modeled below:

$$\hat{x}(2) = K(2)m(2) + L(2)\hat{x}(1) \tag{7.2}$$

The coefficients $K(2)$ and $L(2)$ are best determined by choosing them so that

1. The resultant estimate is unbiased.
2. The variance is minimized.

The error associated with the estimate is

$$\begin{aligned}\tilde{x}(2) &= \hat{x}(2) - x \\ &= K(2)m(2) + L(2)\hat{x}(1) - x \\ &= K(2)[x + n(2)] + L(2)[x + n(1)] - x \\ &= [K(2) + L(2) - 1]x + K(2)n(2) + L(2)n(1)\end{aligned}$$

The first term contributes a bias to the error, whereas the second and third terms have zero-mean value and, so, do not contribute to the bias. Therefore, the new estimate $\hat{x}(2)$ will have zero bias if

$$L(2) = 1 - K(2)$$

With this, Eq. (7.2) can be arranged into the form of Eq. (7.1). By generalizing to the kth measurement, we have

$$\hat{x}(k) = \hat{x}(k-1) + K(k)[m(k) - \hat{x}(k-1)] \tag{7.3}$$

where the new estimate will always be unbiased if the previous one was.

The sequence of gains $K(2)$, $K(3)$, ..., and so on can be determined by requiring the new estimate to have minimum variance. The estimation error is given by

$$\begin{aligned}\tilde{x}(k) &= \hat{x}(k) - x \\ &= \hat{x}(k-1) - x + K(k)[m(k) - \hat{x}(k-1)] \\ &= [1 - K(k)]\tilde{x}(k-1) + K(k)n(k)\end{aligned}$$

The variance is, therefore,

$$\begin{aligned}P(k) &= E[\tilde{x}(k)^2] \\ &= [1 - K(k)]^2 P(k-1) + K(k)^2 \sigma_k^2\end{aligned} \tag{7.4}$$

The cross terms are neglected since the estimation error $\tilde{x}(k-1)$ is unbiased and independent of the noise at the current time $n(k)$. The value $K(k)$ that minimizes $P(k)$ can be found by taking its derivative with respect to $K(k)$ and setting it equal to zero. The resultant is

$$K(k) = \frac{P(k-1)}{P(k-1) + \sigma_k^2} \tag{7.5}$$

After inserting (7.5) into (7.4), the covariance equation becomes

$$P(k) = [1 - K(k)]P(k-1) \tag{7.6}$$

In conclusion, the set of estimation equations are

$$\hat{x}(k) = \hat{x}(k-1) + K(k)[m(k) - \hat{x}(k-1)] \tag{7.7a}$$

$$K(k) = \frac{P(k-1)}{P(k-1) + \sigma_k^2} \tag{7.7b}$$

$$P(k) = [1 - K(k)]P(k-1) \tag{7.7c}$$

This process of combining an old estimate with a new measurement to generate a new estimate is the filtering step in the Kalman filter algorithm.

Next the prediction step is introduced. The preceding example is generalized by allowing the target to move with radial velocity (v_x). For the linear case the true target range can be represented by

$$\mathbf{x}(k+1) = \mathbf{\Phi}\mathbf{x}(k) + \mathbf{q}(k) + \mathbf{f}(k) \tag{7.8}$$

where $\mathbf{x}$ is now a two-dimensional state vector that includes the quantities to be estimated, in this case range and velocity:

$$\mathbf{x}(k) = \begin{vmatrix} x(k) \\ v_x(k) \end{vmatrix} \qquad \mathbf{x}(k+1) = \begin{vmatrix} x(k+1) \\ v_x(k+1) \end{vmatrix} \tag{7.9}$$

and where boldface denotes "vector quantity;" $\mathbf{\Phi}$ is the state transition matrix as given by

$$\mathbf{\Phi} = \begin{vmatrix} 1 & \Delta T \\ 0 & 1 \end{vmatrix} \tag{7.10}$$

for ΔT the time between measurements.

The 2×1 vector $\mathbf{f}(k)$ is a deterministic aiding input used to compensate for known dynamics. It contains terms that subtract out the radar's ownship motion as well as other known dynamical inputs that govern state evolution.

The term $\mathbf{q}(k)$ is zero-mean, white, Gaussian "plant" noise with covariance

$$\mathbf{Q} = E[\mathbf{q}(k)\mathbf{q}^T(k)]$$

In general, $\mathbf{Q}$ may also be a function of k, but the k factor is dropped for notational simplicity.

The process noise $\mathbf{q}(k)$ physically represents all the random disturbances that act to perturb the target from a known trajectory. Examples include trajectory deflections due to maneuvers or due to the inhomogeneous media through which the target travels. It may also represent "perceived" deflections due to computer round-off errors or linearization errors made in implementing the set of dynamical equations.

The measurements are of the form

$$\mathbf{m}(k) = \mathbf{H}\mathbf{x}(k) + \mathbf{n}(k) \tag{7.11}$$

for $\mathbf{H}$ a 1×2 measurement matrix:

$$\mathbf{H} = [1 \quad 0]$$

or, in the case where radial velocity is directly measured, a 2×2 measurement matrix.

$$\mathbf{H} = \begin{vmatrix} 1 & 0 \\ 0 & 1 \end{vmatrix}$$

$\mathbf{n}(k)$ is zero-mean, white, Gaussian measurement noise with covariance $\mathbf{R}(k)$:

$$\mathbf{R}(k) = E[\mathbf{n}(k) * \mathbf{n}(k)]$$

Kalman filtering divides the processing into alternating steps of filtering and prediction. The filtering step is similar to the estimation of a constant parameter, such that the results of (7.7) directly apply. In the prediction step our attention is focused on the next time point. We form an unbiased, minimum variance estimate of $\mathbf{x}$:

$$\bar{\mathbf{x}}(k+1) = \mathbf{\Phi}\hat{\mathbf{x}}(k) + \mathbf{f}(k) \tag{7.12a}$$

based on the current estimate $\hat{\mathbf{x}}(k)$ and our knowledge of $\mathbf{f}(k)$ and $\mathbf{\Phi}$. From (7.8), the prediction estimate has a variance

$$\mathbf{P}(k+1) = \mathbf{\Phi}\mathbf{P}(k)\mathbf{\Phi}^T + \mathbf{Q}(k) \tag{7.12b}$$

With these concepts it is possible to explain the Kalman filter algorithm as shown in Figure 7.1. When the measurement is received, the so-called *residual* is generated as the difference between the new measurement and the predicted measurement, $\mathbf{m} - \mathbf{H}\hat{\mathbf{x}}$. Next the residual is weighted by the Kalman filter and used to update the state estimate. Finally, using the known dynamical model, $\mathbf{\Phi}$, the estimate $\hat{\mathbf{x}}$ is then propagated from one measurement time to the next (prediction).

Mathematically, these steps are summarized as follows:

Filtering:

State variable

$$\hat{\mathbf{x}}(k) = \bar{\mathbf{x}}(k) + \mathbf{K}(k)[m(k) - \mathbf{H}\bar{\mathbf{x}}(k)] \tag{7.13a}$$

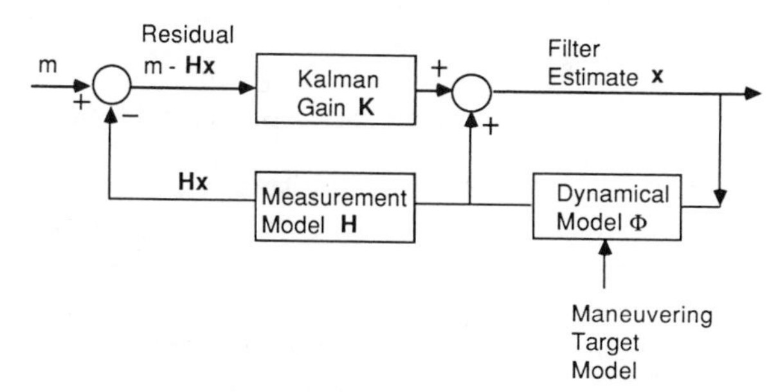

Figure 7.1 Functional elements of the Kalman filter.

where the Kalman gain [in analogy to (7.5)] is

$$\mathbf{K}(k) = \mathbf{P}(k)\mathbf{H}^T(k)[\mathbf{H}(k)\mathbf{P}(k)\mathbf{H}^T(k) + \mathbf{R}(k)]^{-1} \tag{7.13b}$$

Covariance

$$\mathbf{P}^+(k) = [\mathbf{I} - \mathbf{K}(k)\mathbf{H}(k)]\mathbf{P}(k) \tag{7.13c}$$

Prediction:

State variable:

$$\bar{\mathbf{x}}(k+1) = \mathbf{\Phi}\hat{\mathbf{x}}(k) + \mathbf{f}(k) \tag{7.13d}$$

Covariance

$$\mathbf{P}(k+1) = \mathbf{\Phi}\mathbf{P}^+(k)\Phi^T + \mathbf{Q}(k) \tag{7.13e}$$

where $\mathbf{\Phi} \triangleq$ state transition matrix
$\bar{\mathbf{x}} \triangleq$ a priori estimated state variable
$\hat{\mathbf{x}} \triangleq$ a posteriori estimated state variable
$\mathbf{f} \triangleq$ aiding term
$\mathbf{K} \triangleq$ Kalman gain
$\mathbf{P} \triangleq$ a priori Kalman filter covariance matrix
$\mathbf{P}^+ \triangleq$ a posteriori Kalman filter covariance matrix
$\mathbf{H} \triangleq$ measurement matrix
$\mathbf{R} \triangleq$ measurement covariance matrix
$\mathbf{Q} \triangleq$ plant noise covariance matrix

The Kalman filter can be initialized in several different ways. The following illustrates a reasonable set of initial conditions:

$$\hat{\mathbf{x}}(k=1) = \begin{vmatrix} m(k=1) \\ 0 \end{vmatrix} \qquad \mathbf{P}^+(k=1) = \begin{vmatrix} \sigma_1^2 & 0 \\ 0 & \sigma_v^2 \end{vmatrix}$$

where $m(1)$ = first position measurement
σ_1^2 = variance on first position measurement
σ_v^2 = assumed velocity variance based on ensemble statistics.

The off-diagonal elements are zero since the state variables are uncorrelated prior to data filtering. The position state is assumed equal to the first measurement, $m(1)$, and so the associated variance is the variance on measurement noise alone. The velocity state is initialized using *a priori* data (assuming no velocity measurements are available). *A priori*, the target is assumed to possess zero-mean velocity with variance σ_v^2. If additional information is available, such as a nonzero-mean velocity, then this information may be incorporated as well. A realistic value for σ_v^2 should be inserted into the model since too small of a value may desensitize the filter to incoming measurements, whereas too large of a value may create problems during the first gating test. The track file should be able to use the Kalman filter's covariance matrix for gating purposes without further manipulation provided that the filter is properly initialized.

7.2 TARGET ACCELERATION MODELS

7.2.1 White-Noise Model

Equation (7.8) models the acceleration as a Gaussian random variable $\mathbf{q}(k)$ having zero-mean and variance $\mathbf{Q}$. More critically, it is assumed white, such that the acceleration at one measurement time is independent of the acceleration at another measurement time. This set of assumptions is mathematically represented by

$$\begin{aligned} E\{a_x(k)\} &= 0 \\ E\{a_x^2(k)\} &= \sigma_a^2 \\ E\{a_x(k)a_x(j)\} &= 0 \qquad j \neq k \end{aligned} \tag{7.14a}$$

for the acceleration $a_x(k)$ related to the noise vector $\mathbf{q}$ by

$$\mathbf{q}(k) = a_x(k) \begin{vmatrix} \Delta T^2/2 \\ \Delta T \end{vmatrix} \tag{7.14b}$$

and

$$\begin{aligned} \mathbf{Q} &= E[\mathbf{q}(k)\mathbf{q}^T(k)] \\ &= \sigma_a^2 \begin{vmatrix} \dfrac{\Delta T^4}{4} & \dfrac{\Delta T^3}{2} \\ \dfrac{\Delta T^3}{2} & \Delta T^2 \end{vmatrix} \end{aligned} \tag{7.14c}$$

7.2.2 Markov Model

A white-noise model is seldom an accurate description of the target dynamics, as pilot-induced accelerations display a high degree of correlation with time. For aircraft targets typical correlation times vary between 1 min for a slow turn and 10 sec for an evasive maneuver (Singer, 1970). A value of 5 sec was used by Blackman for a highly maneuverable tactical aircraft.

Another example of correlated accelerations occurs in tracking ballistic targets within the earth's atmosphere. In this case aerodynamic drag due to the earth's atmosphere produces an acceleration component that is both correlated and random (due to random atmospheric inhomogeneities). This atmospheric drag can be modeled as a correlated acceleration component that is opposite in direction to the velocity vector.

Either the Markov model or the random-walk model are usually used to model correlated accelerations (Singer, 1970). In the case of the Markov model the acceleration process is modeled as the output of a first-order low-pass filter driven by white noise. Under this model the acceleration satisfies the following continuous-time relation:

$$\dot{a}_x = -\alpha a_x + w \tag{7.15}$$

where $\alpha \triangleq$ inverse acceleration correlation time constant
$w \triangleq$ white noise with zero mean and variance

$$E[w(t)w(t + \tau] = \delta(\tau)2\alpha\sigma_a^2 \tag{7.16}$$

for $\delta(\ )$ the Dirac delta function.

The autocorrelation function for the Markov process is exponential, as illustrated in Figure 7.2; this relationship can be derived by taking the Laplace transform of (7.15), computing the associated power spectral density, and then transforming back into the autocorrelation function (Brown, 1983, p. 186). The importance of finding the autocorrelation function is that it allows the constant

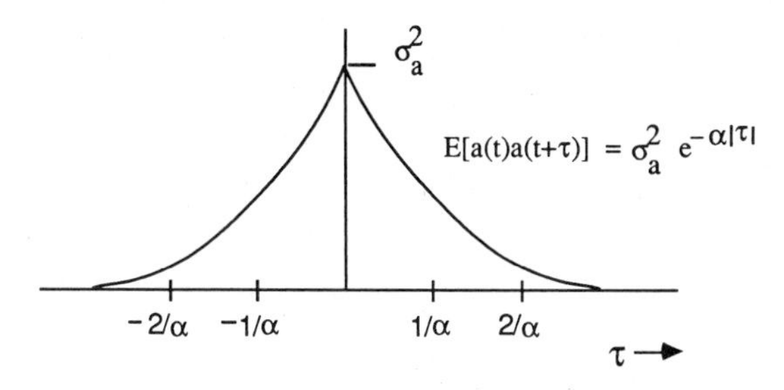

Figure 7.2 Autocorrelation function of Markov process.

σ_a to be identified as the standard deviation of the acceleration process and α to be identified as the inverse of the correlation time constant.

Continuing with the Markov target model, the state vector is

$$\mathbf{x} = \begin{vmatrix} x \\ v_x \\ a_x \end{vmatrix} \tag{7.17}$$

The corresponding continuous-time equation of motion is

$$\dot{\mathbf{x}}(t) = \mathbf{F}\mathbf{x}(t) + \mathbf{q}(t) + \mathbf{f}(t) \tag{7.18}$$

where

$$\mathbf{F} = \begin{vmatrix} 0 & 1 & 0 \\ 0 & 0 & 1 \\ 0 & 0 & -\alpha \end{vmatrix} \tag{7.19}$$

and

$$\mathbf{q}(t) = \begin{vmatrix} 0 \\ 0 \\ w(t) \end{vmatrix} \tag{7.20}$$

Discretization is the process of converting this matrix differential equation into a discrete time difference equation such that the means and variances of the state variables match at the sample points. Following Singer (1970), we begin by converting (7.18) into

$$d/dt[e^{-\mathbf{F}t}\mathbf{x}(t)] = e^{-\mathbf{F}t}[\mathbf{q}(t) + \mathbf{f}(t)] \tag{7.21}$$

After integration over the sampling time ΔT, this equation becomes

$$e^{-\mathbf{F}t}\mathbf{x}(t) = e^{-\mathbf{F}(t-\Delta T)}\mathbf{x}(t - \Delta T) + \int e^{-\mathbf{F}\tau}[\mathbf{q}(\tau) + \mathbf{f}(\tau)]$$

or, after dividing through by $e^{-\mathbf{F}t}$ and defining $\mathbf{x}(k + 1) \triangleq \mathbf{x}(t = k\Delta T)$,

$$\mathbf{x}(k + 1) = \mathbf{\Phi}\mathbf{x}(k) + \mathbf{q}(k) + \mathbf{f}(k) \tag{7.22}$$

where the transition matrix $\mathbf{\Phi}$ is found to be

$$\mathbf{\Phi} = e^{\mathbf{F}\Delta T}$$

$$= \begin{vmatrix} 1 & \Delta T & \frac{1}{\alpha^2}(-1 + \alpha \Delta T + \rho) \\ 0 & 1 & \frac{1}{\alpha}(1 - \rho) \\ 0 & 0 & \rho \end{vmatrix} \tag{7.23}$$

for $\rho \triangleq e^{-\alpha \Delta T}$.

The vector $\mathbf{q}(k)$ is given by

$$\mathbf{q}(k) = \int_{t-\Delta T}^{t} e^{\mathbf{F}(t-\tau)} \mathbf{q}(\tau)\, d\tau \qquad t = k \Delta T$$

which can be shown to have covariance

$$\mathbf{Q} = E[\mathbf{q}(k)\mathbf{q}^T(k)]$$

$$= 2\alpha\sigma_a^2 \begin{vmatrix} q_{11} & q_{12} & q_{13} \\ q_{21} & q_{22} & q_{23} \\ q_{31} & q_{32} & q_{33} \end{vmatrix} \tag{7.24}$$

where

$$q_{11} = \frac{1}{2\alpha^5}\left(1 - \rho^2 + 2\alpha \Delta T + \frac{2\alpha^3 \Delta T^3}{3} - 2\alpha^2 \Delta T^2 - 4\alpha \Delta T \rho\right)$$

$$q_{12} = \frac{1}{2\alpha^4}\left(\rho^2 + 1 - 2\rho + 2\alpha \Delta T \rho - 2\alpha \Delta T + \alpha^2 \Delta T^2\right)$$

$$q_{13} = \frac{1}{2\alpha^3}(1 - \rho^2 - 2\alpha \Delta T \rho)$$

$$q_{22} = \frac{1}{2\alpha^3}(4\rho - 3 - \rho^2 + 2\alpha \Delta T)$$

$$q_{23} = \frac{1}{2\alpha^2}(1 + \rho^2 - 2\rho)$$

$$q_{33} = \frac{1}{2\alpha}(1 - \rho^2)$$

The measurement matrix is of the form

$$\mathbf{H} = [1 \quad 0 \quad 0]$$

or, in the case when radial velocity is directly measured,

$$\mathbf{H} = \begin{vmatrix} 1 & 0 & 0 \\ 0 & 1 & 0 \end{vmatrix}$$

At this stage of the analysis, the simplifying assumption is usually introduced that the sampling interval ΔT is much less than the acceleration time constant ($\alpha \Delta T \ll 1$). This assumption is valid for fixed-rate tracking systems; if it were not, there would be little justification for modeling the acceleration as correlated in the first place. Only in the special case of a variable rate ESA tracker would it be necessary to jointly consider the cases ($\alpha \Delta T \gg 1$) and ($\alpha \Delta T \ll 1$).

Given $\alpha \Delta T \ll 1$, the truncated Taylor series expansion can be substituted in

$$\rho \triangleq e^{-\alpha \Delta T} = 1 - \alpha \Delta T + (\alpha \Delta T^2/2 + \text{HOT} \tag{7.25}$$

so that

$$\mathbf{\Phi} \cong \begin{vmatrix} 1 & \Delta T & \Delta T^2/2 \\ 0 & 1 & \Delta T(1 - \alpha \Delta T/2) \\ 0 & 0 & \rho \end{vmatrix} \tag{7.26}$$

and

$$\mathbf{Q} \cong 2\alpha\sigma_a^2 \begin{vmatrix} \dfrac{\Delta T^5}{20} & \dfrac{\Delta T^4}{8} & \dfrac{\Delta T^3}{6} \\ \dfrac{\Delta T^4}{8} & \dfrac{\Delta T^3}{3} & \dfrac{\Delta T^2}{2} \\ \dfrac{\Delta T^3}{6} & \dfrac{\Delta T^2}{2} & \Delta T \end{vmatrix} \tag{7.27}$$

This form of the Kalman filter is most appropriate whenever the radar's data rate is high enough to support the correlated acceleration assumption.

Similarly, in the limit $\alpha \Delta T \gg 1$, values can be derived for q_{11}, q_{12}, and q_{22} as follows:

$$\mathbf{Q} = 2\sigma_a^2/\alpha \begin{vmatrix} \dfrac{\Delta T^3}{3} & \dfrac{\Delta T^2}{2} \\ \dfrac{\Delta T^2}{2} & \Delta T \end{vmatrix} \tag{7.28}$$

This form for $\mathbf{Q}$ should rightfully replace (7.14c) in the limit $\alpha \Delta T \gg 1$ since it more accurately models the acceleration decorrelation within ΔT (Blackman, 1986).

7.3 SELECTING THE FILTER PARAMETERS

7.3.1 Singer Method

The acceleration standard deviation (σ_a) can be chosen in several different ways. A conservative approach would be to directly take $\sigma_a = A_{max}$. This would correspond to modeling the target as always accelerating at the maximum acceleration limits (A_{max} and $-A_{max}$). Typical values for A_{max} follow:

Target	A_{max} (g)
DC10	0.2–0.5
Missile	15–60
Tactical jet	10
Jeep	0.5–1
Ship	0.2

Singer (1970) proposed a more generalized model wherein the target is assumed to accelerate at the maximum rates (A_{max} and $-A_{max}$), each with probability P_{max}, and have no acceleration with probability P_0. The probability distribution for all other values of acceleration are taken to be uniform in the interval (A_{max}, $-A_{max}$). The height of the uniform distribution is computed to be $(1 - 2P_{max} - P_0)/(2A_{max})$. For this model Singer has shown that the target acceleration variance is

$$\sigma_a^2 = \frac{A_{max}^2}{3}(1 + 4P_{max} - P_0) \tag{7.29}$$

7.3.2 Monte Carlo Method

The Singer model is correct only if the average filter performance is the criteria of interest and the target can be accurately modeled as Markovian. Furthermore, the Singer model only considers the contribution to $\mathbf{Q}$ due to target maneuvers. Depending on the specific application, other sources of error may also be present. These include computer round-off errors, errors in modeling the aiding terms, errors in modeling the acceleration, linearization errors, data miscorrelation errors, and so on.

In general, although the Singer model gives a convenient framework for initial filter design, it is seldom the sole means of filter selection. A more popular method is to use *Monte Carlo simulation* to obtain the exact filter parameters (Maybeck, 1979, p. 337; Farina and Studer, 1985b, Chapter 6). Monte Carlo simulation refers to simulating the set of dynamical equations and various noise inputs, and then averaging over a large number of trials (also called *realizations*). The σ_a and α values are adjusted until the error statistics obtained by averaging are compatible with the predicted values and meet the required accuracy under the conditions of interest.

7.3.3 Mismatch method

In the scalar case (7.22) can be written as

$$a_x(k+1) = \rho a_x(k) + \sqrt{1-\rho^2}\,q(k) \tag{7.30}$$

where $E[q(k)^2] = \sigma_a^2$ and $\rho \triangleq e^{-\alpha \Delta T}$. From this equation it can be seen that the correlation coefficient ρ influences the state estimate in two ways:

1. It "restores" the mean acceleration to zero.
2. It "modulates" the noise variance.

Let us separately denote these two effects as ρ_r (for ρ "restore") and ρ_m (for ρ "modulate"). Thus, in this terminology,

$$a_x(k+1) = \rho_r a_x(k) + \sqrt{1-\rho_m^2}\,q(k) \tag{7.31}$$

The conventional Singer–Markov theory recommends setting $\rho_r = \rho_m$. In practice it sometimes makes more sense to perform independent optimization of ρ_r and ρ_m as a function of time. This idea becomes important in those systems performing variable ESA sampling.

The target's true acceleration profile resembles a Markov process only when viewed macroscopically. When viewed microscopically, it more closely resembles a random-walk process (i.e., no restoring force present). Therefore, a nonunity value for ρ_r produces a short-term bias during periods of constant target acceleration and rapid ESA sampling. The amount of bias can potentially be a large fraction of the acceleration's true value. Conversely, changing ρ_m effectively changes the filter's bandwidth since it changes the **Q** value. In conclusion, independently selecting ρ_r and ρ_m allows the filter's bias and bandwidth to be independently matched to the target's profile and instantaneous update rate.

7.4 EXTENDED KALMAN FILTER

The previous development assumes linear target dynamics. In most cases of actual target dynamics, this linearization assumption is not valid. For instance, when tracking reentry vehicles or ballistic targets, the primary acceleration components are the earth's gravity and aerodynamic drag due to the earth's atmosphere. Aerodynamic drag is known to be a nonlinear function of the target's velocity, altitude, and shape (Mehra, 1971; Chang and Tabaczynski, 1984). Furthermore, the earth's gravity is a nonlinear function of target position. As another example, when tracking highly maneuverable targets, the three-dimensional nature of the maneuver can lead to nonlinear coupling between the states. Finally, the equations of motion may be nonlinear because a polar coordinate frame is utilized in which Newtonian mechanics become nonlinear. An example of this is using a polar coordinate frame to derive range data from angle-only measurements.

It is necessary to introduce the extended Kalman filter whenever the target dynamics are nonlinear. Similar to (7.13), the equations become (Gelb, 1974; Farina and Studer, 1985a, Chapter 3):

Prediction:

State variable

$$\dot{\hat{\mathbf{x}}}(t) = \mathbf{f}[\hat{\mathbf{x}}(t)] \tag{7.32a}$$

Error variable

$$\dot{\tilde{\mathbf{x}}}(t) = \mathbf{F}[\hat{\mathbf{x}}(t)]\tilde{\mathbf{x}}(t) + \mathbf{q}(t) \tag{7.32b}$$

where $\mathbf{F}$ is the Jacobian of $\mathbf{f}[\ ]$:

$$\mathbf{F} = \partial\mathbf{f}[\mathbf{x}(t)]/\partial\mathbf{x}(t)$$

Covariance

$$\dot{\mathbf{P}}(t) = \mathbf{F}\mathbf{P}(t) + \mathbf{P}\mathbf{F}^T + \mathbf{Q}(t) \tag{7.32c}$$

Equation (7.32c) can be derived from (7.32b) by converting into the discrete time case $\tilde{\mathbf{x}}(k+1)$, squaring $\tilde{\mathbf{x}}(k+1)$, taking the expected value, and then reconverting back to the continuous-time case.

This set of prediction equations is evaluated by using numerical integration. For example, the Runge–Kutta numerical integration technique is commonly used for this purpose. Note that the filtering step remains basically the same.

7.5 GLINT STATE

The previous equations modeled the noise as being uncorrelated from measurement to measurement. However, target glint noise does not instantly decorrelate unless frequency agility is employed. Also, deception jamming, errors in ownship velocity, spurious Doppler spectra, and certain other types of noise cannot be modeled as uncorrelated. If the update time ΔT is small enough, then modeling these noise correlations may lead to improved track performance.

Again, a Markov model is proposed to model correlated noise. In discrete form, the equation becomes

$$g(k+1) = \rho_g g(k) + \sqrt{1-\rho_g^2}\,w(k) \tag{7.33}$$

where

$$E[w(k)\cdot w(k)] = \sigma_W^2$$

and where g defines the glint offset state and ρ_g is related to the target's correlation time (see Figure 4.2).

Inserting $g(k)$ into the state variable, we have

$$\mathbf{x} = \begin{vmatrix} x \\ v_x \\ a_x \\ g \end{vmatrix}$$

where the measurement matrix is of the form

$$\mathbf{H} = [1 \quad 0 \quad 0 \quad 1]$$

or, in the case of Doppler glint,

$$\mathbf{H} = \begin{vmatrix} 1 & 0 & 0 & 0 \\ 0 & 1 & 0 & 1 \end{vmatrix}$$

7.6 SQUARE-ROOT FILTERING

The basic form of the Kalman filter described in (7.13) suffers from a numerical stability problem that originates in the recursive formula:

$$\mathbf{P}^+(k) = [\mathbf{I} - \mathbf{K}(k)\mathbf{H}(k)]\mathbf{P}(k) \tag{7.34}$$

Notice that, for a small error in $\mathbf{K}$, denoted as $\delta\mathbf{K}$, the corresponding error in $\mathbf{P}^+$ is $-\delta\mathbf{K}(k)\mathbf{H}(k)\mathbf{P}(k)$. Hence, unless the numerical precision used at every iteration is high enough, the resulting matrix may not be positive definite as required for stable tracking. Furthermore, once the covariance has become nonpositive, all future values of covariance will remain nonpositive provided $\mathbf{Q}$ is small.

A common solution is to increase the size of the $\mathbf{Q}$ matrix to prevent the $\mathbf{P}$ matrix from remaining nonpositive. However, this solution amounts to accepting suboptimal performance in certain cases, especially whenever the true target matrix $\mathbf{Q}$ is small. It is also possible to employ extra precision arithmetic, but in real-time applications this may not be feasible.

To circumvent problems of this kind, alternate recursion relationships have been developed to propagate and update the *square root* of the covariance matrix (Potter, 1963; Bierman, 1977; Maybeck, 1979). Although equivalent algebraically to the conventional Kalman filter, these so-called *square root filters* exhibit improved numerical precision and stability. This advantage is achieved with negligible additional increase in computational complexity.

The most popular methods are Potter's method, Carlson's method, and Bierman's methods. Bierman in particular has developed a numerically favorable version of the square root filter by factoring the covariance matrix into an upper triangular matrix $\mathbf{U}(k)$ with 1's along its main diagonal and a diagonal matrix $\mathbf{D}(n)$ as follows:

$$\mathbf{P}(k) = \mathbf{U}(k)\mathbf{D}(k)\mathbf{U}^T(k) \tag{7.35}$$

The idea is to replace the recursion for the covariance $\mathbf{P}$ with a recursion for the components $\mathbf{U}$ and $\mathbf{D}$. Hence, the nonnegative definiteness of the computed matrix $\mathbf{P}(k)$ is guaranteed. This method is known as the *UD* factorization method.

Equivalently, the factorization may be written as

$$\mathbf{P}(k) = [\mathbf{U}(k)\mathbf{D}^{1/2}(k)][\mathbf{D}^{1/2}(k)\mathbf{U}^T(k)] \tag{7.36}$$

where $\mathbf{D}^{1/2}$ is the square root of $\mathbf{D}$. Accordingly, this factorization is also known as a square root filter even though the computation of square roots is not actually performed.

7.7 SUMMARY

The principles of Kalman and square-root filtering were briefly reviewed. The subjects of one-dimensional two-state, three-state, and four-state filters were discussed in a tutorial fashion.

PROBLEMS

7.1 Let $m(k) = x + n(k)$, for $\sigma^2 = 100$, and $m(1) = 5.3$, $m(2) = 8.1$, $m(3) = 4.2$, and $m(4) = -2.2$. Derive a recursive estimate for x and give values for $\hat{x}(k)$, $K(k)$, and $P^+(k)$ as a function of k.

7.2 Consider the following measurement model:

$$m(k) = x + n(k)$$

where $\{n\}$ are i.i.d. Gaussian noise with variance σ^2, and x is a constant random variable with mean a, variance b, and assumed Gaussian statistics. Find a recursive estimate for x by using a Kalman filter, and state the initial conditions on the state and covariance.

7.3 Any least-squares problem can be posed as a Kalman filtering problem by setting $\mathbf{Q} = 0$. A classic example of this is the problem of fitting a straight line through a set of measurements to minimize the sum of the square of the error. Show that this problem can be posed as a Kalman filtering problem and find the corresponding values of $\mathbf{H}$, $\mathbf{\Phi}$, and $\hat{\mathbf{x}}(k)$.

7.4 Our work in Chapter 3 could have been approached from the viewpoint of a Kalman filter problem. Let us restate (3.1) as follows: given an assumed target frequency ω_T, then estimate **V**. Define appropriate values for **H**, **Φ**, $\hat{\mathbf{x}}(k)$, **Q**, and **R** in this case. Why do we not estimate **V** and ω_T simultaneously?

7.5 Derive the autocorrelation function $R(t, \tau)$ associated with the first-order Markov process.

7.6 Prove that the residual $\tilde{\mathbf{y}} = \mathbf{m} - \mathbf{H}\hat{\mathbf{x}}$ obeys the relation

$$E[\tilde{\mathbf{y}}^T\mathbf{S}^{-1}\tilde{\mathbf{y}}] = M$$

where M is the dimensionality of the measurement and **S** is (by definition) the residual covariance matrix. (Hint: use the factorization $\mathbf{S}^{-1} = \mathbf{A}^T\mathbf{A}$, for **S** any positive definite matrix, and then define $\mathbf{z} = \mathbf{A}\tilde{\mathbf{y}}$.)

7.7 Show that, by proper normalization of the two-state Kalman filter set of equations, the performance can be compactly parameterized by means of a single parameter $\sigma_a \Delta T^2/\sigma_k$. Repeat this example for the three-state case and define any new parameter(s) necessary to describe filter performance.

7.8 Assume there is a bias in the measurement equation as given by $E(n) = b$. Consider two cases:

(1) b is a known constant.

(2) b is unknown and time-varying with time constant τ_b and variance σ_b^2.

State the most efficient way to remove this bias for the two cases.

7.9 The Kalman gain and covariance update equations where previously given in (7.13) as

$$\mathbf{K}(k) = \mathbf{P}(k)\mathbf{H}^T(k)[\mathbf{H}(k)\mathbf{P}(k)\mathbf{H}^T(k) + \mathbf{R}(k)]^{-1} \qquad \text{(P7.1a)}$$

$$\mathbf{P}^+(k) = [\mathbf{I} - \mathbf{K}(k)\mathbf{H}(k)]\mathbf{P}(k) \qquad \text{(P7.1b)}$$

Fill in the steps necessary to derive the following alternate formulation:

$$\mathbf{P}^+(k)^{-1} = \mathbf{P}(k)^{-1} + \mathbf{H}^T(k)\mathbf{R}(k)^{-1}\mathbf{H}(k) \qquad \text{(P7.2a)}$$

$$\mathbf{K}(k) = \mathbf{P}^+(k)\mathbf{H}^T(k)\mathbf{R}(k)^{-1} \qquad \text{(P7.2b)}$$

[Hint: substitute (P7.1a) into (P7.1b) and then prove that the result is equivalent to (P7.2a). Next substitute (P7.2a) into (P7.2b) and prove that the result is equivalent to (P7.1a).] These equations are known as the *information* form of the Kalman filter, whereas the previous form is known as the *covariance* form. This formulation is sometimes used in the literature as it is convenient when deriving theoretical results. The disadvantage of the information formulation is the higher computational burden (since a larger matrix must be inverted).

7.10 Similar to (7.4) show that, for an arbitary gain $\mathbf{K}$, the covariance update equation can be written as

$$\mathbf{P}^{+}(k) = [\mathbf{I} - \mathbf{K}(k)\mathbf{H}(k)]\mathbf{P}(k)[\mathbf{I} - \mathbf{K}(k)\mathbf{H}(k)]^{T} + \mathbf{K}(k)\mathbf{R}(k)\mathbf{K}(k)^{T}$$

This equation is known as the *Joseph* form of the covariance equation (Gelb, 1974). The Joseph formulation is more general than the standard Kalman formulation since no assumption is made about the explicit value of $\mathbf{K}$. It also exhibits improved numerical stability and so is sometimes used when stability is the key issue. In fact, show that for a small error in $\mathbf{K}$, denoted as $\delta\mathbf{K}$, to first order the corresponding error in $\mathbf{P}$ is zero. The disadvantage of the Joseph formulation is the considerably higher computational burden. This can be seen by inspection since the computation of $[\mathbf{I} - \mathbf{K}(k)\mathbf{H}(k)]\mathbf{P}(k)$ is common to both formulations.

7.11 In Section 7.2.2 the acceleration a_x was modeled as a Markov process. As an exercise, let us instead model the position x as Markovian:

$$\dot{x} = -\alpha x + w$$

where $\alpha \triangleq$ inverse position time constant
$w \triangleq$ white noise with zero mean and variance

$$E[w(t)w(t+\tau] = \delta(\tau)N_0/2$$

for $\delta(\,)$ the Dirac delta function. Find the corresponding expressions for Φ, $\hat{x}(k)$, and Q in this case. Comment on when such a model might be appropriate.

7.12 Steady-state solutions often provide some theoretical insight into the operation of the Kalman filter. Consider the following scalar system of dynamical equations:

$$x(k+1) = x(k) + q(k)$$

$$m(k+1) = x(k+1) + n(k+1)$$

where $E[n(k) * n(k)] = \sigma_k^2$ and $E[q(k) * q(k)] = Q$. The Kalman filter system of equations become in the scalar case

$$P(k+1) = P^{+}(k) + Q$$

$$K(k+1) = \frac{P(k+1)}{P(k+1) + \sigma_k^2}$$

$$P^{+}(k+1) = [1 - K(k+1)]P(k+1)$$

Using the fact that $P^+(k+1) = P^+(k)$ in the steady state, solve the system of difference equations for the steady-state covariance and gain.

7.13 A process $w(t)$ with autocorrelation:

$$\mathbf{R}(\mathbf{t}, \tau) = E[w(t)w(t+\tau]$$
$$= \delta(\tau)N_0/2$$

has corresponding power spectral density:

$$S_w(f) = N_0/2$$

Let us model target motion as the output of a lowpass filter driven by this white-noise process. Consider the following candidate filters:

(a) $H(f) = \begin{cases} 1 & -B < f < B \\ 0 & \text{otherwise} \end{cases}$ (Ideal lowpass)

(b) $H(f) = \dfrac{1}{1 + j\omega}$ (first-order lowpass)

(c) $H(f) = \dfrac{1}{j\omega}$ (integration)

Determine the corresponding autocorrelation function $R(t, \tau)$ in each case. [Hint: the autocorrelations for (a) and (b) are easier to derive by first deriving the power spectral density and then transforming back to the time domain, whereas the autocorrelation of (c) is easier to derive by beginning in the time domain.]

7.14 A shell is shot into the air by a mortar. Denote x to be the horizontal position of the shell and z to be the vertical position. The equations of motion are of the following nonlinear type:

$$\dot{v}_x = -(\alpha\rho V/2)v_x + w_x$$

and

$$\dot{v}_z = -(\alpha\rho V/2)v_z - g + w_z$$

where α = atmospheric drag coefficient (its reciprocal is referred to as the ballistic coefficient)
ρ = air density
g = earth's gravity (assumed constant in this experiment)
V = target velocity magnitude $= \sqrt{v_x^2 + v_z^2}$
w_x, w_z = white noise

Derive the $\mathbf{F}$ matrix, the Jacobian of $f[\ ]$, in the matrix equation (7.32c).

CHAPTER 8

ADAPTIVE KALMAN FILTERING

The Kalman filter requires that precise models be postulated for both the target dynamics and the radar measurement process. If these underlying models are not correct, the Kalman filter will not perform optimally. The approach most often taken in this case is to adaptively "match" the filter's state and parameters to the operational environment. This approach is discussed next using a series of evolutionary examples.

8.1 COVARIANCE ADJUSTMENT

Covariance adjustment is the most straightforward and extensively used method of adaptive Kalman filtering. It involves monitoring the performance of the Kalman filter residual and iteratively modifying one or more parameters of the Kalman filter covariance until satisfactory performance is obtained. Due to performance as well as computational reasons, only a simplified introduction to covariance adjustment is outlined in this section. The interested reader is referred to the literature (Jazwinski, 1970; Mehra, 1970, 1972; Carew and Belanger, 1973; Athans and Chang, 1976; Chin, 1979; Farina and Pardini, 1980; Castella, 1980; Maybeck, chapter 10, 1979; Blackman, 1986, p. 40; Bar-Shalom and Fortmann, 1988).

The Kalman filter's residual $\tilde{\mathbf{y}}(k)$ is given by

$$\tilde{\mathbf{y}}(k) = \mathbf{m}(k) - \mathbf{H}\bar{\mathbf{x}}(k) \tag{8.1}$$

where $\mathbf{H}$ is the measurement matrix and $\mathbf{m}$ is the measurement vector. The residual $\tilde{\mathbf{y}}(k)$ is a zero-mean white random variable with covariance $\mathbf{S}$ (Gelb, 1974):

$$\mathbf{S} = \mathbf{HPH}^T + \mathbf{R} \tag{8.2}$$

assuming the Kalman filter models are optimally matched to the target and measurement characteristics. Hence, any deviation from this state of affairs is an indication that these models are not correct. By comparing the residual's statistics (mean, variance, autocorrelation function) to preset threshold(s), the presence of a model mismatch can be inferred and corrected.

Averaging may be used to obtain a more well-behaved measure (at a cost of an added delay in detecting the mismatch). A convenient way to average the residuals is to form a sliding window average. For example, defining the normalized residual d^2 to be

$$d^2 = \tilde{\mathbf{y}}(k)^T \mathbf{S}^{-1} \tilde{\mathbf{y}}(k) \tag{8.3}$$

then a sliding window average over the normalized variance is

$$Z(k) = aZ(k-1) + d^2(k) \tag{8.4}$$

In the absence of a mismatch, $Z(k)$ will have a steady-state average value of

$$\begin{aligned} E(Z) &= aE[Z] + E[d^2(k)] \\ &= aE[Z] + M \end{aligned}$$

where M is the dimensionality of the measurement (see Problem 7.6). Therefore,

$$E[Z] = M/(1-a) \tag{8.5}$$

Similarly, the variance can be found:

$$\begin{aligned} E[Z - E(Z)]^2 &= a^2 E[Z - E(Z)]^2 + 2M \\ &= 2M/(1-a^2) \end{aligned} \tag{8.6}$$

By matching an (assumed) chi-squared statistic to this mean and variance, $Z(k)$ can be identified as (approximately) chi-squared with dimensionality $M(1+a)/(1-a)$ (Bar-Shalom and Fortmann, 1988, p. 299). Declaration of a mismatch is performed by comparing $Z(k)$ to a threshold:

$$Z(k) \gtrless T_h \tag{8.7}$$

where T_h is determined from the assumed chi-squared statistics and the required P_{FA}. Additional tests may also be performed, for example, a simple procedure is to monitor the sign of the last N residuals to determine the presence of any mean component to the error (Farina and Studer, 1985a, p. 247). Also, more than one value of N and a may be used to improve responsiveness.

If a model mismatch is declared, one or more parameters of the Kalman filter are modified (preferably before the Kalman gain is computed on that measurement). One method is to vary the σ_a parameter in the $\mathbf{Q}$ matrix (Castella, 1980). This method is intuitively appealing since the $\mathbf{Q}$ matrix is dynamic and is often blamed for the filter mismatch in the first place. On the other hand, sometimes a better method is to directly increment the covariance matrix $\mathbf{P}$ in addition to incrementing σ_a (Farina and Studer, 1985a, p. 250). This method is better if the model mismatch remains undetected for a significant number of measurements since a substantial buildup in error can occur that is not immediately remedied by increasing σ_a.

The larger the $Z(k)$ threshold crossing, the larger the applied covariance correction. Likewise, the covariance may be reduced whenever a low value is consistently observed in $Z(k)$. A unique solution describing this relationship cannot be determined since σ_a^2 and $\mathbf{R}$ are indeterminant in the steady state (Fitzgerald, 1980) and the time of the error is generally unknown in any case. Therefore, an assumption must be made about the relative relationships of these parameters and the algorithm tuned to ensure convergence. Generally speaking, the filter parameters are adjusted to account for any discrepancy between the observed value of $Z(k)$ and the expected value and to force the Kalman filter's value of covariance to coincide with the actual value of covariance.

8.2 BATCH PROCESSING

Targets are often characterized by straight-line segments between discrete heading changes. A Kalman filter that is optimized for straight flight will be unresponsive if the target begins to maneuver. The approach most often taken in this case is to first detect the maneuver using (8.7) and to subsequently correct the state or covariance estimate in some manner.

Covariance adjustment may be used for this purpose. However, covariance adjustment does not utilize past measurement data in correcting for the maneuver event, and thus loses information at a time when a responsive correction is needed.

One proposed solution is to store all the old measurements up to $t - s\Delta T$ seconds, and then to reinitialize a new filter $s\Delta T$ seconds in the past based on the assumed maneuver event (Bar-Shalom and Birmiwal, 1982). The resulting reinitialized filter can be assumed to be based only on data relevant to the maneuver event and so is automatically corrected for the maneuver event.

State reinitialization is an attractive solution from the standpoint that it makes almost no assumptions about the nature of the maneuver. On the other hand, a perceived disadvantage of batch processing is that it produces a sharp discontinuity in the computational load at the precise instant of maneuver detection and batch correction. This is critical if the length of the observation window s is large and the computational resources do not permit a sharp discontinuity in the processing.

For high data rate tracking systems, one way to implement batch processing is to first prefilter the measurement data to reduce the storage rate. Given that the assumed target dynamics are not critical over short time segments, then it is valid to prefilter the measurements assuming straight-line motion over short segments $m\,\Delta T$. The output of this prefilter forms a "condensed measurement vector" that is representative of the target's position, velocity, and acceleration at time $t - (m/2)\,\Delta T$. These condensed measurement vectors are then used to update the current Kalman filter, as well as being stored for possible use in later batch processing. The disadvantage of prefiltering, of course, is that the assumption of quasi-stationarity may not be valid over a time period sufficiently long to provide any computational savings. Prefiltering also introduces an average time delay in the estimate on the order of $(m/2)\,\Delta T$.

8.3 PARALLEL KALMAN FILTERS

Given that the discussion is limited to real-time recursive filters, and given that the time delay associated with covariance adjustment cannot be tolerated, then a common technique is to form multiple recursive filters. As Magill proposed as early as 1965, a bank of N parallel Kalman filters is formed. Each filter utilizes a different target model. The "correct" filter is that filter observed to perform best as judged by the performance of its residuals. Alternately, a weighted average over all N filters may be taken.

In its simplest form the processing is as follows. Two parallel Kalman filters are set up. The "slow" filter uses a small $\mathbf{Q}$ value based on quiescent target dynamics. The "fast" filter uses a large $\mathbf{Q}$ value based on maneuvering target dynamics. The fast filter is periodically reinitialized from the slow filter. However, whenever a maneuver is suspected, the fast filter is not reinitialized. Whenever a meneuver is confirmed, the states of the slow filter are reinitialized to the states of the fast filter, and tracking proceeds as before.

Multiple-filter processing requires more numerical computations than single-filter processing. Additionally, these computations must be performed all the time (recursively). Nevertheless, this type of solution has become more and more popular as processing costs go down.

8.4 WEIGHTED LEAST-SQUARES ESTIMATION

The least-squares approach was originally studied by Friedland (1969) and later by Chan, Hu, and Plant (1979). Its main advantage is that a second Kalman filter gain computation is avoided; instead, statistics are accumulated that carry the information needed to recompute a filter correction at any future date. The least-squares approach was not originally stated in this manner, rather, the emphasis was on decoupling the state and acceleration estimates (Friedland, 1969).

Again, a "slow" Kalman filter is set up. If the target initiates a maneuver, the residuals of the slow filter begin to display a characteristic divergence pattern. Reinitialization is accomplished by a least-squares estimator that forms a new estimate of the state over a predefined observation window using accumulated statistics.

To begin, let us define an observation window of length s. The observation window begins at time $n\,\Delta T$ in the past and extends over s measurements to the present time, $(n + s - 1)\,\Delta T$. The integer s is the total number of measurements used by the least-squares estimator (i.e., s is the observation window length). Choosing a proper value for s is critical; s needs to be large enough to supply a statistically well-behaved estimate yet small enough that the data contained in the observation window is relevant to the maneuver event.

At the start of the observation window, there is available to us an estimate of the target's state $\hat{\mathbf{x}}_q(n)$ and covariance $\mathbf{P}_q$ from the quiescent (slow) Kalman filter. The target's true state is perturbed from this estimate by an amount, $\Delta\mathbf{x}(n)$, due to the presence of the prior maneuver event at some unknown time in the past. The problem statement is to design a recursive least-squares estimate of the target's true state at the start of the observation window, denoted here as $\hat{\mathbf{x}}(n)$, based only on data relevant to the maneuver.

A Priori Data. First let us incorporate any *a priori* data that is available to the estimator. Assuming a sustained target maneuver, we know that $\Delta\mathbf{x}(n)$ must be Newtonian in form:

$$\Delta\mathbf{x}(n) = \begin{vmatrix} a_x t^2/2 \\ a_x t \\ a_x \end{vmatrix} \tag{8.8}$$

where a_x is the random target acceleration and t is the initiation time of the maneuver relative to the start of the observation window. Therefore, the expected value of $\Delta\hat{\mathbf{x}}(n)$ is zero, and its variance is given by

$$\Delta\mathbf{P} \cong \sigma_a^2 \begin{vmatrix} \dfrac{\Delta s^4}{20} & \dfrac{\Delta s^3}{8} & \dfrac{\Delta s^2}{6} \\ \dfrac{\Delta s^3}{8} & \dfrac{\Delta s^2}{3} & \dfrac{\Delta s}{2} \\ \dfrac{\Delta s^2}{6} & \dfrac{\Delta s}{2} & 1 \end{vmatrix} \tag{8.9}$$

where $E\{a_x^2(k)\} \triangleq \sigma_a^2$ and t is modeled as being uniformly distributed between $(0, \Delta s)$.

In conclusion, the *a priori* information available to us is

$$\hat{\mathbf{x}}(n) = \hat{\mathbf{x}}_q(n) \tag{8.10}$$

with covariance $\mathbf{P}_q + \Delta\mathbf{P}$. The covariance $\mathbf{P}_q$ is the predicted covariance from the quiescent Kalman filter, and $\Delta\mathbf{P}$ models any buildup in error that occurs before the start of the observation window. Note that the start of the observation window cannot coincide with the start of the maneuver in order to guarantee relevance to the unknown maneuver event.

Measurement Data. Subsequent measurements and variances are observed as given by

$$\mathbf{m}(k) = \mathbf{H}\mathbf{x}(k) + \mathbf{n}(k) \qquad \text{for } k \geqslant n \tag{8.11}$$

where $E[\mathbf{n}(k) \cdot \mathbf{n}(k)^T] = \mathbf{R}$ and $\mathbf{H} = [1 \quad 0 \quad 0]$.

Next consider the process model as given by (7.8):

$$\mathbf{x}(i + 1) = \mathbf{\Phi}(i + 1)\mathbf{x}(i) \tag{8.12}$$

It is convenient to define

$$\mathbf{\Phi}(k:n) = \sum_{i=n+1}^{k} \mathbf{\Phi}(i) \tag{8.13}$$

where, for a linear target model,

$$\mathbf{\Phi}(k:n) = \begin{vmatrix} 1 & (k-n)\Delta T & (k-n)^2 \Delta T^2/2 \\ 0 & 1 & (k-n)\Delta T \\ 0 & 0 & 1 \end{vmatrix}$$

The process is then

$$\mathbf{x}(k) = \mathbf{\Phi}(k:n)\mathbf{x}(n) \tag{8.14}$$

In general, the problem becomes one of estimating $\mathbf{x}(n)$ given the known prior data $\hat{\mathbf{x}}(n) = \hat{\mathbf{x}}_q(n)$ with covariance $\mathbf{P}_q + \Delta\mathbf{P}$, the measurements $\mathbf{m}(k)$ with covariance $\mathbf{R}$, and the following linearized process model:

$$\mathbf{m}(k) = \mathbf{H}'(k)\mathbf{x}(n) + \mathbf{n}(k) \tag{8.15}$$

where $\mathbf{H}'(k) \triangleq \mathbf{H\Phi}(k:n)$ and $\mathbf{n}(k)$ is the measurement noise.

When implemented in batch form, the least-squares estimate is expressed as

$$\hat{\mathbf{x}}(n) = \left[\sum_{i=n-1}^{k} \mathbf{H}'^{T}(i)\mathbf{R}^{-1}\mathbf{H}'(i) \right]^{-1} \left[\sum_{i=n-1}^{k} \mathbf{H}'^{T}(i)\mathbf{R}^{-1}\mathbf{m}(i) \right] \tag{8.16}$$

where the term $i = n - 1$ is included to allow the a priori information from (8.10) to be incorporated.

The weighted least-squares estimate can be viewed as a special case of the Kalman filter estimate when all the measurement data are processed in batch. This can be derived by examining the Kalman filter gain:

$$\begin{aligned}\mathbf{K}(k) &= \mathbf{P}(k)\mathbf{H}^{T}(k)[\mathbf{H}(k)\mathbf{P}(k)\mathbf{H}^{T}(k) + \mathbf{R}(k)]^{-1} \\ &= [\mathbf{P}(k)^{-1} + \mathbf{H}^{T}\mathbf{R}^{-1}\mathbf{H}]^{-1}\mathbf{H}^{T}\mathbf{R}^{-1}\end{aligned}$$

which, in the limit $\mathbf{P} \to \infty$, becomes the same as the weighted least-squares estimate that appears in (8.16):

$$\mathbf{K}(k) = (\mathbf{H}^{T}\mathbf{R}^{-1}\mathbf{H})^{-1}\mathbf{H}^{T}\mathbf{R}^{-1}$$

The additions and summations that appear in (8.16) may be computed continuously or in batch. At any rate the inversion is only performed if a maneuver is detected and always in batch. Since, in the absence of a maneuver, only the **R** covariance matrix must be inverted, the least-squares formulation reduces the computational burden relative to a second Kalman filter.

The corrected state at time k is the projection of $\hat{\mathbf{x}}(n)$ onto k,

$$\hat{\mathbf{x}}_{\text{corrected}}(k) = \mathbf{\Phi}(k:n)\hat{\mathbf{x}}(n) \tag{8.17}$$

Since the state correction affects a change in the state, a new covariance estimate must also be computed. The new covariance matrix is of the form

$$\mathbf{P}_{\text{corrected}}(k) = \mathbf{\Phi}(k:n)\mathbf{P}_{x}\mathbf{\Phi}(k:n)^{T} \tag{8.18}$$

where $\mathbf{P}_x$ is the error due to the imperfect estimation of $\hat{\mathbf{x}}(n)$. Inserting (8.15) into (8.16), the state correction error can be written as

$$\hat{\mathbf{x}}(n) - \mathbf{x}(n) = \left[\sum_{i=n-1}^{k} \mathbf{H}'^{T}(i)\mathbf{R}^{-1}\mathbf{H}'(i) \right]^{-1} \left[\sum_{i=n-1}^{k} \mathbf{H}'^{T}(i)\mathbf{R}^{-1}\mathbf{n}(i) \right]$$

Squaring both sides results in

$$\mathbf{P}_{x} = \left[\sum_{i=n-1}^{k} \mathbf{H}'^{T}(i)\mathbf{R}^{-1}\mathbf{H}'(i) \right]^{-1} \tag{8.19}$$

The length of the observation window s should be long enough to reduce $\mathbf{P}_x$ to fractional levels (especially in the acceleration state).

8.4.1 Simulation Results

An example was simulated consisting of a two-dimensional (x, y) tracker with $\Delta T = 10\,\text{sec}$, $\mathbf{R} = \text{diag}\{10^4\ \text{m}^2, 10^4\ \text{m}^2\}$, and $\mathbf{Q} = 0$. This is the same example as previously simulated by Chan, Hu, and Plant (1979) and by Bar-Shalom and Birmiwal (1982). By repeating this example, confidence is placed in the similarity of the various approaches despite the basic differences in implementation.

The simulated target maintained a constant course and speed until time $t = 400\,\text{sec}$ after which it commenced a 90° turn with acceleration $0.075\,\text{m/sec}^2$. The maneuver algorithm used a window length $s = 50\,\text{sec}$ and $\sigma_a = 0.1\ \text{m/sec}^2$. Figure 8.1a–c shows the average estimation errors after Monte Carlo averaging. As expected, the performance of the maneuver corrector was virtually the same as that found in the prior literature. Also shown in dots is the case $\Delta s = 20\,\text{sec}$, which exhibited substantially the same performance as the case $\Delta s = 0$.

Our primary aim was to investigate the sensitivity of the algorithm to the assumed time of the maneuver event. Relaxing the accuracy requirements on the assumed maneuver time allows a reduction in the computational complexity by applying coarser quantization steps. To analyze the effect of uncertainty in the maneuver time, the starting time of the window was assumed to lag the starting time of the maneuver by 20 sec. As Figure 8.2a–c shows, the filter experienced

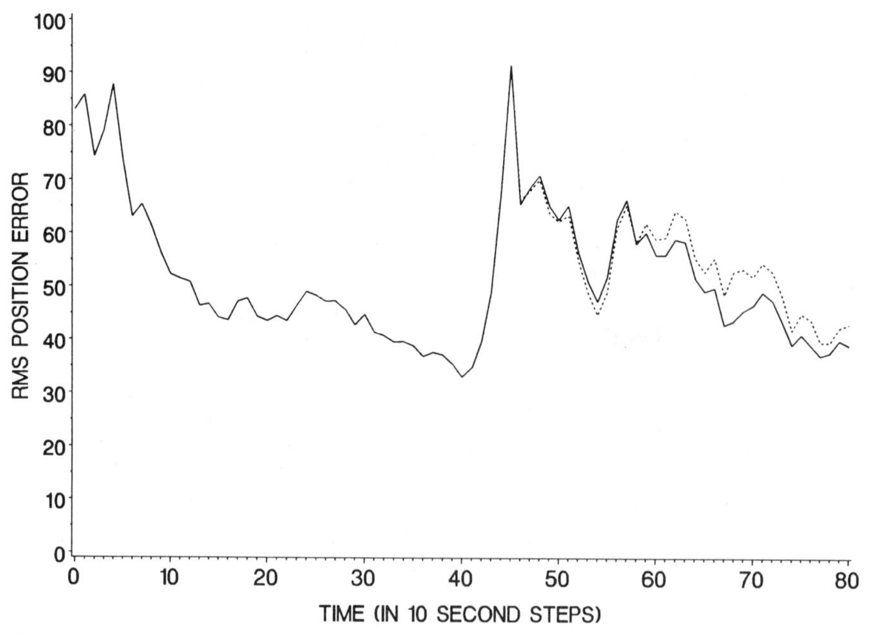

Figure 8.1 Correcting a target maneuver with perfect a priori information. Solid $\Delta S = 0$, Dots $\Delta S = 20\,\text{sec}$.

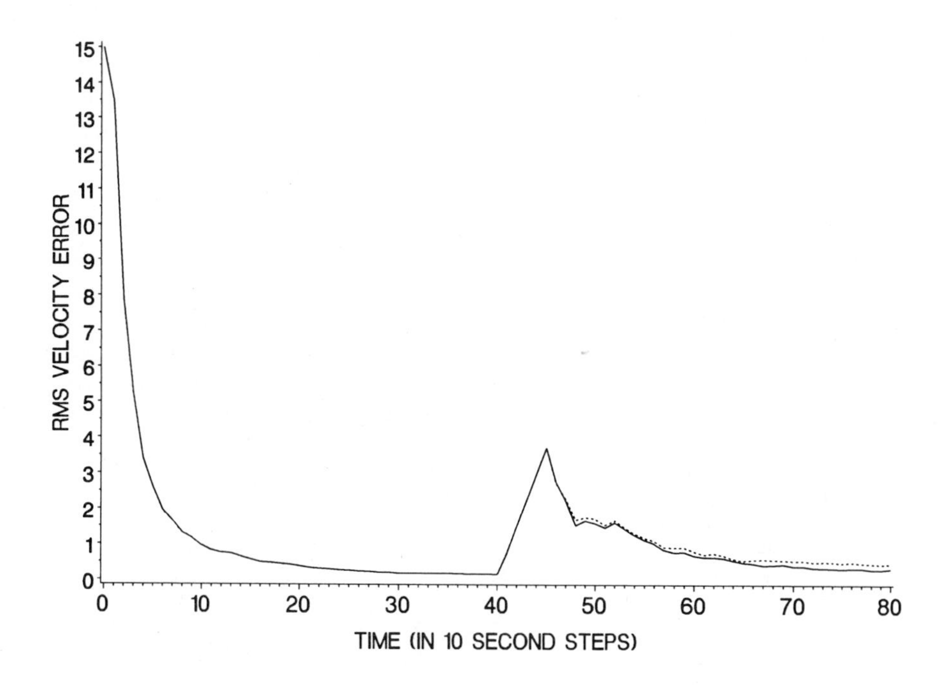

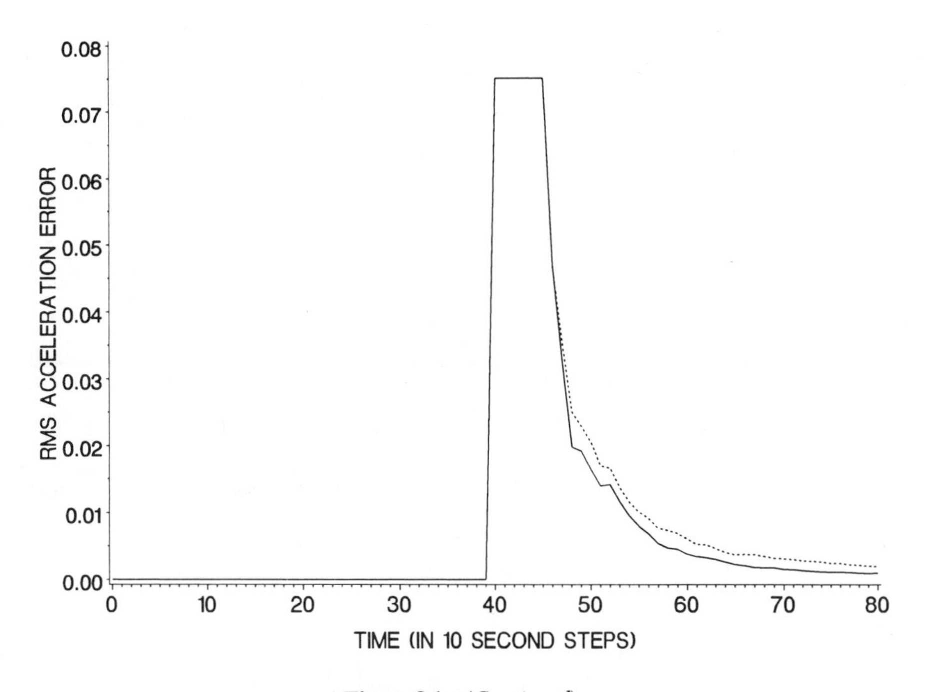

Figure 8.1 (*Continued*)

significant divergence in this case due to difficulties in separating out the velocity and acceleration estimates. By modeling this phenomenon (i.e., by setting $\Delta s = 20\,\text{sec}$ in our model) filter convergence was reestablished (see dotted line). Thus, Figure 8.2 confirms the validity of our model and the utility of incorporating *a priori* data into the least-square estimator.

Figure 8.3 graphs the performance of $\sigma_a = 0.1\,\text{m/sec}^2$ (in solid) versus $\sigma_a = 0.2\,\text{m/sec}^2$ (in dots) for the same case considered in Figure 8.1. This figure shows that the previous statements are not contingent on the assumed σ_a parameter. Figure 8.4 graphs the performance of $\Delta s = 0$ (in solid) versus $\Delta s = 40$ seconds (in dots) for the case considered in Figure 8.2. The case $\Delta s = 40\,\text{sec}$ is interesting because it shows a slight performance improvement over the case $\Delta s = 20\,\text{sec}$. This performance improvement is a consequence of the better damping provided by the adjusted covariance and the better matching to a uniform pdf.

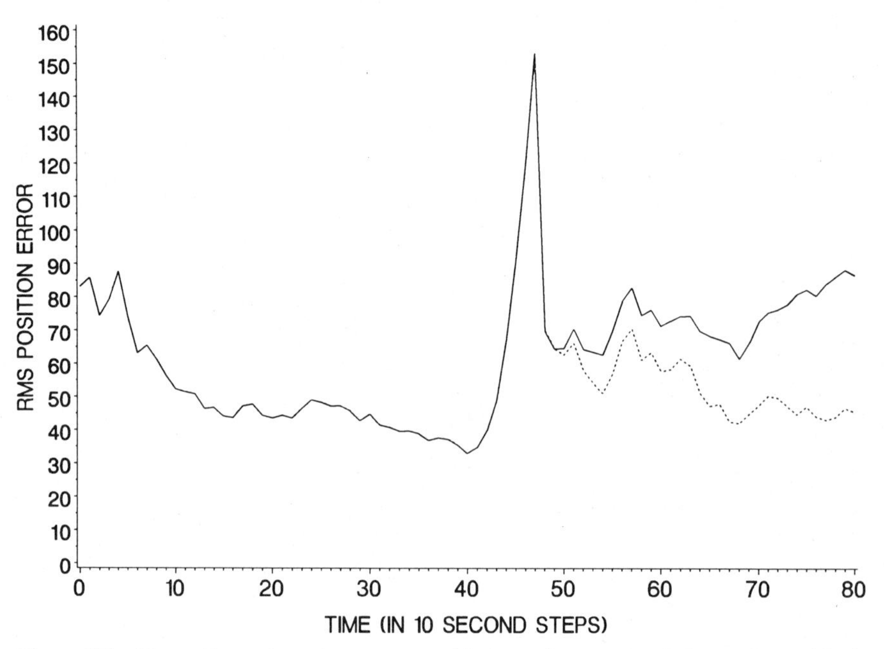

Figure 8.2 Correcting a target maneuver with imperfect *a priori* information, with the assumed maneuver event lagging the true maneuver by 20 sec. Solid $\Delta S = 0$; Dots $\Delta S = 20$ sec.

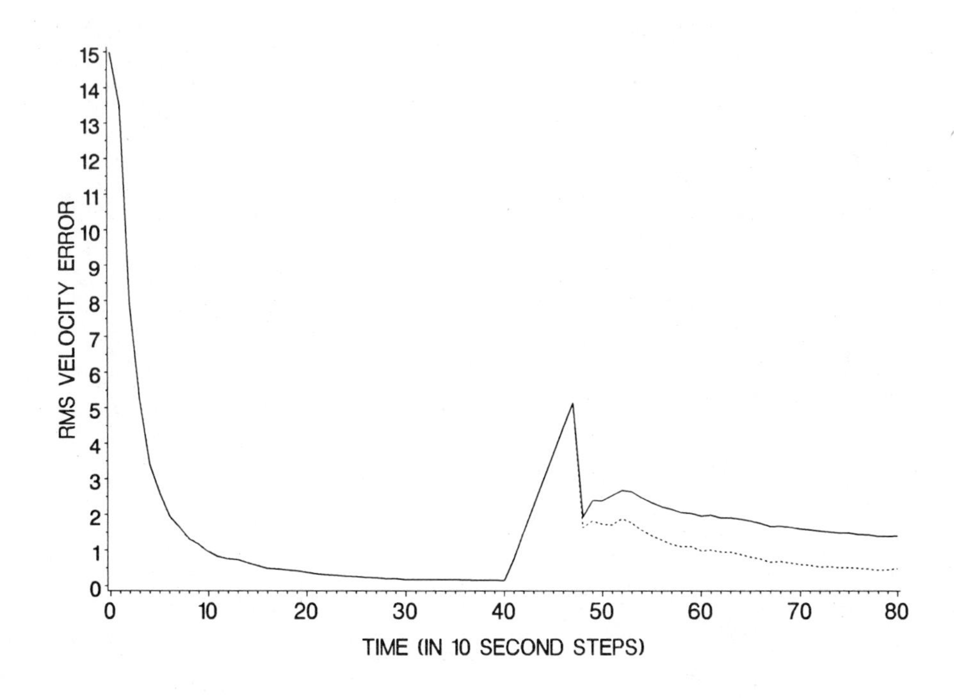

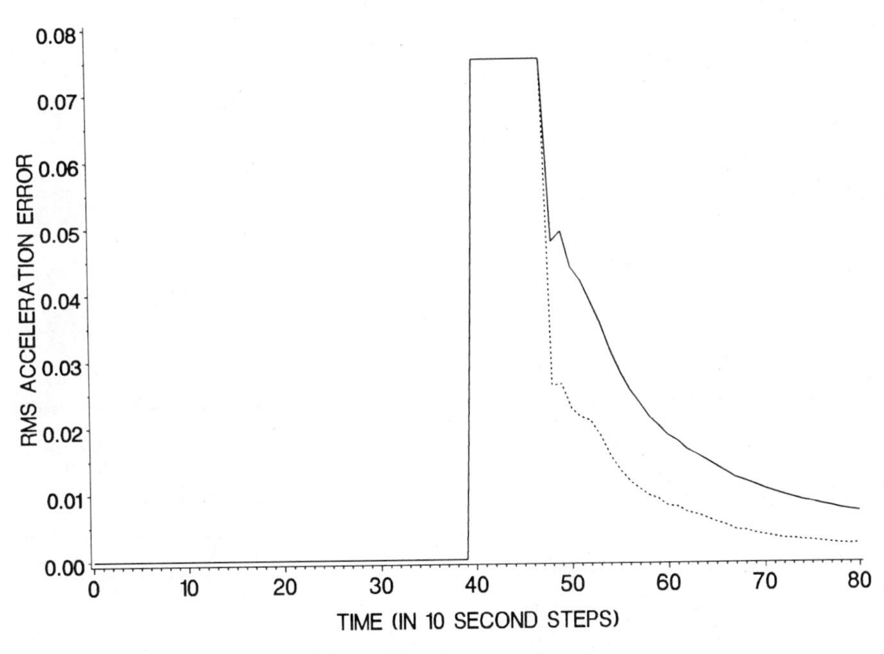

Figure 8.2 (*Continued*)

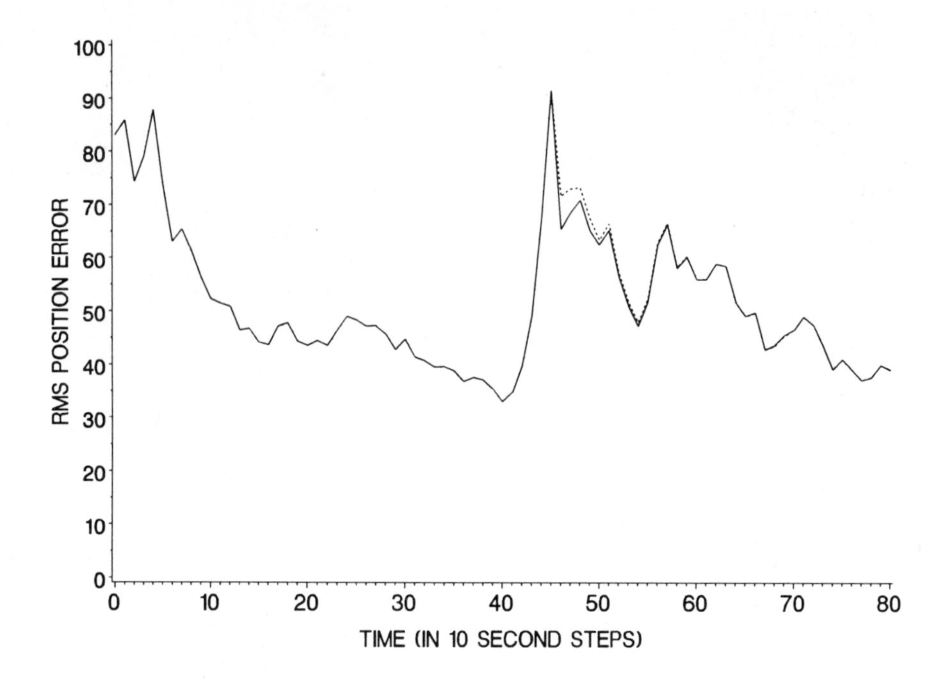

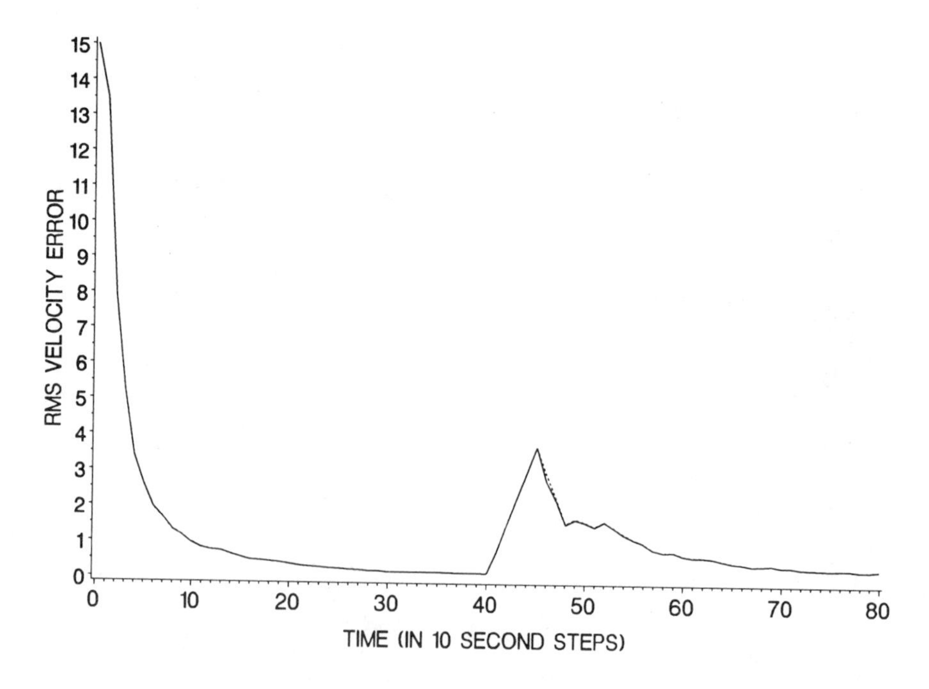

Figure 8.3 Perturbing the assumed acceleration statistics. Solid $\sigma_a = 0.1\ \text{m/sec}^2$, Dots $\sigma_a = 0.2\ \text{m/sec}^2$, $\Delta S = 0$.

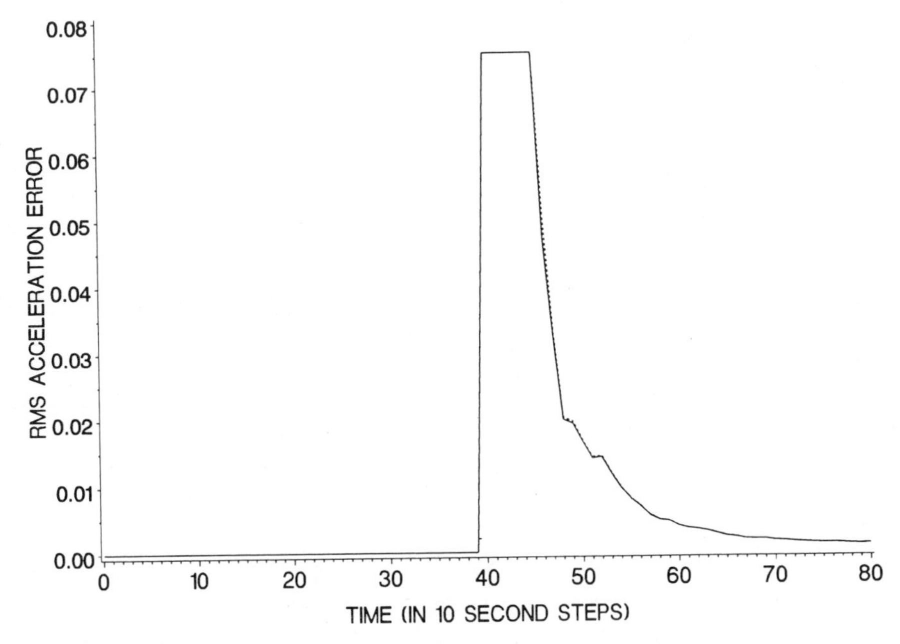

Figure 8.3 (*Continued*)

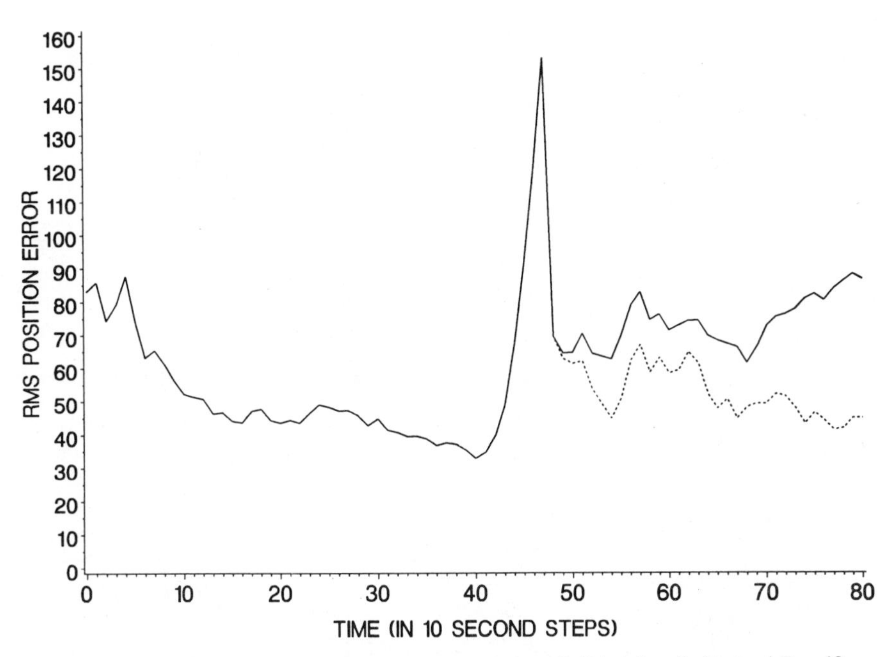

Figure 8.4 Perturbing the assumed time statistics. Solid $\Delta S = 0$, Dots $\Delta S = 40$ sec, $\sigma_a = 0.1$ m/sec^2.

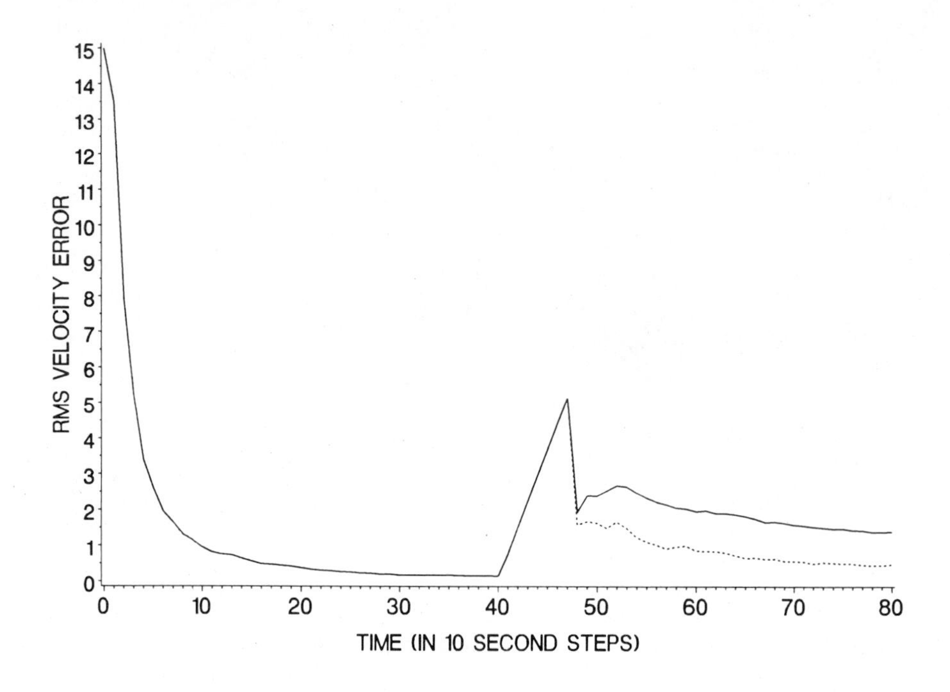

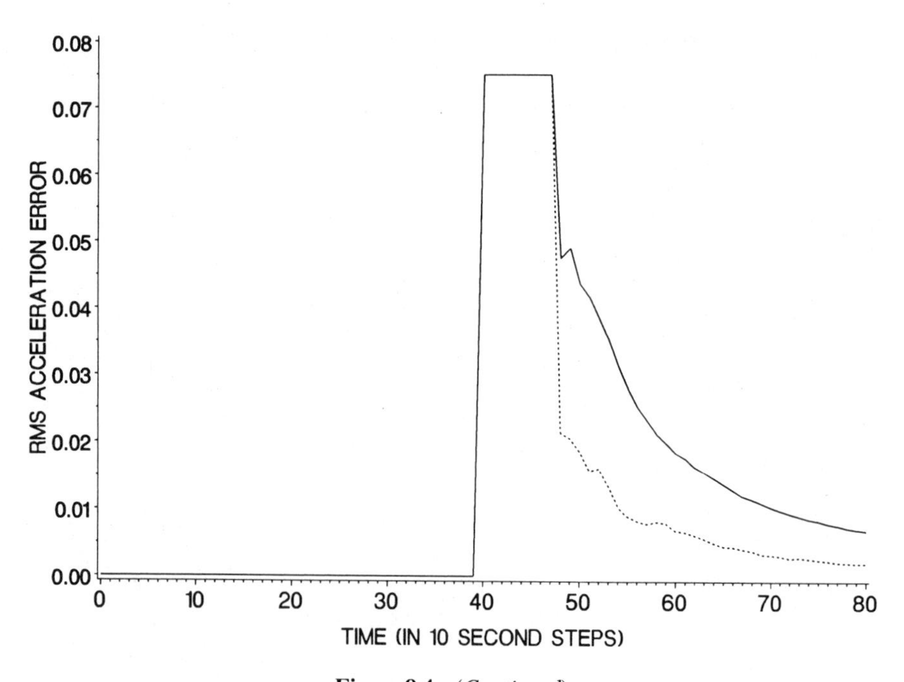

Figure 8.4 (*Continued*)

8.5 DOPPLER MEASUREMENTS

Accurate knowledge of the target's Doppler is of great benefit in determining the strength of the maneuver in the range dimension and the time of the maneuver's initiation prior to any buildup in position error. However, it would be incorrect to state that maneuvers do not pose a problem in those systems incorporating Doppler measurements. To understand this apparent contradiction, it is necessary to understand the phenomenon of Doppler spectra (DS).

Doppler spectra are a problem when tracking targets in Doppler. Spurious Doppler spectra returns can be caused by the target's propeller blades, jet engine turbine blades, helicopter blades, tank treads, or wheel modulations. DS returns have the same azimuth, elevation, and range as the target but possess a different Doppler frequency from the true return (or "skin").

If the track is incorrectly updated with a DS return, the Doppler measurement is corrupted by the amount of the DS offset. Furthermore, once a track has been updated by a DS return, it often will not be associated with future skin returns. When this occurs, the range and range rate track predictions will diverge greatly from the true range and range rate. Eventually, unless the track is corrected, the target will be lost.

The presence of an offset in the Doppler measurement can be corrected using the procedures outlined previously. Let us define g to be the DS offset state. Inserting $g(k)$ into the state variable, we have

$$\mathbf{x}(n) = \begin{vmatrix} x \\ v_x \\ a_x \\ g \end{vmatrix} \tag{8.20}$$

where the measurement matrix is of the form:

$$\mathbf{H} = \begin{vmatrix} 1 & 0 & 0 & 0 \\ 0 & 1 & 0 & 1 \end{vmatrix} \tag{8.21}$$

The initial values (x, v_x, a_x) are known at the start of the observation window $n\Delta T$ with covariance $\mathbf{P}$ (since there is little uncertainty in establishing the time of the offset when tracking the target in Doppler). The offset g is modeled as a zero-mean random variable with variance σ_g.

Subsequent measurements and variances are observed as given by

$$\mathbf{m}(k) = \mathbf{H}\mathbf{x}(k) + \mathbf{n}(k) \qquad \text{for } k \geqslant n \tag{8.22}$$

where $E[\mathbf{n}(k) \cdot \mathbf{n}(k)^T] = \mathbf{R}$. The process model is

$$\mathbf{x}(k) = \mathbf{\Phi}(k:n)\mathbf{x}(n) \tag{8.23}$$

where

$$\mathbf{\Phi}(k:n) = \begin{vmatrix} 1 & (k-n)\Delta T & (k-n)^2 \Delta T^2/2 & 0 \\ 0 & 1 & (k-n)\Delta T & 0 \\ 0 & 0 & 1 & 0 \\ 0 & 0 & 0 & 1 \end{vmatrix}$$

Least-squares estimation proceeds, as outlined previously, to correct the Kalman filter estimate if it is later decided that the Kalman filter has falsely locked onto a DS line. The estimated value of g is used to reset the track onto the target skin line.

8.6 SUMMARY

Four subjects seem to dominate the theory of adaptive filters: covariance adjustment, batch processing, multiple recursive filters, and least-squares estimation (either recursive or batch). The first approach suffers delay problems that prevent it from attaining the performance of the latter three approaches although covariance adjustment does suffice to follow the target under benign maneuver conditions. The last three approaches perform similarly except for minor differences; it is believed that their primary distinctions are in implementation. The reader is referred to the books by Bar-Shalom and Fortmann, Farina and Studer, and Blackman for further study.

CHAPTER 9

COORDINATE SYSTEMS

In order to describe the target's position in three dimensions, a set of coordinate frames must first be specified. Probably no other area of tracking has received as much attention as this subject. This is because reducing the computational complexity is a primary consideration in the design of most tracking systems. Typically this involves selecting a coordinate frame that allows the filter to be at least partially decoupled.

For convenience, we shall ignore the topic of computational reduction in this chapter. Rather, some of the salient background material is presented to aid the reader in general theoretical understanding of radar coordinate frames.

9.1 MEASUREMENT COORDINATES

The raw three-dimensional measurements constitute the following quantities:

$$\mathbf{m} = \begin{vmatrix} r_m \\ \dot{r}_m \\ \zeta_{a_m} \\ \zeta_{e_m} \end{vmatrix} \tag{9.1}$$

where r_m is the range measurement, $\dot{r}_m$ is the range rate or Doppler measurement, and ζ_e, ζ_a are the direction cosines referenced to the monopulse phase centers, elevation (e) and azimuth (a), respectively. This "measurement" coordinate frame is the most direct coordinate frame and has been used in many past applications (Wishner et al., 1970; Mehra, 1971; Fitzgerald, 1974b; Bath et al., 1980). However, its use entails working with a nonlinear measurement vector.

9.2 ANTENNA COORDINATES

An alternate and simpler approach is to reformulate the measurement into a Cartesian coordinate frame. It can be shown that this approach results in negligible performance loss provided some simplifying assumptions can be made. Namely, the underlying assumption is that the measurement error comprises a linear error ellipsoid around a predictable three-dimensional Cartesian vector.

To this end, let us introduce the *antenna* (or *face*) coordinate frame. The antenna coordinate frame is that Cartesian coordinate frame that coincides with the electrical axis of the antenna. To be precise, the b, or boresight, axis coincides with the antenna electrical boresight. The e and a axes coincide with the monopulse elevation and azimuth axes. [Note that the *electrical* axis (b, a, e) may differ slightly from the actual *mechanical* axis of the antenna face.]

Ignoring for now the Doppler measurement, the new measurement vector is defined as*

$$\mathbf{m}_{bae} \triangleq \begin{vmatrix} \zeta_{b_m} r_m \\ \zeta_{a_m} r_m \\ \zeta_{e_m} r_m \end{vmatrix} \triangleq \begin{vmatrix} x_{b_m} \\ x_{a_m} \\ x_{e_m} \end{vmatrix} . \tag{9.2}$$

where

$$(\zeta_a, \zeta_e) = \text{true direction cosines}$$

$$(\zeta_{a_m}, \zeta_{e_m}) = \text{measured direction cosines}$$

$$\zeta_{b_m} \triangleq \sqrt{1 - \zeta_{a_m}^2 - \zeta_{e_m}^2}$$

$$(x_b, x_a, x_e) = \text{target position in } (b, a, e) \text{ coordinates}$$

$$(x_{b_m}, x_{a_m}, x_{e_m}) = \text{measured target position in } (b, a, e) \text{ coordinates}$$

The corresponding covariance matrix is

$$\mathbf{R}_{bae} = \begin{vmatrix} r_{11} & r_{12} & r_{13} \\ r_{12} & r_{22} & r_{23} \\ r_{13} & r_{23} & r_{33} \end{vmatrix} \tag{9.3}$$

*The (b, a, e) notation is a fusion of Mehra's (u, v, w) notation and Pearson's (r, e, d) notation. This notation is used because it is equally compatible with the STT, TWS, and ESA tracking applications.

where

$$
\begin{aligned}
r_{11} &= \sigma_r^2 \bar{\zeta}_b^2 + r^2 \sigma_{\zeta_e}^2 \bar{\zeta}_e^2 / \bar{\zeta}_b^2 + r^2 \sigma_{\zeta_a}^2 \bar{\zeta}_a^2 / \bar{\zeta}_b^2 \\
r_{22} &= r^2 \sigma_{\zeta_a}^2 + \sigma_r^2 \bar{\zeta}_a^2 \\
r_{33} &= r^2 \sigma_{\zeta_e}^2 + \sigma_r^2 \bar{\zeta}_e^2 \\
r_{23} &= \sigma_r^2 \bar{\zeta}_e \bar{\zeta}_a \\
r_{13} &= \sigma_r^2 \bar{\zeta}_b \bar{\zeta}_e - r^2 \sigma_{\zeta_e}^2 \bar{\zeta}_e / \bar{\zeta}_b \\
r_{12} &= \sigma_r^2 \bar{\zeta}_b \bar{\zeta}_a - r^2 \sigma_{\zeta_a}^2 \bar{\zeta}_a / \bar{\zeta}_b
\end{aligned}
$$

and where

$$
\begin{aligned}
\sigma_r^2 &= \text{range measurement variance} \\
\sigma_{\zeta_e}^2 &= \text{elevation measurement variance} \\
\sigma_{\zeta_a}^2 &= \text{azimuth measurement variance}
\end{aligned}
$$

The inherent assumption of (9.3) is that the Gaussian error ellipsoid in measurement coordinates can be modeled as a Gaussian error ellipsoid in antenna coordinates. This assumption is equivalent to truncating the expansion of the ellipsoid to the first partial derivative. It is also assumed that the point of linearization is equal to the predicted target position. Both of these assumptions may be only approximate for the application of interest. Alternate formulations and some insight into the problem may be found in (Bar-Shalom and Fortmann, 1988, pp. 106–121).

9.3 COVARIANCE COORDINATES

In general the covariance matrix in (9.3) contains off-diagonal elements if the target is off electrical boresight. A covariance coordinate is (by definition) any coordinate frame that diagonalizes **R**. The required transformation is

$$\mathbf{m}^* = \{*, bae\}\mathbf{m}_{bae} \tag{9.4}$$

where, in order to diagonalize **R**, the transformation matrix $\{*, bae\}$ must obey

$$\{*, bae\}\mathbf{R}_{bae}\{*, bae\}^T = \text{diagonal} \tag{9.5}$$

There are several reasons for wishing to diagonalize **R**. For one, a diagonal **R** matrix prevents an accurate measurement in one direction (such as range) from being corrupted due to computer round-off errors introduced by a less accurate measurement in another direction (Daum and Fitzgerald, 1983). Second, the measurements can be more easily processed one at a time by processing the appropriate rows in the **H** matrix, and thus there is no longer any need to compute the matrix inverse of the $M \times M$ matrices $\mathbf{HPH}^T$ and **R** when computing the Kalman filter gain. A third benefit is that the transformation $\{*, bae\}$ can also be used to simplify the gating and correlation processes discussed in Chapter 11. Diagonalizing **R** also facilitates filter decoupling of the type discussed in Baheti (1986) and Daum and Fitzgerald (1983). Finally, it simplifies the introduction of square-root filtering and adaptive filtering techniques (Maybeck, 1979).

9.3.1 RPCC Coordinates

Equation (9.5) does not uniquely define a single coordinate frame. Various authors (Brown et al., 1977) have suggested choosing $\{*, bae\}$ such as to correspond to a Cartesian coordinate frame that is aligned with the predicted target line-of-sight vector $\mathbf{i}_r$,

$$\mathbf{i}_r = (\bar{\zeta}_b \mathbf{i}_b + \bar{\zeta}_a \mathbf{i}_a + \bar{\zeta}_e \mathbf{i}_e)$$

As Figure 9.1 shows, the other axes are oriented to align with a spherical coordinate frame. This coordinate frame is denoted as the radar principal Cartesian coordinates (RPCC), and the letters (r, θ, ϕ) are used for subscripting.

The spherical transformation obeys (Hayt, 1974, p. 25)

$$\{r\theta\phi, bae\} = \begin{vmatrix} \bar{\zeta}_b & \bar{\zeta}_a & \bar{\zeta}_e \\ \bar{\zeta}_a & -\bar{\zeta}_a\bar{\zeta}_b/\bar{\zeta}_c & -\bar{\zeta}_e\bar{\zeta}_b/\bar{\zeta}_c \\ 0 & \bar{\zeta}_e/\bar{\zeta}_c & -\bar{\zeta}_a/\bar{\zeta}_c \end{vmatrix} \tag{9.6}$$

where

$$\bar{\zeta}_c \triangleq \sqrt{\bar{\zeta}_e^2 + \bar{\zeta}_a^2} \triangleq \sqrt{1 - \bar{\zeta}_b^2}$$

and where the overbar denotes predicted quantities.

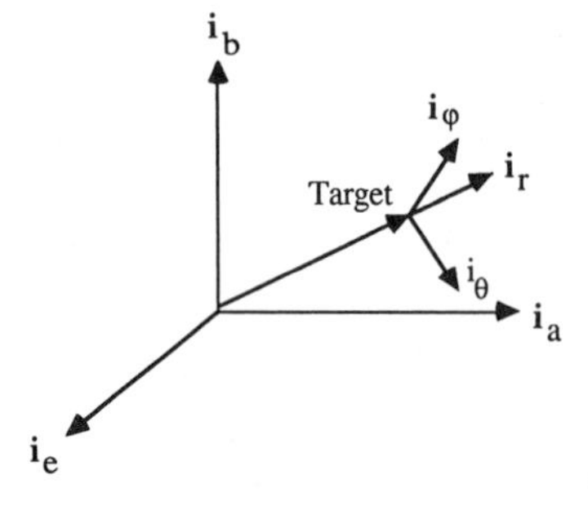

Figure 9.1 Radar principal Cartesian coordinates (r, θ, ϕ) superimposed on antenna (b, a, e) coordinates.

The new measurement vector is approximately

$$\mathbf{m}_{r\theta\phi} \cong \begin{vmatrix} r_m \\ -r_m[\bar{\zeta}_e(\zeta_{e_m} - \bar{\zeta}_e) + \bar{\zeta}_a(\zeta_{a_m} - \bar{\zeta}_a)]/(\bar{\zeta}_c\bar{\zeta}_b) \\ r_m[\bar{\zeta}_e(\zeta_{a_m} - \bar{\zeta}_a) - \bar{\zeta}_a(\zeta_{e_m} - \bar{\zeta}_e)]/\bar{\zeta}_c \end{vmatrix} \cong \begin{vmatrix} x_{r_m} \\ x_{\theta_m} \\ x_{\phi_m} \end{vmatrix} \tag{9.7}$$

Equation (9.7) is derived by substituting (9.6) and (9.2) into (9.4). The ζ_{b_m} term is then expanded in a Taylor series about $\bar{\zeta}_b$, and the higher-order angular error terms (ζ_a, ζ_e) are dropped. The accuracy of this approximation can be improved by employing multiple passes (Gelb, 1974) or by ordering the sequence of measurements to process the angle measurements first (Miller and Leskiw, 1982).

Given (9.7), the corresponding covariance matrix is

$$\mathbf{R}_{r\theta\phi} = \begin{vmatrix} \sigma_r^2 & 0 & 0 \\ 0 & r^2\sigma_\zeta^2/\bar{\zeta}_b^2 & 0 \\ 0 & 0 & r^2\sigma_\zeta^2 \end{vmatrix} \tag{9.8}$$

provided $\sigma_{\zeta_e}^2 \cong \sigma_{\zeta_a}^2 = \sigma_\zeta^2$.

Incorporating the Doppler measurement at this stage is straightforward. Recognizing that the Doppler measurement corresponds to the target's projected velocity in the true line-of-sight direction, the measurement becomes

$$\mathbf{m}_{r\theta\phi} \cong \begin{vmatrix} r_m \\ -r_m[\bar{\zeta}_e(\zeta_{e_m} - \bar{\zeta}_e) + \bar{\zeta}_a(\zeta_{a_m} - \bar{\zeta}_a)]/(\bar{\zeta}_c\bar{\zeta}_b) \\ r_m[\bar{\zeta}_e(\zeta_{a_m} - \bar{\zeta}_a) - \bar{\zeta}_a(\zeta_{e_m} - \bar{\zeta}_e)]/\bar{\zeta}_c \\ \dot{r}_m \end{vmatrix} \cong \begin{vmatrix} x_{r_m} \\ x_{\theta_m} \\ x_{\phi_m} \\ (v_r + x_\theta v_\theta/x_r \\ \quad + x_\phi v_\phi/x_r)_m \end{vmatrix} \tag{9.9}$$

with associated covariance

$$\mathbf{R}_{r\theta\phi} = \begin{vmatrix} \sigma_r^2 & 0 & 0 & 0 \\ 0 & r^2\sigma_\zeta^2/\bar{\zeta}_b & 0 & 0 \\ 0 & 0 & r^2\sigma_\zeta^2 & 0 \\ 0 & 0 & 0 & \sigma_{\dot{r}}^2 \end{vmatrix} \tag{9.10}$$

and with $\mathbf{H}$ given by

$$\mathbf{H}_{r\theta\phi} \cong \begin{vmatrix} 1 & 0 & 0 & 0 & 0 & 0 & 0 & 0 & 0 \\ 0 & 0 & 0 & 1 & 0 & 0 & 0 & 0 & 0 \\ 0 & 0 & 0 & 0 & 0 & 0 & 1 & 0 & 0 \\ 0 & 1 & 0 & v_\theta/x_r & 0 & 0 & v_\phi/x_r & 0 & 0 \end{vmatrix} \tag{9.11}$$

To first order, the Doppler measurement is equal to the predicted line-of-sight velocity, $\dot{r} \cong v_r$. However, the measurement is also a function of the cross-range position errors x_θ and x_ϕ. Thus, incorporation of the Doppler measurement has the effect of coupling the three coordinate axes together. In those situations where the predicted position is reasonably well known, this coupling can be ignored, thus simplifying the gain computation. The resulting covariance in this case is

$$\mathbf{R}_{r\theta\phi} = \begin{vmatrix} \sigma_r^2 & 0 & 0 & 0 \\ 0 & r^2\sigma_\zeta^2/\overline{\zeta}_b & 0 & 0 \\ 0 & 0 & r^2\sigma_\zeta^2 & 0 \\ 0 & 0 & 0 & r_{44} \end{vmatrix} \tag{9.12}$$

for

$$r_{44} = \sigma_{\dot{r}}^2 + \sigma_{x_\theta}^2 v_\theta^2/x_r^2 + \sigma_{x_\phi}^2 v_\phi^2/x_r^2 \tag{9.13}$$

9.4 RANGE–VELOCITY COORDINATES

One interesting result of the previous analysis was that the Doppler measurement couples the predicted line-of-sight velocity, v_r, and the cross-range position errors x_θ and x_ϕ. If we transform to a new coordinate system aligned with the range and velocity vectors, however, this coupling can be partially reduced, thus simplifying the measurement updating process (Daum and Fitzgerald, 1983, p. 278).

Denote x_r the target position along the predicted range vector, denote x_v the target position normal to x_r in the range–velocity plane, and denote x_l the target position normal to that plane. The required transformation matrix is

$$\{rvl, r\theta\phi\} = \begin{vmatrix} 1 & 0 & 0 \\ 0 & v_\theta/v_v & v_\phi/v_v \\ 0 & -v_\phi/v_v & v_\theta/v_v \end{vmatrix} \tag{9.14}$$

for

$$v_v^2 \triangleq v_\phi^2 + v_\theta^2$$

Since $x_v = x_\theta(v_\theta/v_v) + x_\phi(v_\phi/v_v)$, the transformation results in the following partially decoupled Doppler measurement:

$$\dot{r}_m = (v_r + x_v v_v/x_r)_m \tag{9.15}$$

This coordinate system is referred to as the range–velocity Cartesian coordinate (RVCC) system (Brown et al., 1977).

9.5 NORTH, EAST, AND DOWN COORDINATES

State extrapolation is most easily carried out in an *inertial* Cartesian coordinate frame. Inertial means nonaccelerating and nonrotating. Use of the term *inertial* must be taken in context since no coordinate frame is inertial in an absolute sense (i.e., a fixed object is actually rotating with the earth, the earth is actually rotating around the sun, and the sun is accelerating with our galaxy). In this context an *inertial* coordinate frame refers to any coordinate frame where the rotations and accelerations are small relative to the target accelerations of interest.

For tracking over short distances, the most popular inertial coordinate frame is the north, east, down (NED) coordinate frame. NED coordinates are oriented to align with local north, east, down, and so are invariant to changes in the antenna's instantaneous pointing angle or the platform's instantaneous yaw, pitch, and roll angle. However, not all NED coordinates are invariant to gross platform motion; indeed, most applications peg the origin of the NED coordinate frame to the platform (platform-pegged NED coordinates).

In NED coordinates the state vector is given by

$$\mathbf{x}_{\text{NED}} = \begin{vmatrix} x_N \\ v_N \\ x_E \\ v_E \\ x_D \\ v_D \end{vmatrix} \tag{9.16a}$$

or, in those cases where acceleration is incorporated as a state,

$$\mathbf{x}_{\text{NED}} = \begin{vmatrix} x_N \\ v_N \\ a_N \\ x_E \\ v_E \\ a_E \\ x_D \\ v_D \\ a_D \end{vmatrix} \tag{9.16b}$$

9.5.1 Coordinate Transformations

A coordinate transformation must be performed to convert the measurement vector $\mathbf{m}_{r\theta\phi}$ to the state vector $\mathbf{x}_{\text{NED}}$. Let us first consider the transformation from antenna coordinates to NED coordinates. This may require the use of an

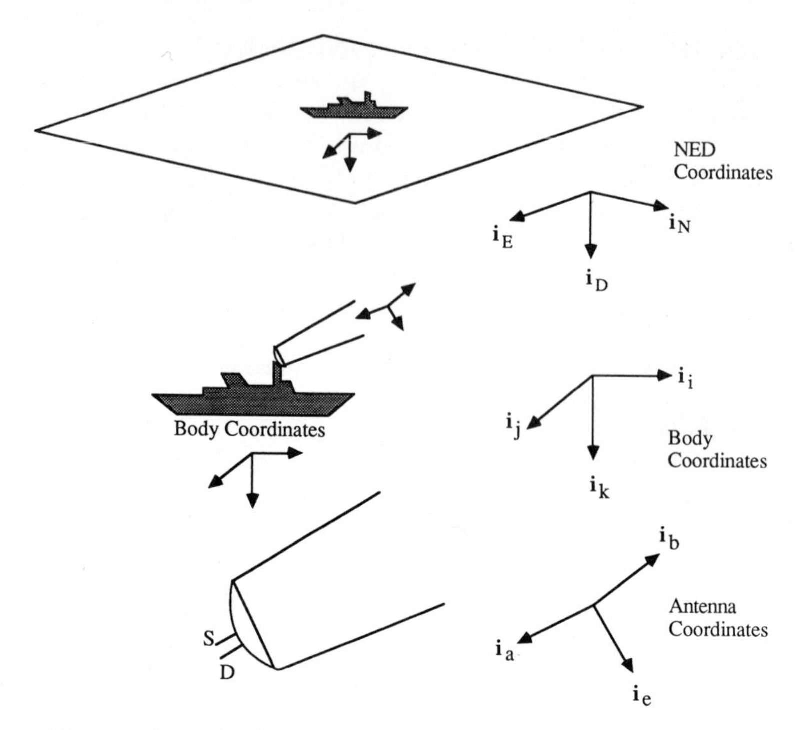

Figure 9.2 Transformation between NED and antenna coordinates in general requires the use of body coordinates.

intermediary coordinate frame called *body coordinates*. The geometry to which this applies is illustrated by Figure 9.2.

Let us define coordinate vectors as follows:

NED: Cartesian coordinate vectors aligned with inertial north, east, down.

- $\mathbf{i}_N$ North
- $\mathbf{i}_E$ East
- $\mathbf{i}_D$ Down

Body: Noninertial Cartesian coordinate vectors aligned with the platform's body.

- $\mathbf{i}_i$ Vector coincident with $\mathbf{i}_b$ when the gimbal angles are zero
- $\mathbf{i}_j$ Vector orthogonal to $\mathbf{i}_i$ and $\mathbf{i}_k$, and coincident with $\mathbf{i}_a$ when the gimbal angles are zero
- $\mathbf{i}_k$ Vector orthogonal to $\mathbf{i}_i$ and $\mathbf{i}_j$, and coincident with $\mathbf{i}_e$ when the gimbal angles are zero

Antenna: Noninertial Cartesian coordinate vectors aligned with antenna's electrical boresight.

- $\mathbf{i}_b$ Vector coincident with the antenna's electrical boresight

$\mathbf{i}_a$ Vector in the plane of the antenna and coincident with the azimuth monopulse channel

$\mathbf{i}_e$ Vector in the plane of the antenna and coincident with the elevation monopulse channel

Notice that the antenna and body coordinate vectors coincide when the gimbal angles are zero (by definition). However, the gimbal angles change if the antenna mechanically scans.

Each coordinate rotation is attained by multiplying the original input vectors with various combinations of the matrices defined by

$$\{1, \delta\} = \begin{vmatrix} 1 & 0 & 0 \\ 0 & \cos(\delta) & \sin(\delta) \\ 0 & -\sin(\delta) & \cos(\delta) \end{vmatrix} \tag{9.17a}$$

$$\{2, \delta\} = \begin{vmatrix} \cos(\delta) & 0 & -\sin(\delta) \\ 0 & 1 & 0 \\ \sin(\delta) & 0 & \cos(\delta) \end{vmatrix} \tag{9.17b}$$

$$\{3, \delta\} = \begin{vmatrix} \cos(\delta) & \sin(\delta) & 0 \\ -\sin(\delta) & \cos(\delta) & 0 \\ 0 & 0 & 1 \end{vmatrix} \tag{9.17c}$$

for δ a dummy variable.

The transform operation is as follows:

$$\begin{vmatrix} \mathbf{i}_b \\ \mathbf{i}_a \\ \mathbf{i}_e \end{vmatrix}_{\text{antenna}} = \{2, v\}\{3, \eta\}\{1, \phi\}\{2, \delta\}\{3, \psi\} \begin{vmatrix} \mathbf{i}_N \\ \mathbf{i}_E \\ \mathbf{i}_D \end{vmatrix}_{\text{NED}} \tag{9.18a}$$

where the coordinate rotation utilizes the known antenna gimbal angles, η and v, and body angles, ϕ, δ, and ψ, corresponding to roll, pitch, and heading, respectively.

Likewise,

$$\begin{vmatrix} \mathbf{i}_N \\ \mathbf{i}_E \\ \mathbf{i}_D \end{vmatrix}_{\text{NED}} = \{3, -\psi\}\{2, -\delta\}\{1, -\phi\}\{3, -\eta\}\{2, -v\} \begin{vmatrix} \mathbf{i}_b \\ \mathbf{i}_a \\ \mathbf{i}_e \end{vmatrix}_{\text{Antenna}} \tag{9.18b}$$

The antenna-to-RPCC transformation from (9.6) is

$$\begin{vmatrix} \mathbf{i}_r \\ \mathbf{i}_\theta \\ \mathbf{i}_\phi \end{vmatrix}_{\text{RPCC}} = \begin{vmatrix} \bar{\zeta}_b & \bar{\zeta}_a & \bar{\zeta}_e \\ \bar{\zeta}_a & -\bar{\zeta}_a\bar{\zeta}_b/\bar{\zeta} & -\bar{\zeta}_e\bar{\zeta}_b/\bar{\zeta}_c \\ 0 & \bar{\zeta}_e/\bar{\zeta} & -\bar{\zeta}_a/\bar{\zeta} \end{vmatrix} \begin{vmatrix} \mathbf{i}_b \\ \mathbf{i}_a \\ \mathbf{i}_e \end{vmatrix}_{\text{Antenna}}$$

The residual is then

$$\text{Residual}_{r\theta\phi} = \mathbf{m}_{r\theta\phi} - \mathbf{H}_{r\theta\phi}\bar{\mathbf{x}}_{r\theta\phi}$$

$$\cong \begin{vmatrix} r_m - \bar{r} \\ -r_m[\bar{\zeta}_e(\zeta_{e_m} - \bar{\zeta}_e) + \bar{\zeta}_a(\zeta_{a_m} - \bar{\zeta}_a)]/(\bar{\zeta}_c\bar{\zeta}_b) \\ r_m[\bar{\zeta}_e(\zeta_{a_m} - \bar{\zeta}_a) - \bar{\zeta}_a(\zeta_{e_m} - \bar{\zeta}_e)]/\bar{\zeta}_c \\ \dot{r}_m - \bar{\dot{r}} \end{vmatrix} \tag{9.19a}$$

The Kalman gain in RPCC is given by

$$\mathbf{K}(k)_{r\theta\phi} = \mathbf{P}(k)_{r\theta\phi}\mathbf{H}^T_{r\theta\phi}(k)[\mathbf{H}_{r\theta\phi}(k)\mathbf{P}_{r\theta\phi}(k)\mathbf{H}^T_{r\theta\phi}(k) + \mathbf{R}_{r\theta\phi}(k)]^{-1}$$

with **H** given by

$$\mathbf{H}_{r\theta\phi} \cong \begin{vmatrix} 1 & 0 & 0 & 0 & 0 & 0 & 0 & 0 & 0 \\ 0 & 0 & 0 & 1 & 0 & 0 & 0 & 0 & 0 \\ 0 & 0 & 0 & 0 & 0 & 0 & 1 & 0 & 0 \\ 0 & 1 & 0 & v_\theta/x_r & 0 & 0 & v_\phi/x_r & 0 & 0 \end{vmatrix} \tag{9.19b}$$

The new state is computed by multiplying the residual by the gain in RPCC ($r\theta\phi$) and then transforming back to NED coordinates for state extrapolation.

9.6 SITE COORDINATES

Site coordinates (also called *earth-pegged NED* or *navigation coordinates*) refers to any NED coordinate frame whose origin is pegged (or fixed) to a specific location on the earth's surface while the radar is moving. Those applications that use site coordinates are characterized by a local target area under constant surveillance by a moving platform. A specific example includes an airborne radar in a racetrack flight pattern. The z axis always points down, and the x-y plane lies on the surface of the earth. In the case of surface targets, cartographic data may be used to transform the Doppler measurement into a 3-D target velocity in site coordinates.

9.7 EARTH-CENTERED COORDINATES

Earth-centered (EC) coordinates (also called *geocentric coordinates*) is that Cartesian coordinate frame centered at the earth's center. As depicted in Figure 9.3, the x-y plane is in the earth's equatorial plane and the z axis points toward Polaris. The x axis may be either rotating with the earth (ECR) or inertial (ECI). In the latter case the x axis points in the direction of the vernal equinox (e.g., Escobal, 1965, p. 134).

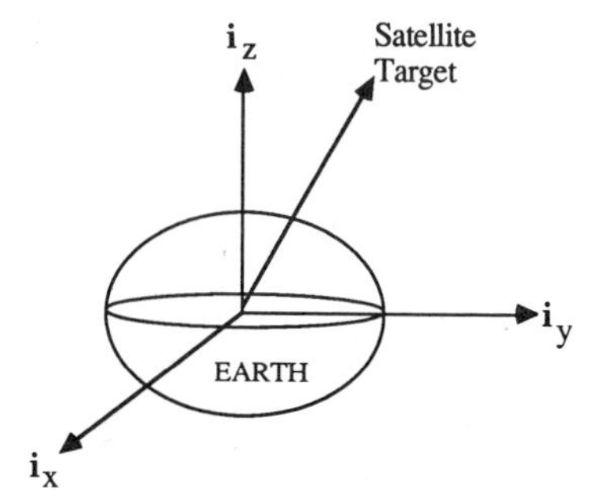

Figure 9.3 Earth-centered (EC) coordinates.

Earth-centered coordinates are useful in tracking satellites and space reentry vehicles. In this application the primary advantage of EC coordinates is that state extrapolation can be more accurately modeled since the earth's gravity is generally considered to be symmetric about its equatorial plane. Earth-centered coordinates are also used when tracking aircraft targets if the track length is a significant fraction of the earth's circumference. EC coordinates may also find a potential use as an intermediary coordinate frame in future netted ATC radar systems. In any case the use of earth-centered coordinates is perhaps most appropriate whenever the computational resources are available to perform the additional transformation.

9.8 STEREOGRAPHIC PROJECTION

Many applications are characterized by the target flying at nearly constant altitude relative to sea level. To state this more precisely, the target's dynamical properties are anisotropic with respect to the earth's altitude direction. Most large passenger aircraft can be described by this statement. When the length of the extrapolated track becomes significant with respect to the earth's curvature, then modeling this dynamical property leads to improved track performance.

The coordinate system of choice in this case is one in which all targets with the same (latitude, longitude) are mapped into the same (x, y) coordinate space. Altitude is then handled separately as the z coordinate. Provided that tracking in a linear Cartesian coordinate system is desired, then the target's position in (latitude, longitude) must be mapped into an appropriate Cartesian coordinate surface (x, y). This is essentially the same as projecting the earth's elliptical surface onto a flat plane (stereographic projection).

An existing application uses a coordinate system based on this principle (the National Airspace System) and several other long-range tracking applications are currently being studied. A variety of techniques are available for performing the required stereographic projection (Shank, 1986; Mulholland and Stout, 1979; 1982). Generally speaking, the drawback of these techniques is the inevitable kinematic distortion that occurs after nonlinear transformation. This requires that the transformation be carefully selected with respect to the particular application at hand.

9.9 VELOCITY COORDINATES

When tracking highly maneuvering targets, it is appropriate to consider the use of velocity coordinates. Velocity coordinates, as the name implies, comprise any coordinate system oriented to align with the target's own velocity vector. Velocity coordinates will be considered in some detail in the section to follow. This is primarily because, except for a few exceptions (Scheder, 1976; Berg, 1983; Maybeck, 1985), the use of velocity coordinates has not been adequately treated in the available literature.

For simplicity, the assumption made in this text is that the target's acceleration can be modeled as a decoupled Markovian relation in any velocity coordinate frame:

$$\begin{aligned} \dot{a}_V &= -\alpha a_V + w_V \\ \dot{a}_\theta &= -\alpha a_\theta + w_\theta \\ \dot{a}_\psi &= -\alpha a_\psi + w_\psi \end{aligned} \tag{9.20}$$

where $a_V \triangleq$ acceleration along the velocity vector
$a_\theta, a_\psi \triangleq$ acceleration components normal to the velocity vector
$\alpha \triangleq$ inverse correlation time constant
$w_V, w_\theta, w_\psi \triangleq$ white noise with zero mean and variances:

$$\begin{aligned} E[w_V(t)w_V(t+\tau)] &= \delta(\tau)2\alpha\sigma_V^2 \\ E[w_\theta(t)w_\theta(t+\tau)] &= \delta(\tau)2\alpha\sigma_\theta^2 \\ E[w_\psi(t)w_\psi(t+\tau)] &= \delta(\tau)2\alpha\sigma_\psi^2 \end{aligned} \tag{9.21}$$

In all other coordinate systems, such as NED coordinates, the accelerations must be modeled as being fully coupled. This defines our problem statement.

Note that generally $\sigma_V \ll (\sigma_\theta, \sigma_\psi)$ in the case of aircraft targets since most pilot-induced maneuvers consist of a constant turn without any increase in speed. In other applications, however, σ_V can be the dominant acceleration component. For instance, when tracking reentry vehicles, the primary unknown acceleration consists of atmospheric drag opposite to the velocity vector.

The velocity vector is given as (see Figure 9.4):

$$\begin{aligned} v_N &= v\sin(\theta)\cos(\psi) \\ v_E &= v\sin(\theta)\sin(\psi) \\ v_D &= v\cos(\theta) \end{aligned}$$

for $v = |\mathbf{v}|$ the velocity magnitude.

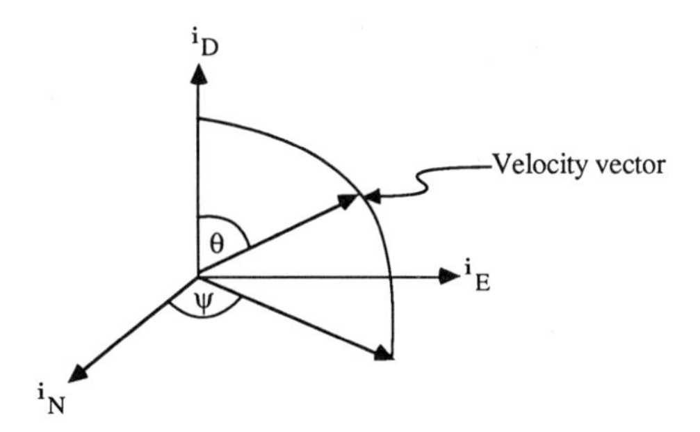

Figure 9.4 Definition of angles θ, ψ.

Taking the derivative of v_N, v_E, v_D, we obtain

$$\begin{aligned}
a_N &= \dot{v}_N \\
&= a_V \sin(\theta)\cos(\psi) + a_\theta \cos(\theta)\cos(\psi) - a_\psi \sin(\psi) \\
a_E &= \dot{v}_E \\
&= a_V \sin(\theta)\sin(\psi) + a_\theta \cos(\theta)\sin(\psi) + a_\psi \cos(\psi) \\
a_D &= \dot{v}_D \\
&= a_V \cos(\theta) - a_\theta \sin(\theta)
\end{aligned} \tag{9.22}$$

for

$$a_V \triangleq \dot{v} \qquad a_\theta \triangleq v\dot{\theta} \qquad a_\psi \triangleq v\dot{\psi}\sin(\theta) \tag{9.23}$$

The derivative of a_N is

$$\begin{aligned}
\dot{a}_N = \dot{a}_V \sin(\theta)\cos(\psi) + \dot{a}_\theta \cos(\theta)\cos(\psi) - \dot{a}_\psi \sin(\psi) + a_V\dot{\theta}\cos(\theta)\cos(\psi) \\
- a_V\dot{\psi}\sin(\theta)\sin(\psi) - a_\theta\dot{\theta}\sin(\theta)\cos(\psi) - a_\theta\dot{\psi}\cos(\theta)\sin(\psi) - a_\psi\dot{\psi}\cos(\psi)
\end{aligned}$$

Substituting in (9.20) through (9.23),

$$\begin{aligned}
\dot{a}_N = &-\alpha a_N + w_N - (\dot{\theta}^2 + \dot{\psi}^2)v_N - \dot{\theta}\dot{\psi}\frac{v_D v_E}{\sqrt{v_N^2 + v_E^2}} \\
&- \frac{\dot{\psi} a_V}{v} v_E + \frac{\dot{\theta} a_V}{v}\frac{v_D v_N}{\sqrt{v_N^2 + v_E^2}}
\end{aligned} \tag{9.24}$$

Likewise for a_E, a_D,

$$\begin{aligned}
\dot{a}_E = &-\alpha a_E + w_E - (\dot{\theta}^2 + \dot{\psi}^2)v_E + \dot{\theta}\dot{\psi}\frac{v_D v_N}{\sqrt{v_N^2 + v_E^2}} \\
&+ \frac{\dot{\psi} a_V}{v} v_N + \frac{\dot{\theta} a_V}{v}\frac{v_D v_E}{\sqrt{v_N^2 + v_E^2}} \\
\dot{a}_D = &-\alpha a_D + w_D - \dot{\theta}^2 v_D - \frac{\dot{\psi} a_V}{v}\sqrt{v_N^2 + v_E^2}
\end{aligned}$$

It is reasonable to drop those cross terms containing both a_V and turn rate $(\dot{\theta}, \dot{\psi})$. The expression thus simplifies to

$$\dot{a}_N = -\alpha a_N + w_N - (\dot{\theta}^2 + \dot{\psi}^2)v_N - \dot{\theta}\dot{\psi}\frac{v_D v_E}{\sqrt{v_N^2 + v_E^2}}$$

$$\dot{a}_E = -\alpha a_E + w_E - (\dot{\theta}^2 + \dot{\psi}^2)v_E + \dot{\theta}\dot{\psi}\frac{v_D v_N}{\sqrt{v_N^2 + v_E^2}} \tag{9.25}$$

$$\dot{a}_D = -\alpha a_D + w_D - \dot{\theta}^2 v_D$$

This represents the matrix of fully coupled Markov equations in NED coordinates. The turn rate parameters $(\dot{\theta}, \dot{\psi})$ may be estimated off-line and inserted into the transition matrix.

Also, note that the noise quantities w_N, w_E, and w_D are now velocity dependent and correlated:

$$w_N \triangleq w_V \frac{v_N}{v} + w_\theta \frac{v_D v_N}{v\sqrt{v_N^2 + v_E^2}} - w_\psi \frac{v_E}{\sqrt{v_N^2 + v_E^2}}$$

$$w_E \triangleq w_V \frac{v_E}{v} + w_\theta \frac{v_D v_E}{v\sqrt{v_N^2 + v_E^2}} + w_\psi \frac{v_N}{\sqrt{v_N^2 + v_E^2}} \tag{9.26}$$

$$w_D \triangleq w_V \frac{v_D}{v} - w_\theta \frac{\sqrt{v_N^2 + v_E^2}}{v}$$

By modeling this target property, an accurate measurement of range/Doppler can couple into the cross-range (angle) direction. Improved performance is therefore obtained during target accelerations if the maneuver is in the plane of the range vector and the velocity vector.

Discretization is the process of converting the matrix differential equation into a discrete time difference equation. Treating the terms $\dot{\theta}$ and $\dot{\psi}$ as aiding terms with known dynamics considerably simplifies the discretization development. We thereby obtain

$$\mathbf{x}(k+1) = \mathbf{\Phi}\mathbf{x}(k) + \mathbf{q}(k) + \mathbf{f}(k) \tag{9.27}$$

where the transition matrix is block diagonalized:

$$\mathbf{\Phi}_{3D} = \begin{vmatrix} \mathbf{\Phi}_{1D} & 0 & 0 \\ 0 & \mathbf{\Phi}_{1D} & 0 \\ 0 & 0 & \mathbf{\Phi}_{1D} \end{vmatrix} \tag{9.28a}$$

for

$$\mathbf{\Phi}_{1D} = \begin{vmatrix} 1 & \Delta T & \frac{1}{\alpha^2}(-1 + \alpha\Delta T + \rho) \\ 0 & 1 & \frac{1}{\alpha}(1 - \rho) \\ 0 & 0 & \rho \end{vmatrix} \tag{9.28b}$$

The vector $\mathbf{f}(k)$ contains the turn rate terms:

$$\mathbf{f}(k) = \int_{t-\Delta T}^{t} e^{\mathbf{F}(t-\tau)}\mathbf{f}(\tau)\,d\tau \qquad t = k\,\Delta T \tag{9.29}$$

for

$$\mathbf{f}(\tau) = \begin{vmatrix} 0 \\ 0 \\ -(\dot{\theta}^2 + \dot{\psi}^2)v_N - \dot{\theta}\dot{\psi}\dfrac{v_D v_E}{\sqrt{v_N^2 + v_E^2}} \\ 0 \\ 0 \\ -(\dot{\theta}^2 + \dot{\psi}^2)v_E + \dot{\theta}\dot{\psi}\dfrac{v_D v_N}{\sqrt{v_N^2 + v_E^2}} \\ 0 \\ 0 \\ -\dot{\theta}^2 v_D \end{vmatrix}$$

where (9.29) is numerically integrated. The aiding vector (9.29) should be included in all equations in Chapters 7 and 8.

A lower bound on the covariance is given by

$$\begin{aligned} \mathbf{Q}_{3D} &= E[\mathbf{q}(k)\mathbf{q}^T(k)] \\ &= \begin{vmatrix} \mathbf{Q}_{1D}E(w_N^2) & \mathbf{Q}_{1D}E(w_N w_E) & \mathbf{Q}_{1D}E(w_N w_D) \\ \mathbf{Q}_{1D}E(w_N w_E) & \mathbf{Q}_{1D}E(w_E^2) & \mathbf{Q}_{1D}E(w_E w_D) \\ \mathbf{Q}_{1D}E(w_N w_D) & \mathbf{Q}_{1D}E(w_E w_D) & \mathbf{Q}_{1D}E(w_D^2) \end{vmatrix} \end{aligned} \tag{9.30}$$

for

$$\mathbf{Q}_{1D} = \begin{vmatrix} q_{11} & q_{12} & q_{13} \\ q_{21} & q_{22} & q_{23} \\ q_{31} & q_{32} & q_{33} \end{vmatrix}$$

and q_{ij} given by (7.24). The value for $E(w_N^2)$ is given by

$$E(w_N^2) = 2\alpha\sigma_V^2 \frac{v_N^2}{v^2} + 2\alpha\sigma_\theta^2 \frac{v_D^2 v_N^2}{v^2(v_N^2 + v_E^2)} + 2\alpha\sigma_\psi^2 \frac{v_E^2}{v_N^2 + v_E^2}$$

The other parameters may be found by extension.

9.10 MEASUREMENT ACCURACIES

In the case of phase comparison monopulse, typical angular accuracies are of the form

$$\sigma_\zeta^2 = \frac{\theta_{\mathrm{HP}}^2}{2k_m^2[\mathrm{SNR} + 1]} \tag{9.31}$$

for θ_{HP} the angular beamwidth, SNR the sum-channel signal-to-noise ratio, and $k_m \cong 1.35$ on boresight.

The range measurement variance is conventionally written as

$$\sigma_r^2 = \frac{(c\tau/2)^2}{4\,SNR_p} \tag{9.32}$$

where SNR_p refers to the peak detection SNR in the sum channel (e.g., see Barton, 1964, p. 41).

Referencing the range accuracy expression to the theoretical *peak* value of SNR, SNR_p, is reasonable from an analysis viewpoint. However, it is also a potential source of confusion in our application. Recall that the SNR must be estimated in real time in order to complete the Kalman filter loop. Therefore, either the SNR must be measured (sampled) at the peak, or the measured value of SNR must be properly transformed before inserting into (9.32), or (9.32) must be rederived with the SNR normalized to a nonpeak (measured) quantity.

The accuracy of the Doppler measurement cannot be simply stated since it varies as a function of filter weighting, filter spacing, number of filters used in the measurement (typically greater than two), noise correlation between filters, and the precise definition of SNR. It can be argued that the accuracy of the Doppler measurement is best evaluated via computer simulation for the particular filter parameters at hand.

Additional terms may be added under special conditions. For instance, if multipath is corrupting the angle measurement, then an additional term may be added to stabilize the Kalman filter. Likewise if Doppler spectra (DS) is corrupting the Doppler measurement.

9.11 SUMMARY

The choice of coordinate systems is a complex design question that depends on the application and the computational resources available. To provide a convenient framework for discussion, two simplifications were made. First, the discussion was limited to Cartesian coordinate systems and, second, it was assumed that computational considerations were not a major factor in selecting the coordinate system. Relative to these two restrictions, a reasonably complete discussion was presented of the coordinate systems needed for efficient three-dimensional radar tracking.

The measurement process characterizes a signal in terms of range and two angles (or, in the case of an ESA, two direction cosines). Although it is theoretically possible to update the target using the raw measurements themselves, it is usually more convenient to immediately convert the measurements into an equivalent Cartesian coordinate system. The first of these Cartesian coordinate systems is the so-called *antenna coordinates* (also called *face coordinates*). Antenna coordinates are all that is necessary for simple applications such as those where the target is close to mechanical boresight. For tracking targets off boresight, however, additional coordinate systems may be introduced.

Target updating is most efficiently carried out in a coordinate system aligned with the measurement covariance matrix. This is especially true if the covariance matrix is highly anisotropic, such as happens when the range measurement accuracy is much better (or much worse) than the angular measurement accuracy. To date, covariance coordinates have been mainly used in ESA applications and whenever fine range resolution is provided.

Target prediction is most easily carried out in an inertial coordinate frame. The most popular "inertial" coordinate system is the north, east, down (NED) coordinate system. NED coordinates are primarily used whenever a limited area is under surveillance. Its chief deficiency is that it is pegged to the radar's platform and, thus, is not truly inertial when highly accurate tracking must be considered. More stable coordinate frames include site coordinates, earth-centered coordinates, navigation coordinates, and stereographic projection coordinates. Applications that use these types of coordinate systems are those where (a) the target travels an appreciable fraction of the earth's curvature, or (b) the radar must fuse data from multiple sensors or navigation devices, or (c) the target's dynamics are anisotropic with respect to the earth. Examples of the latter include working with stereographic coordinates to track large passenger airliners, or extrapolating ground-based targets along the surface of the earth.

Velocity coordinates are primarily used in military applications where the target is highly maneuvering. The underlying concept is to replace the conventional Singer–Markov target model with a target dynamical model oriented with the target's velocity vector and (sometimes) with its acceleration vector. A simple example was given in the chapter to illustrate the concept.

CHAPTER 10

A REPRESENTATIVE STT SYSTEM

Single-target tracking (STT) avoids many of the problems associated with other tracking systems (i.e., no data miscorrelation problems and no antenna scanning issues). Nevertheless, some of the filtering and waveform selection issues can be complex. Therefore, after some brief introductory comments, the majority of the discussion focuses on these issues.

10.1 THE RADAR HARDWARE

The basic STT philosophy is to optimize the sensor and processing for the purpose of achieving very accurate tracking of a single target (or target cluster). Hence, STT typically uses a narrow pencil beam in the interest of achieving a small angular resolution and associated high gain. A mechanical servomechanism is used to keep the primary target continuously on antenna boresight (called *on-null tracking*). Associated targets, such as guided missiles in the fire control application, are tracked off boresight (*off-null tracking*).

10.2 ACQUISITION

Before zeroing in on a single target, the STT system must first *acquire* the target. This typically entails cooperating with a separate *search* radar so that an approximate estimate of target location can be obtained. Or, the same radar may perform both search and track functions via two distinct modes of operation.

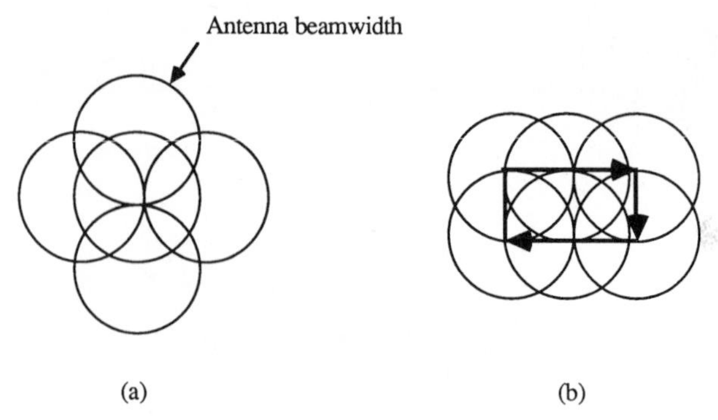

Figure 10.1 (a) Spiral acquisition pattern. (b) Raster-scan acquisition pattern.

If the beam is narrow, however, finding the target may still pose a problem even if the approximate target location is known. Thus, most STT radars possess some limited search capability. This process is called *acquisition* and essentially comprises a minisearch centered on the expected target location, as shown in Figure 10.1. Thus, in sequential order, the three primary modes of operation are search, acquisition, and track. Acquisition differs from true search in that only a limited window of range/Doppler/angle space is examined for target detection purposes, with the size of the window based on prior covariance analysis as gained from search.

10.3 CLOSED-LOOP TRACK

After acquisition, closed-loop tracking is initiated. This is typically decomposed into semiindependent angle, range, and Doppler tracking loops. The functional diagram of each loop is configured as shown in Figure 10.2.

The first block comprises the radar sensor and signal processing. The output of this block is a single set of measurements and associated statistics. The second block is the data filter that smooths the measurement data. The controller is the third block in the processing flow; its function is to develop commands to keep

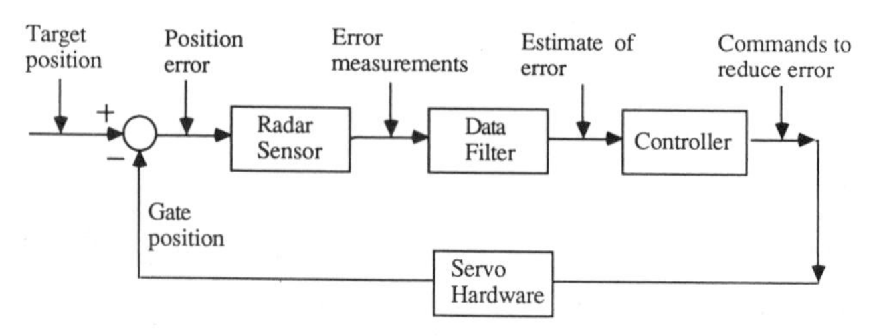

Figure 10.2 STT tracking is typically configured in a closed-loop fashion.

the track "gates" centered on the target. All returns outside the gate region are ignored for detection purposes. The width of the gate region is on the order of a beamwidth in angle, pulse width in range, and Doppler bin size in velocity. The radar is said to be "locked" onto the target if the target return is continuously maintained within its gate region. The fourth block consists of a servo device that accepts the commands from the controller and interfaces with the hardware. After the fourth block completes its task, the whole sequence repeats.

10.4 DATA FILTERING

As mentioned previously, STT data filtering is traditionally decomposed into semiindependent angle, range, and Doppler tracking loops. Some commonly cited advantages of decoupled range/Doppler loops include:

1. The quantities of real interest can be more directly estimated. In STT the quantities of real interest are the pointing errors to the target, since it is the pointing errors themselves that determine the gate settings, antenna gimbal commands, navigation commands, weapon delivery, and so on. Therefore, it is advisable to directly filter these error quantities.
2. Elimination of the need to transform the raw measurements into an inertial NED coordinate frame at the high data rates commensurate with STT.
3. Reduction of error cross coupling. For example, the target can be tracked in angle and illuminated by the beam despite the presence of severe range errors. Sources of unexpected errors include electronic countermeasures (ECM), Doppler spectra (DS), or any uncalibrated measurement bias error (such as gimbal offset).
4. Decoupling facilitates the successful application of simple precomputed filter gains such as those of the $\alpha\beta\gamma$ type.
5. Range/angle decoupling permits angle-only measurements to be more easily accommodated while avoiding the well-known phenomenon of range covariance collapsing.
6. Range/Doppler decoupling permits greater responsiveness to maneuvers by independently monitoring the range/Doppler residual.

10.4.1 A Rrepresentative Filter Structure

This section describes a representative STT digital filter. A good part of the theory of STT digital filters as outlined here was originated by J. B. Pearson and associates of Hughes Aircraft (Pearson, 1970). The liberty has been taken, however, to rephrase Dr. Pearson's ideas to maintain continuity with the rest of this book.

Let us write

$$\mathbf{R} = \mathbf{i}_r r \tag{10.1}$$

where $\mathbf{R}$ is the vector range to the target, r is the scalar range to the target, and $\mathbf{i}_r$ is a unit vector in the direction of $\mathbf{R}$. From these definitions, it follows that

$$\mathbf{i}_r r = \mathbf{x}_T - \mathbf{x}_A \tag{10.2}$$

where $\mathbf{x}_T$ and $\mathbf{x}_A$ are the position vectors of the target and antenna, respectively. Taking the first derivative of this equation gives

$$\dot{\mathbf{i}}_r r + \mathbf{i}_r \dot{r} = \mathbf{v}_T - \mathbf{v}_A \tag{10.3}$$

and taking the second derivative

$$\ddot{\mathbf{i}}_r r + 2\dot{\mathbf{i}}_r \dot{r} + \mathbf{i}_r \ddot{r} = \mathbf{a}_T - \mathbf{a}_A \tag{10.4}$$

In terms of the line-of-sight rate vector, $\boldsymbol{\omega}$, this equation is

$$\dot{\boldsymbol{\omega}} \times \mathbf{i}_r r + \boldsymbol{\omega} \times (\boldsymbol{\omega} \times \mathbf{i}_r) r + 2(\boldsymbol{\omega} \times \mathbf{i}_r)\dot{r} + \mathbf{i}_r \ddot{r} = \mathbf{a}_T - \mathbf{a}_A \tag{10.5}$$

The term $\mathbf{a}_T$ is the inertial acceleration of the target, and $\mathbf{a}_A$ is the inertial acceleration of the radar's antenna.

At this stag' :t is convenient to introduce the concept of modified antenna coordinates. By definition, *modified antenna coordinates* (b', a', e') are antenna coordinates as rotated to align with the predicted line-of-sight vector $\mathbf{i}_r$. Modified antenna coordinates are convenient to use in STT since the target is close to mechanical boresight anyway and, hence, almost aligned with the natural coordinate frame of the antenna.

It is relatively easy to verify that the second and fourth term of (10.5) lies along the range vector $\mathbf{i}_r$. Thus,

$$-|\boldsymbol{\omega}|^2 r + \ddot{r} = a_{T_{b'}} - a_{A_{b'}} \tag{10.6}$$

where

$$|\boldsymbol{\omega}|^2 = \omega_{e'}^2 + \omega_{a'}^2$$

and where the fictional term $|\boldsymbol{\omega}|^2 r$ is called the *centripetal* acceleration. Likewise, for the remaining terms,

$$\begin{aligned} \dot{\omega}_{e'} - \omega_{b'}\omega_{a'} + (2\dot{r}/r)\omega_{e'} &= -(1/r)(a_{T_{a'}} - a_{A_{a'}}) \\ \dot{\omega}_{a'} + \omega_{b'}\omega_{e'} + (2\dot{r}/r)\omega_{a'} &= +(1/r)(a_{T_{e'}} - a_{A_{e'}}) \end{aligned}$$

In many cases of practical interest, $\omega_{b'}$ is either zero (i.e., the antenna is roll stabilized) or $\omega_{b'}$ can be adequately handled by means of an external aiding term. Hence, dropping $\omega_{b'}$ and arranging the remaining terms into matrix notation, we have

$$d/dt \begin{vmatrix} r \\ \dot{r} \\ \xi_{a'} \\ \omega_{e'} \\ \xi_{e'} \\ \omega_{a'} \end{vmatrix} = \begin{vmatrix} \dot{r} \\ |\boldsymbol{\omega}|^2 r + a_{T_{b'}} \\ -\omega_{e'} \\ -(2\dot{r}/r)\omega_{e'} - (1/r)a_{T_{a'}} \\ \omega_{a'} \\ -(2\dot{r}/r)\omega_{a'} + (1/r)a_{T_{e'}} \end{vmatrix} + \begin{vmatrix} 0 \\ -a_{A_{b'}} \\ \omega_{A_{e'}} \\ +(1/r)a_{A_{e'}} \\ -\omega_{A_{a'}} \\ -(1/r)a_{A_{e'}} \end{vmatrix} \tag{10.7}$$

where the known acceleration terms a_A, ω_A have been segregated from the unknown terms. The term ω_A represents the known angular rate of the noninertial (b', a', e') coordinate axes.

10.4.2 Decoupled Filters

Strictly speaking, the "extended" Kalman filter should be introduced to implement the data filtering, based on the nonlinear state equation given previously. However, in normal STT, the quantities r, $\dot{r}$, and ω are *usually* known to high accuracy. In this case it is legitimate to treat these quantities as known parameters thus simplifying the filtering.

The form of the filter then reverts to the standard linear Kalman filter:

$$d/dt \begin{vmatrix} r \\ \dot{r} \end{vmatrix} = \begin{vmatrix} 0 & 1 & 0 \\ |\boldsymbol{\omega}|^2 & 0 & 1 \end{vmatrix} \begin{vmatrix} r \\ \dot{r} \\ a_{T_{b'}} \end{vmatrix} - \begin{vmatrix} 0 \\ a_{A_{b'}} \end{vmatrix} \tag{10.8}$$

where the term $|\boldsymbol{\omega}|^2$ is supplied by the angle filters. Likewise,

$$d/dt \begin{vmatrix} \xi_{e'} \\ \omega_{a'} \end{vmatrix} = \begin{vmatrix} 0 & 1 & 0 \\ 0 & -(2\dot{r}/r) & (1/r) \end{vmatrix} \begin{vmatrix} \xi_{e'} \\ \omega_{a'} \\ a_{T_{e'}} \end{vmatrix} - \begin{vmatrix} \omega_{A_{a'}} \\ (1/r)a_{A_{e'}} \end{vmatrix}$$

The terms r, $\dot{r}$ are supplied by the range filter. The azimuth case is parallel in structure to the elevation case. This semiuncoupled configuration is graphically portrayed in Figure 10.3.

The line-of-sight accelerations can be approximated as a decoupled Markov process,

$$\begin{aligned} \dot{a}_{T_{b'}} &= -\alpha a_{T_{b'}} + w_{b'} \\ \dot{a}_{T_{e'}} &= -\alpha a_{T_{e'}} + w_{e'} \\ \dot{a}_{T_{a'}} &= -\alpha a_{T_{a'}} + w_{a'} \end{aligned} \tag{10.9}$$

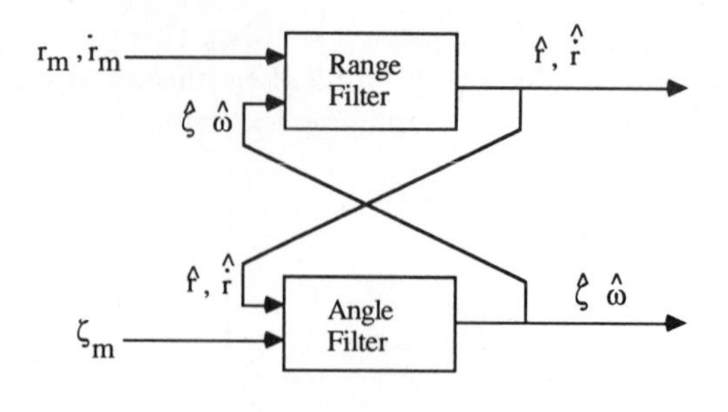

^ = "best estimate of"

m = "best measurement of"

r = range

$\dot{r}$ = range rate

ζ = angle off-boresight

ω = angular rate

Figure 10.3 Semiuncoupled filter strategy.

where w is white noise and α is the correlation time constant. Thus, the net differential equations are

$$d/dt \begin{bmatrix} r \\ \dot{r} \\ a_{T_{b'}} \end{bmatrix} = \begin{bmatrix} 0 & 1 & 0 \\ |\omega|^2 & 0 & 1 \\ 0 & 0 & -\alpha \end{bmatrix} \begin{bmatrix} r \\ \dot{r} \\ a_{T_{b'}} \end{bmatrix} - \begin{bmatrix} 0 \\ a_{A_{b'}} \\ 0 \end{bmatrix} \tag{10.10a}$$

and

$$d/dt \begin{bmatrix} \xi_{e'} \\ \omega_{a'} \\ a_{T_{e'}} \end{bmatrix} = \begin{bmatrix} 0 & 1 & 0 \\ 0 & -(2\dot{r}/r) & (1/r) \\ 0 & 0 & -\alpha \end{bmatrix} \begin{bmatrix} \xi_{e'} \\ \omega_{a'} \\ a_{T_{e'}} \end{bmatrix} - \begin{bmatrix} \omega_{A_{a'}} \\ (1/r)a_{A_{e'}} \\ 0 \end{bmatrix}$$

Exact discretization is unnecessarily complicated. A close approximation follows:

$$\begin{bmatrix} \hat{r}(k+1) \\ \hat{\dot{r}}(k+1) \\ \hat{a}_{T_{b'}}(k+1) \end{bmatrix} = \begin{bmatrix} 1 & \Delta T & (\Delta T)^2/2 \\ |\omega|^2 \Delta T & 1 & \Delta T \\ 0 & 0 & e^{-\alpha \Delta T} \end{bmatrix} \begin{bmatrix} \hat{r}(k) \\ \hat{\dot{r}}(k) \\ \hat{a}_{T_{b'}}(k) \end{bmatrix} - \begin{bmatrix} 0 \\ \Delta T a_{A_{b'}}(k) \\ 0 \end{bmatrix}$$

and

$$\begin{vmatrix} \hat{\xi}_{e'}(k+1) \\ \hat{\omega}_{a'}(k+1) \\ \hat{a}_{T_{e'}}(k+1) \end{vmatrix} = \begin{vmatrix} 1 & \Delta T & 0 \\ 0 & 1-\Delta T(2\dot{r}/r) & \Delta T/r \\ 0 & 0 & e^{-\alpha \Delta T} \end{vmatrix} \begin{vmatrix} \hat{\xi}_{e'}(k) \\ \hat{\omega}_{a'}(k) \\ \hat{a}_{T_{e'}}(k) \end{vmatrix} - \begin{vmatrix} \Delta T \omega_{A_{a'}} \\ (\Delta T/r) a_{A_{e'}} \\ 0 \end{vmatrix}$$

Finally, all of the turn parameters from (9.25) can be lumped into a single variable, thereby preserving the desired decoupled property:

$$\begin{vmatrix} \hat{r}(k+1) \\ \hat{\dot{r}}(k+1) \\ \hat{a}_{T_{b'}}(k+1) \end{vmatrix} = \begin{vmatrix} 1 & \Delta T & (\Delta T)^2/2 \\ |\boldsymbol{\omega}|^2 \Delta T & 1 & \Delta T \\ 0 & -\dot{\psi}^2 \Delta T & e^{-\alpha \Delta T} \end{vmatrix} \begin{vmatrix} \hat{r}(k) \\ \hat{\dot{r}}(k) \\ \hat{a}_{T_{b'}}(k) \end{vmatrix} - \begin{vmatrix} 0 \\ \Delta T a_{A_{b'}}(k) \\ 0 \end{vmatrix} \tag{10.11a}$$

and

$$\begin{vmatrix} \hat{\xi}_{e'}(k+1) \\ \hat{\omega}_{a'}(k+1) \\ \hat{a}_{T_{e'}}(k+1) \end{vmatrix} = \begin{vmatrix} 1 & \Delta T & 0 \\ 0 & 1-\Delta T(2\dot{r}/r) & \Delta T/r \\ 0 & -\dot{\psi}^2 r \Delta T & e^{-\alpha \Delta T} \end{vmatrix} \begin{vmatrix} \hat{\xi}_{e'}(k) \\ \hat{\omega}_{a'}(k) \\ \hat{a}_{T_{e'}}(k) \end{vmatrix} - \begin{vmatrix} \Delta T \omega_{A_{a'}} \\ (\Delta T/r) a_{A_{e'}} \\ 0 \end{vmatrix} \tag{10.11b}$$

This represents a system of semiuncoupled angle-range filter equations in modified antenna coordinates.

The radar measurements comprise

$$\mathbf{x}_m = \begin{vmatrix} r_m \\ \dot{r}_m \\ \zeta_{e_m} \\ \zeta_{a_m} \end{vmatrix} \tag{10.12}$$

where r_m is the range measurement, $\dot{r}_m$ is the range rate or Doppler measurement, and ζ_e, ζ_a are the direction cosines referenced to the monopulse phase centers, elevation (e) and azimuth (d), respectively.

The residuals become

$$\text{Residual}_e = \zeta_{e_m} - \hat{\xi}_e \tag{10.13a}$$

$$\text{Residual}_a = \zeta_{a_m} - \hat{\xi}_a \tag{10.13b}$$

$$\text{Residual}r = r_m - \hat{r} \tag{10.13c}$$

$$\text{Residual}_{\dot{r}} = \dot{r}_m - \hat{\dot{r}} \tag{10.13d}$$

The range and Doppler tracking filters may be further decoupled as a pair of redundant filters. Each filter estimates the same states and is constructed with the same state transition matrices (10.11a). The only difference is the use of (10.13c) in the case of range and the use of (10.13d) in the case of Doppler. The purpose of range/Doppler decoupling is to allow cross checks to be made between the range filter's estimate of range rate (differentiated range) and the Doppler filter's estimate of velocity. Any diverence between the two is taken as an indication that the radar has locked onto a false return such as a DS line. The filter considered in error is then reset.

Of course, the optimal arrangement would be to provide three independent filters: one based on range measurements alone, one based on Doppler measurements alone, and a third based on both range and Doppler measurements. The computational complexity of this arrangement, however, precludes it from being implemented in most situations.

10.4.3 Coupled Filters

Coupled angle-range filters must be used if range data is to be derived from angle-only measurements of a moving target. Returning to (10.7):

$$d/dt \begin{vmatrix} r \\ \dot{r} \\ \xi_{a'} \\ \omega_{e'} \\ \xi_{e'} \\ \omega_{a'} \end{vmatrix} = \begin{vmatrix} \dot{r} \\ |\boldsymbol{\omega}|^2 r + a_{T_{b'}} \\ -\omega_{e'} \\ -(2\dot{r}/r)\omega_{e'} - (1/r)a_{T_{a'}} \\ \omega_{a'} \\ -(2\dot{r}/r)\omega_{a'} + (1/r)a_{T_{e'}} \end{vmatrix} + \begin{vmatrix} 0 \\ -a_{A_{b'}} \\ \omega_{A_{e'}} \\ +(1/r)a_{A_{a'}} \\ -\omega_{A_{a'}} \\ -(1/r)a_{A_{e'}} \end{vmatrix} \tag{10.14}$$

In particular, focus on the following equations:

$$\dot{\omega}_{e'} = -(2\dot{r}/r)\omega_{e'} - (1/r)a_{T_{a'}} + (1/r)a_{A_{e'}} \tag{10.15}$$

In the absence of any maneuvers, the quantity $\dot{r}/r$ is directly observable from differentiated angle estimates alone. That is to say,

$$\dot{r}/r = -\dot{\omega}_{e'}/(2\omega_{e'}) \qquad \text{given } a_{T_{a'}} = 0 \text{ and } a_{A_{a'}} = 0 \tag{10.16}$$

Thus, the quantities and $\dot{\omega}$, ω, and $(\dot{r}/r)$ are all completely observable in the absence of maneuvers and valid range measurements. The quantity $(1/r)$ remains unobservable until a maneuver is initiated. As such, these four quantities are good candidates for state variables in the absence of accurate range measurements. This idea was originally conceived by K. R. Brown and significantly developed by H. D. Hoelzer and co-workers while at IBM (Hoelzer et al., 1978).

Let us begin by defining the quantities

$$\dot{u} \triangleq (\dot{r}/r) \qquad s \triangleq (1/r) \tag{10.17}$$

for $\dot{u}$ the inverse time-to-go.

Equation (10.15) may be rewritten as

$$\dot{\omega}_{e'} = -2\dot{u}\omega_{e'} - sa_{T_{a'}} + sa_{A_{a'}} \tag{10.18}$$

The reciprocal of range is an efficient choice for the fourth coordinate since $1/r = s$ enters linearly in the equations during maneuvers. This is important since equations that are not linear must be linearized about a state estimate for the propagation of covariance (Hoelzer et al., 1978).

The range equation may also be rewritten in terms of $\dot{u}$ by noting that

$$\begin{aligned} \ddot{u} &= \frac{d}{dt}\frac{\dot{r}}{r} \\ &= \frac{\ddot{r}}{r} - \left(\frac{\dot{r}}{r}\right)^2 \\ &= |\boldsymbol{\omega}|^2 - \dot{u}^2 + sa_{T_{b'}} - sa_{A_{b'}} \end{aligned} \tag{10.19}$$

The selection of state variables is now complete, resulting in

$$d/dt \begin{vmatrix} s \\ \dot{u} \\ \xi_{a'} \\ \omega_{e'} \\ \xi_{e'} \\ \omega_{a'} \end{vmatrix} = \begin{vmatrix} -\dot{u}s \\ |\boldsymbol{\omega}|^2 - \dot{u}^2 + sa_{T_{b'}} \\ -\omega_{e'} \\ -2\dot{u}\omega_{e'} - sa_{T_{a'}} \\ \omega_{a'} \\ -2\dot{u}\omega_{a'} + sa_{T_{e'}} \end{vmatrix} + \begin{vmatrix} 0 \\ -sa_{A_{b'}} \\ \omega_{A_{e'}} \\ sa_{A_{a'}} \\ -\omega_{A_{a'}} \\ -sa_{A_{e'}} \end{vmatrix} \tag{10.20}$$

In TWS the solution is similar except for the need to predict the estimates forward in Cartesian coordinates for the sake of accuracy and simplicity.

It is fairly easy to see that range can be estimated from angle-only measurements if the antenna acceleration a_A is known. The unknown target acceleration a_T is either treated as pure noise or as a fixed-plane turn. For the problem to be solved, the radar must "out maneuver" the target during the observation period; that is, a_A must be much larger than the random noise component of a_T in order to derive accurate range information from angle-only measurements.

10.5 WAVEFORM SELECTION

Although dynamic waveform selection is very complicated in practice, the basic idea can be stated simply. The controller acts to continuously center the target return at mid-PRF and mid-PRI. When the target changes its position and velocity, the controller changes the PRF accordingly. This, in effect, causes the PRF to adaptively change to maximize the target signal energy received.

The compressed pulsewidth is typically matched to the estimated target range extent, speed, and aspect angle. A larger pulsewidth is commanded whenever the target return is not corrupted by clutter; this ensures a higher SNR when the competing interference is thermal noise. A smaller compressed pulsewidth is commanded whenever clutter competes with the target's Doppler; this ensures a higher SCR when the competing interference is clutter.

The filter weights may also be adjusted as a function of the known separation from clutter. This includes both the Doppler filter weights (clutter rejection in frequency) and the antenna weights (clutter rejection in angle). For example, sidelobes with sharp skirts may be formed to reject nearby clutter; decaying sidelobes are formed to reject far clutter.

10.6 CONTROL LAWS

In early analog versions of STT, the design of the servomechanism was essentially viewed as synonymous with the design of the data filter. However, beginning with J. B. Pearson, a number of engineers began to realize that designing an efficient servomechanism is at odds with designing an efficient data filter. Specifically, the data filter needs to optimize track accuracy under all target conditions. The servomechanism must follow the target under worst-case target conditions. Also, the filter gains should be automatically adjusted according to the measured SNR to avoid small signal suppression.

The question then becomes "what is a good design for the control law?" In the approach outlined in Pearson and Steer (1974), the commanded servo rates are taken to be linearly proportional to a combination of the pointing error estimates as expressed by

$$\begin{aligned} \omega_{C_a} &= \hat{\omega}_a + k\hat{\xi}_e \\ \omega_{C_e} &= \hat{\omega}_e - k\hat{\xi}_a \\ r_C &= \hat{r} + \Delta T\hat{\dot{r}} \end{aligned} \tag{10.21}$$

where ω_C = antenna servo command
r_C = range servo command
k = antenna servo gain (normally k is slightly less than the inverse of the loop iteration time, $k \lesssim 1/\Delta T$).

If the iteration time ΔT is small enough, this method of servo control is effective.

Finally, in the event that the target passes within a region of mainlobe clutter and cannot be detected above the CFAR threshold, the controller may command the track gates to be repositioned at both sides of the mainlobe clutter region in expectation of target reentry.

10.6.1 Off-Null Tracking

Off-null tracking may be invoked whenever doing so attenuates the level of competing clutter relative to the target signal. Moving the antenna up (positive elevation) reduces the mean level of clutter in the beam and also helps control multipath. In an airborne radar, moving it sideways (azimuth) affects the clutter's mean spectral distribution (as per Figure 4.8).

The new antenna stabilization commands are

$$
\begin{aligned}
\omega_{C_a} &= \hat{\omega}_a + k(\hat{\xi}_e - \mathrm{BIAS}_e) \\
\omega_{C_e} &= \hat{\omega}_e - k(\hat{\xi}_a - \mathrm{BIAS}_a)
\end{aligned}
\tag{10.22}
$$

where k = antenna servo gain

BIAS_e, BIAS_a = off-null bias angles

10.7 ANTENNA SERVOMECHANISM

The antenna servomechanism is typically composed of four parts. The gimbal mechanism (or rotary joint) supports the antenna. The motor and associated drive provide the driving force. The gyro senses the antenna pointing angles and provides an input for feedback control. The stabilization loop ensures proper feedback control of the lower frequency components.

In the ideal case, which is almost always approximated in practice, the servomechanism stabilizes the antenna pointing angle such that only residual "jitter" errors are present. This jitter is negligible relative to the antenna beamwidth and, hence, is often neglected in evaluating the overall performance of the STT system.

10.8 ELECTRONIC COUNTERMEASURES (ECM)

Military radars may be subjected to interference deliberately designed to degrade the measurement process. Such interference is known as *electronic countermeasures* (ECM). STT military radars in particular are frequently the victim of ECM since STT operation is often an immediate precursor to weapon delivery.

The various forms of ECM include noise jamming, deception jamming, chaff, and decoys. A noise jammer, for instance, attempts to deny detection by injecting a noiselike interference into the victim system and so raise the CFAR threshold above the signal level. Noise jamming is generally considered less effective against an STT radar for several reasons. First, guard processing can be used to classify the jammer as being in the mainbeam or in the sidelobes. Jammers in the sidelobes are nulled out, or a new RF is commanded until an open RF is discovered. Jammers in the mainlobe are tracked in angle. Noise jamming does deny valid range measurements; however, approximate range can still be determined from coupled angle-range filtering as outlined in this chapter.

Deception countermeasures (DECM) are generally considered more effective against STT radars. Deception jammers contrive to "deceive" the radar by injecting a modified replica of the real signal into the victim system. The end goal is to "break lock."

Closed-loop tracking maximizes the signal and minimizes the noise in the absence of deception jamming. But, it offers the opportunity for deception jammers to disrupt operation by *breaking lock*. This involves falsely simulating a target maneuver that pulls the track gates away from the real target. A classic example of this is range gate stealing (RGS). The RGS jammer generates a false target return by storing the received radar signals and retransmitting at a slightly later time. This is accomplished by means of a repeater that initially repeats the received radar pulse with a minimum time delay and a stronger power than the original signal return. The radar's AGC and CFAR circuitry eventually adjust to the stronger jammer signal; in effect, desensitizing the radar to the true signal. Then, the repeater begins to introduce increasing amounts of time delay into the repeated signal. The range loop tracks the stronger jammer signal, and so the gates are gradually drawn away from the true target return.

In this manner the RGS jammer simulates an apparent maneuver and "steals" the track gates away from the real target. Then, the repeater may be turned off at some distance from the true target return. The radar must then revert to acquisition mode or lose the target. The end result is to confuse the radar and to occupy its time in continually reacquiring the target.

Range deception may be combined with some form of angle deception. A classic example of angle deception is *terrain bounce* jamming. In this type of jamming, the jammer bounces a transmission off of the earth's surface and so simulates a false multipath condition. There are numerous other angle deception measures of this type; they all attempt to distort the wavefront arriving at the radar.

Some DECM systems may also attempt Doppler deception as well using a velocity gate stealer (VGS). Coordinated VGS and RGS operation can prevent the radar from making a consistency check between the Doppler estimate and differentiated range estimate to determine the presence of a DECM jammer. If all three loops are successfully disrupted, then the STT radar must reacquire the target in angle, range, and Doppler simultaneously after break-lock.

10.9 SUMMARY

The basic elements of an STT system were outlined, with particular emphasis placed on closed-loop tracking. A semiuncoupled filter strategy was presented to implement the set of dynamical equations. Both more sophisticated and more simplistic systems have been implemented in the past, ranging from simple analog designs to elaborate digital designs in coupled Cartesian coordinates. The important topic of adaptive waveform selection and the ECM aspects of the tracking theory were briefly discussed.

CHAPTER 11

DATA CORRELATION LOGIC

In a multiple-target environment it is necessary to first pair the correct radar measurement with the correct target track. This procedure is known as *data correlation* and is conventionally comprised of two steps called *gating* and *association.*

11.1 GATING

Gating is a coarse test that screens the measurements and classifies them according to one of two ways:

1. The measurement is tentatively assigned to one or more existing tracks.
2. The measurement is not assigned to any existing track. In this case the measurement either represents a new target detection, false alarm, or new Doppler spectra (DS) detection.

Data association is introduced only if gating fails to produce an unambiguous pairing. An ambiguous pairing occurs if more than one measurement is assigned to the same track or if a measurement is assigned to more than one track.

Data association represents a more formal procedure for pairing measurements to tracks. Alone, it is sufficient to perform the entire data correlation task. As a practical matter, however, gating is almost always performed first, primarily due to the higher computational burden involved in data association. Gating is also sometimes used to simplify the data association algorithm by eliminating unlikely pairings.

Gating and data association are both performed by calculating a normalized *distance* d^2 related to the residual given by

$$\tilde{\mathbf{y}}(k) = \mathbf{m}(k) - \mathbf{H}\bar{\mathbf{x}}(k) \tag{11.1}$$

which has residual covariance matrix

$$\mathbf{S} = \mathbf{H}\mathbf{P}\mathbf{H}^T + \mathbf{R} \tag{11.2}$$

The relationship is as follows:

$$d^2 = \tilde{\mathbf{y}}(k)^T\mathbf{S}^{-1}\tilde{\mathbf{y}}(k) \tag{11.3}$$

Gating is then performed by comparing d^2 to a threshold,

$$d^2 \gtrless G \tag{11.4}$$

If the distance d^2 is less than the *track gate* G, it is tentatively assigned to the track file. If the distance exceeds the gate G, it is assumed that the measurement came from some other source. This process is repeated for each track-to-measurement pairing.

Figure 11.1 illustrates a spatial diagram of the gating process. The predicted values $\mathbf{H}\bar{\mathbf{x}}(k)$ define the center of the gate region. The size of the gate region is directly proportional to the size of the gate threshold G and the normalizing covariance $\mathbf{S}$.

The gate threshold G is typically computed by requiring that the probability of the target falling inside the gate region (defined as P_G) be a fixed constant. The relationship between G and P_G is shown in Figure 11.2. This type of gating test is equivalent to a Neyman–Pearson test with a fixed P_G, and hence the gate is denoted G_{NP}. Given Gaussian statistics the probability density of d^2 is chi-squared with dimensionality M (χ_M^2), so that G_{NP} can be determined from the required P_G and the known pdf of d^2.

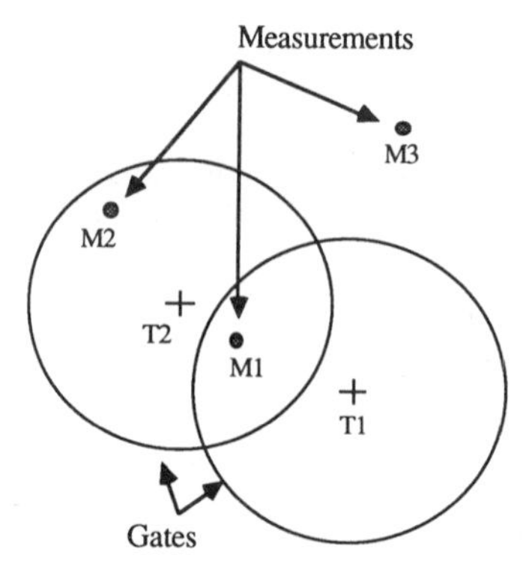

Figure 11.1 Example of gating with two targets and three measurements.

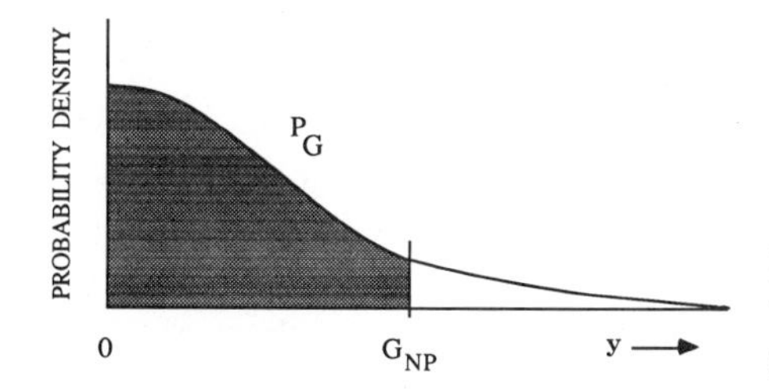

Figure 11.2 P_G is the probability of a valid target measurement falling in the gate region.

11.1.1 Rectangular Gating

A rectangular gate is (by definition) a preliminary gating test of the form

$$|\tilde{\mathbf{y}}_i(k)|^2 \gtrless G_i S_{ii} \qquad \text{for } 1 \leqslant i \leqslant M \tag{11.5}$$

where S_{ii} is the iith component of $\mathbf{S}$. This type of gating is suboptimal since it approximates an ellipsoid gate with a rectangular region. More importantly, it ignores the information contained in the off-diagonal components of the covariance matrix $\mathbf{S} = \mathbf{HPH}^T + \mathbf{R}$.

The use of rectangular gates as the sole means of gating is only justified if $\mathbf{HPH}^T$ and $\mathbf{R}$ are both diagonal. Strictly speaking, this condition never occurs. However it is often approximately valid, particularly if (1) $\mathbf{R}$ has been previously diagonalized, (2) the target line-of-sight rate is slow, such that diagonalizing $\mathbf{R}$ implies that $\mathbf{HPH}^T$ is also diagonalized, and (3) Doppler is not measured. If Doppler is one of the measurements, $\mathbf{HPH}^T$ is highly nondiagonal regardless of the coordinate system, and so rectangular gating is not valid. This is easily seen since a positive range-rate residual should imply a positive range residual. Modeling this correlation reduces the probability of a mistaken DS correlation.

11.1.2 Nonkinematic Decision Features

Normally the measurement set $\mathbf{m}(k)$ comprises the kinematic measurements including two angles, range, and radial velocity. Thus, M has a nominal value of 3 or 4. It is often suggested that $\mathbf{m}(k)$ should also incorporate additional decision features such as observed signal magnitude, harmonic signal properties, or target identity (ID) declaration. This argument states that these additional decision features can potentially help differentiate signal returns from spurious returns with similar kinematics.

Theoretically, the incorporation of these additional decision features is straightforward since the theory remains basically the same. The measurement $\mathbf{m}$ is simply augmented by the new decision features (Blackman, 1986). The drawback is that each new decision feature must possess a probability density and measurement covariance $\mathbf{R}$ that are reasonably well known. This objection has limited the prior use of this technique to certain specialized cases (e.g., see Blackman, 1986, pp. 376–380).

11.2 MULTIPLE GATES

The previous theory assumed that the Kalman filter prediction covariance, **P**, gives an accurate representation of the target's true covariance error. However, the two can deviate significantly upon occasion, such as in the presence of a target maneuver or data miscorrelation.

If the tracker is required to follow the target under all possible conditions, then gating must be based on worst-case target conditions. On the other hand, for computational reasons, it is usually desired to reduce the size of the gate. This reduces the number of false alarms that fall within the gate region and hence reduces the computational load of the subsequent data association task.

This dilemma is conventionally overcome by employing multiple gates (e.g., Trunk, 1983, p. 82). The first gate, called the *small gate* or *covariance gate*, takes **P** from the filter covariance matrix. If no measurements appear in the small gate, a larger gate region is introduced. The size of the large gate region is determined by the worst-case target conditions including maneuver and data miscorrelations.

In this context the gating test classifies the measurements in one of three ways:

1. The measurement falls within the small gate region of an existing track, and so it is tentatively assigned to that track.
2. The measurement falls within the large gate region of an existing track, and so it is tentatively assigned to that track provided no other targets appear inside its small gate.
3. The measurement falls outside of all gates, and therefore it is not assigned to any existing track.

The presence of a sustained maneuver induces an additional error in the x direction, as given by

$$P_x = (a_x \Delta T^2/2)^2 \tag{11.6}$$

or, in velocity,

$$P_{v_x} = (a_x \Delta T)^2$$

This value of covariance **P** replaces the Kalman filter prediction covariance in the large gating test.

The calculation of a_x may be based on a target dynamical model wherein the target initiates a fixed-plane turn maneuver. The corresponding acceleration a_x is given by

$$a_x = a_{\text{MAX}} \frac{|\mathbf{v} \times \mathbf{i}_x|}{|\mathbf{v}|} \tag{11.7}$$

where a_x = projection of a_{MAX} upon the i_x direction
a_{MAX} = maximum acceleration limits of the target
$\mathbf{v}$ = target velocity vector
$\mathbf{i}_x$ = unit vector in the direction of the gating test

and where it is implicitly assumed that rectangular gating is employed and the direction of a_{MAX} is orthogonal to $\mathbf{v}$ (fixed-plane turn assumption).

11.3 SIMPLE ASSOCIATION RULES

Data association takes the output of the gating algorithm and makes final measurement-to-track associations. In those cases where a single measurement is gated to a single track, an assignment can be immediately made. However, for closely spaced targets, it is more likely that conflict situations will arise. Conflict situations arise when multiple measurements fall within a single gate, or when a single measurement falls within the gates of more than one track.

The data association algorithm attempts to resolve these conflicts using probabilistic methods, of which there are several ways for doing this of interest. The simplest is the so-called nearest-neighbor (NN) approach. The NN approach looks through the gated measurements and pairs the minimum distance d^2 with the track file under consideration.

Assuming Gaussian statistics, the probability of observing the residual $\tilde{\mathbf{y}}$ is

$$g(\tilde{\mathbf{y}}) = \frac{e^{-d^2/2}}{(2\pi)^{M/2}|\mathbf{S}|^{1/2}} \tag{11.8}$$

per incremental area, where M is the measurement dimension and $|\mathbf{S}|$ is the determinant of $\mathbf{S}$. Hence, minimizing d^2 is equivalent to minimizing $g(\tilde{\mathbf{y}})$, and so NN is the optimum procedure if the decision is limited to a single assignment.

The classical NN sequence is as follows:

1. For each track, find that measurement with the closest distance d^2 and make that assignment.
2. Repeat rule 1 for the next track.

The above NN approach does not guarantee feasibility. A track-to-measurement pairing is said to be *feasible* if each measurement is assigned to at most one track. To generate only feasible assignments, each measurement may be removed from further consideration after assignment. Therefore, an "improved" NN approach may be summarized as follows:

1. For each track find that measurement with the closest distance d^2 and make that assignment.

2. Remove the assigned measurement from the pool of measurements to be assigned and repeat rule 1 for the next track.

The chief deficiency with this approach is that the outcome depends on the order in which the measurements are processed. Thus, for example, an early bad assignment can prevent a later good assignment from being made, in effect stealing the measurement from the pool. The NN approach can be improved by ordering the sequence of measurements according to minimum distance before processing.

Finally, the sequence of steps for the NN solution becomes:

1. Tracks that gate with a single measurement are immediately assigned. Measurements that gate with a single track T_i invalidate any multiply gated measurements that gate with the same track T_i.
2. Order the remaining track measurement pairings according to minimum distance d^2.
3. Starting with those pairings with minimum distance d^2, assign measurements to tracks. Remove the assigned measurements from the pool.
4. Repeat rule 3 until all feasible assignments are made.

The first rule equates to giving preference to those measurements that are singly gated (i.e., gated to a single track). Then, for those tracks that are multiply gated, track measurement pairings are made in order of increasing distance d^2.

In the absence of any universally accepted convention, this algorithm can serve as a realistic definition of an NN approach as applied to the problem of data association. However, the reader should be warned that there is actually no single NN approach, and properly speaking NN refers to any one of a number of algorithms that use a minimum distance criteria.

11.4 GENERAL EVALUATION STATISTIC

A more sophisticated data association algorithm would select a combination of pairings to satisfy some overall distance criteria. In other words, provided other measurements are observed in the vicinity of the track file, it is reasonable to expect them to change the outcome of the data assignments.

The deficiency of the NN approach can be seen by examining Figure 11.1. The nearest neighbor approach would pair measurement M1 with track T2, despite the presence of a second measurement M2 near T2, and no other measurement near T1. Consequently, it is clear that a more optimal association algorithm would minimize the sum of the distances, rather than minimize each distance when viewed in isolation.

Assuming Gaussian statistics, and defining $\tilde{\mathbf{y}}_{ij}$ to be the residual for the ith measurement and jth track, then the probability of observing m independent residual errors is

$$P_{\mathrm{ER}} = \prod_{i,j=1}^{m} g(\tilde{\mathbf{y}}_{ij}) \tag{11.9}$$

where m is the total number of pairings. Therefore, the quantity to minimize is proportional to the sum of the normalized distance squared:

$$\sum_{i,j=1}^{m} \frac{d_{ij}^2}{2} - \ln\left[\frac{1}{(2\pi)^{M/2}|\mathbf{S}_{ij}|^{1/2}}\right] \tag{11.10}$$

This logarithmic quantity is called the probability *score*. Logarithmic quantities are used to prevent floating-point overflow.

A more general criteria would incorporate the probability of observing a particular detection sequence as given by the probability of detection and false alarm. For example, when tracking a weak target, rather than assigning a measurement to the track that barely falls within its large gate, it may be more probable that the target was undetected on a particular scan.

To this end, let us define P_D to be the probability of updating a track and $(1 - P_D)$ to be the probability of not updating the track. The resulting score function is (Blackman, 1986, p. 275)

$$\text{Score} = \sum_{i,j=1}^{m} a_{ij} \tag{11.11}$$

where

$$a_{ij} = \begin{cases} \dfrac{d_{ij}^2}{2} - \ln\left[\dfrac{P_D}{(2\pi)^{M/2}|\mathbf{S}|^{1/2}}\right] & \text{(if track is updated)} \\ -\ln(1 - P_D) & \text{(if track is not updated)} \end{cases}$$

The value P_D may be based on the track's predicted signal magnitude, as per (3.23), as well as its position in the scan volume.

Next define β_{FA} and β_{NT} to be the probability of observing a false alarm and probability of observing a new target, respectively, per incremental area. The score associated with observing a particular sequence of new targets, false alarms, and updated tracks is

$$a_{ij} = \begin{cases} -\ln(\beta_{\mathrm{NT}}) & \text{if the measurement is used to initiate a new target track} \\ -\ln(\beta_{\mathrm{FA}}) & \text{if the measurement is declared a false alarm} \\ \dfrac{d_{ij}^2}{2} - \ln\left[\dfrac{P_D}{(2\pi)^{M/2}|\mathbf{S}|^{1/2}}\right] & \text{if the measurement updates a track} \\ -\ln(1 - P_D) & \text{if track is not updated} \end{cases}$$

It should be noted that all of these quantities are normalized to incremental area, and so have units of $\mathrm{m}^{-3}\,(\mathrm{m/sec})^{-1}$. The values β_{FA} and β_{NT} may be based on the observed amplitude of the measurement, previous scan history, scan boundaries, and possible sources of targets in the scan volume (airports).

Finally, the various hypotheses can be conveniently represented by means of an *assignment matrix* of marginal scores (Blackman, 1986, p. 276) as illustrated for the 3×3 case:

$$\left|\begin{array}{cccccc} a_{11} & a_{12} & a_{13} & a_{14} & \infty & \infty \\ a_{21} & a_{22} & a_{23} & \infty & a_{25} & \infty \\ a_{31} & a_{32} & a_{33} & \infty & \infty & a_{36} \\ a_{41} & \infty & \infty & & & \\ \infty & a_{52} & \infty & & 0 & \\ \infty & \infty & a_{63} & & & \end{array}\right|$$

Again, the objective is to assign measurements to tracks, false alarms, and to new tracks such as to minimize the overall score.

11.5 OPTIMAL SEQUENTIAL DATA ASSOCIATION ALGORITHMS

The most straightforward way to determine the optimum solution is the *enumeration method*, wherein all feasible assignments are listed, the sum of the scores are computed for each, and then that assignment selected that achieves minimum score. Enumeration is appropriate when solving small assignment matrices. For large assignment matrices, on the other hand, total enumeration becomes extremely undesirable. To illustrate this point, consider the case of pairing 12 measurements to 12 tracks. Assuming all 12 measurements gate to all 12 tracks, then there exists at least 12! feasible paths through the assignment matrix. To enumerate all 12! paths would be a prohibitive computational burden. Fortunately, there exists several strategies for finding the optimal path without resorting to total enumeration.

General Remark. The objective can be mathematically phrased as follows. For each row of the assignment matrix, select a unique column such that the total score is minimized:

$$\text{MINIMIZE} \sum_{i=1}^{m} a_{ij_i}$$

Most optimal data association methods are based on the following basic principle: the assignment matrix a_{ij} can be altered by adding and subtracting constant row and column vectors without changing the ranking among the various assignment paths (Kuhn, 1955). That is, since every assignment path must contain an entry in each column and each row, the minimal path is not altered by adding or subtracting out constant vectors.

To illustrate the use of this principle, consider the following 3×3 assignment matrix

	T_1	T_2	T_3
M_1	0	7	4
M_2	∞	3	5
M_3	4	1	2

where infinity (∞) represents an inadmissible pairing (i.e., a pairing that falls outside the track gate region). Consider adding 2.5 to the second column and subtracting 2.5 from the second row. The new matrix is

$$\begin{vmatrix} 0 & 9.5 & 4 \\ \infty & 3 & 2.5 \\ 4 & 3.5 & 2 \end{vmatrix}$$

This matrix is said to be *equivalent* to the first one. The optimal solution to this equivalent matrix can be easily identified as the same as the NN solution (i.e., T_1 is assigned M_1, T_2 is assigned M_2, and T_3 is assigned M_3).

In general, by adding and subtracting constant column and row vectors, an equivalent matrix can always be found for which the optimal solution can be found by inspection. The strategy is in knowing what vectors to add and subtract. This is usually done through a series of approximations. As many passes through the data as necessary are applied until an optimal solution is found. This is in contrast to the NN algorithm that uses one, or at most two, passes through the data before declaring a solution.

The complexity of the earlier example was reduced by considering only the partial problem of assigning measurements to tracks. A more general treatment would consider the problem of assigning measurements to tracks, false alarms β_{FA}, and new tracks β_{NT}, as well as considering the possibility of not updating a track on that scan (Blackman, 1986, p. 279). "Fleshing out" the assignment matrix, the new matrix becomes

	T_1	T_2	T_3	NS_1	NS_2	NS_3
M_1	0	7	4	6	∞	∞
M_2	∞	3	5	∞	5	∞
M_3	4	1	2	∞	∞	4
NM_1	7	∞	∞			
NM_2	∞	9	∞		0	
NM_3	∞	∞	9			

Each of the measurements can associate with one of the tracks or form a new source (NS). Each of the tracks can associate with one of the measurements or not be assigned with any measurement (NM).

The last three rows can be eliminated. First, normalize the last three rows to a constant value:

$$\begin{vmatrix} 2 & 7 & 4 & 15 & \infty & \infty \\ \infty & 3 & 5 & \infty & 14 & \infty \\ 6 & 1 & 2 & \infty & \infty & 13 \\ 9 & \infty & \infty & & & \\ \infty & 9 & \infty & & 9 & \\ \infty & \infty & 9 & & & \end{vmatrix}$$

Since the last rows contain no new information, they can be eliminated, and the problem thus reduces to a rectangular matrix:

$$\begin{vmatrix} 2 & 7 & 4 & 15 & \infty & \infty \\ \infty & 3 & 5 & \infty & 14 & \infty \\ 6 & 1 & 2 & \infty & \infty & 13 \end{vmatrix}$$

Any tracks not assigned a measurement are assumed to be undetected on that scan by default.

We next discuss two algorithms that are based on the aforementioned remarks: the auction algorithm and Munkres algorithm. These two algorithms were chosen for presentation because they represent different ways of solving the problem, and thus serve as two representative benchmarks. The reader should be warned that this list is certainly not exhaustive, and numerous other approaches may be found by consulting any textbook on operations research (e.g., Rockafellar, 1984). The auction algorithm is presented next since it is the simpler of the two to describe and the most recent.

11.5.1 Auction Algorithm

Bertsekas proposed an coordinate-descent assignment algorithm where measurements bid for assignments and where each track is assigned to the highest bidder (Bertsekas, 1979, 1985, 1988). Constant vectors are then added/subtracted from the assignment matrix to discourage further bidding on the assigned measurement. To paraphrase Bertsekas: Objects are awarded to the highest bidder . . . thereby raising their price.

The algorithm consists of two phases per iteration:

Bidding Phase. Each unassigned measurement independently determines its best assignment according to a local distance criteria (similar to NN). It also computes a *bid* for that assignment that approximately equals the difference between the best score and the second best score. This bid represents the worth of the assignment based on local coordinate information.

Assignment Phase. The track is tentatively assigned to the measurement that makes the highest bid (if any). Once an assignment is awarded, the relative scores are changed to discourage further bidding on that object.

These two steps are best illustrated by means of an example. Consider the previous assignment matrix:

	T_1	T_2	T_3	NS_1	NS_2	NS_3
M_1	2	7	4	15	∞	∞
M_2	∞	3	5	∞	14	∞
M_3	6	1	2	∞	∞	13

During the bidding phase, M_1 bids for T_1 with a bid of 2 ($2 = 4 - 2$), M_2 bids for T_2 with a bid of 2 ($2 = 5 - 3$), and M_3 bids for T_2 with a bid of 1 ($1 = 2 - 1$). Both M_1 and M_2 win their bids. At this point the assignment matrix is changed to discourage further bidding. For example, one procedure is to add a constant column vector that forces the best score and the second best score to be equivalent for the measurement just awarded the bid. Under this rule the new assignment matrix becomes

$$\begin{vmatrix} 4^* & 9 & 4 & 15 & \infty & \infty \\ \infty & 5^* & 5 & \infty & 14 & \infty \\ 8 & 3 & 2 & \infty & \infty & 13 \end{vmatrix}$$

where the asterisk (*) denotes a tentative assignment.

On the next iteration, M_3 bids 1 for T_3 and is awarded the track. The new assignment is

$$\begin{vmatrix} 4^* & 9 & 5 & 15 & \infty & \infty \\ \infty & 5^* & 6 & \infty & 14 & \infty \\ 8 & 3 & 3^* & \infty & \infty & 13 \end{vmatrix}$$

The algorithm terminates at this stage since all the measurements have been awarded tracks.

This description omits several important details needed for actual implementation, such as how to resolve conflicts between tie bids. The interested reader is referred to the references for more detailed information of this type (Bertsekas, 1979, 1985, 1988).

11.5.2 Munkres Algorithm

In 1955, Kuhn developed a steepest-descent method for solving the assignment problem based on an earlier proof due to the Hungarian mathematician Egervary (Kuhn, 1955). Munkres developed an efficient computer implementation for the Hungarian method in the case of square matrices $m = n$ (Munkres, 1957). Bourgeois and LaSalle generalized the Munkres algorithm for nonsquare matrices without adding extra rows or columns (Bourgeois and LaSalle, 1971). Stephans and Krupa extended the Bourgeois–LaSalle algorithm to efficiently process large sparse matrices (Stephans and Krupa, 1979). The classical Bourgeois–LaSalle version of the Munkres algorithm is summarized next.

Central to the Munkres algorithm is the dual problem of creating independent zeros. An element is said to be *independent* if no similar element occupies the same row or column. For example, consider the 2×2 matrix of values,

$$\begin{array}{c|cc|} & T_1 & T_2 \\ M_1 & 2 & 4 \\ M_2 & 1 & 1 \end{array}$$

Two independent zeros can be created by subtracting 2 from the first row and subtracting 1 from the second row:

$$\begin{vmatrix} 0^* & 2 \\ 0 & 0^* \end{vmatrix}$$

The two independent zeros are denoted by an asterisk (*).

The objective of the Munkres algorithm can be stated as follows. Find a matrix equivalent to the original one with all elements positive and containing the maximum number of independent zeros (or asterisks). This is done through a sequence of finding lines containing independent zeros, subtracting constant vectors, followed by transferring stars until the maximum number of independent zeros are created.

Two additional constructs need to be introduced. Any zero without an asterisk may be distinguished by a prime. Basically a primed zero is a candidate for an asterisk. Some rows and columns are distinguished too; they are said to be *covered*. A zero is said to be covered if it is contained in a covered line.

With these constructs the Munkres algorithm can be summarized as follows:

Preliminary. Subtract the smallest element from each row in the $m \times n$ assignment matrix.

Step 1. Place an asterisk with each independent zero. Go to step 2.

Step 2. Cover every column containing a zero with an asterisk. If m columns are covered, then the zeros with asterisks form the desired set. Otherwise go to step 3.

Step 3. Choose an uncovered zero and prime it ($0'$). Then consider the row containing it. If there is no zero with an asterisk in this row, go to step 4. If there is a zero with an asterisk in this row, cover this row and uncover the column containing the asterisk. Repeat until all the zeros are covered. Go to step 5.

Step 4. There is a sequence of zeros with alternating asterisks and primes that is created as follows: let z_0 denote the uncovered zero prime $0'$. Let z_1 denote the 0^* in z_0's column (if any). Let z_2 denote the $0'$ in z_1's row. Continue in a similar way until the sequence stops at a $0'$ that has no 0^* in its column. Remove the asterisk from each zero with an asterisk in the sequence, and place an asterisk on each primed zero of the sequence. Erase all primes and uncover every line. Return to step 2.

Step 5. Let h denote the smallest uncovered element of the matrix; h is positive. Add h to each covered row and subtract h from each uncovered column. Return to step 3 without altering any asterisks, primes, or covered lines.

The algorithm is best illustrated by means of an example. Again consider our running example:

$$\begin{bmatrix} 2 & 7 & 4 & 15 & \infty & \infty \\ \infty & 3 & 5 & \infty & 14 & \infty \\ 6 & 1 & 2 & \infty & \infty & 13 \end{bmatrix}$$

The sequence of steps to reduce this matrix to three independent zeros is constructed as follows:

Preliminary step

$$\begin{vmatrix} 0 & 5 & 2 & 13 & \infty & \infty \\ \infty & 0 & 2 & \infty & 11 & \infty \\ 5 & 0 & 1 & \infty & \infty & 12 \end{vmatrix}$$

Steps 1, 2, and 3

$$\begin{vmatrix} 0^* & 5 & 2 & 13 & \infty & \infty \\ \infty & 0^* & 2 & \infty & 11 & \infty \\ 5 & 0 & 1 & \infty & \infty & 12 \end{vmatrix}$$

Step 5 ($h = 1$)

$$\begin{vmatrix} 0^* & 5 & 1 & 12 & \infty & \infty \\ \infty & 0^* & 1 & \infty & 10 & \infty \\ 5 & 0 & 0 & \infty & \infty & 11 \end{vmatrix}$$

Steps 3, 4, and 2

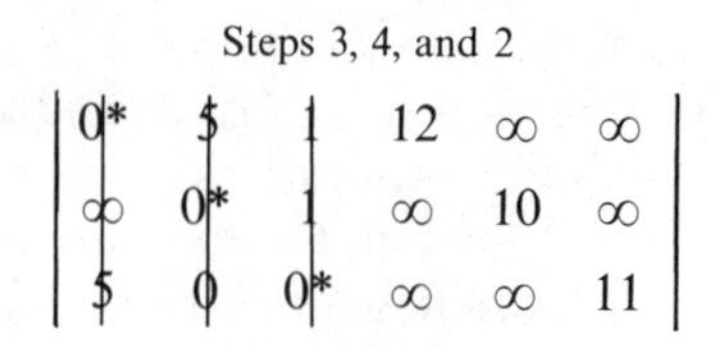

The algorithm terminates since three independent zeros are found.

11.6 MAXIMUM-LIKELIHOOD GATING

The gating test can be improved upon by incorporating the previous probabilistic concepts. One drawback of the standard Neyman–Pearson gate is that it ignores the surrounding new target β_{NT} and false alarm environment β_{FA}, and so an extrapolating track with a large value of **S** can potentially be gated with multiple false alarms or data miscorrelations. If the error of the track grows to the point where there is a high probability of it being gated with multiple false alarms, it is better to limit (bound) the size of the gate region to prevent unnecessary processing of data.

Consider the following maximum-likelihood (ML) gate test (Blackman, 1986, pp. 190–191, 265–269),

$$P_D g(\tilde{\mathbf{y}}_{ij}) \gtrless (\beta_{NT} + \beta_{FA})(1 - P_D) \tag{11.12}$$

As Figure 11.3 shows, the above ML gate test eliminates a pairing if the probability that the pairing is valid $[P_D g(\tilde{\mathbf{y}}_{ij})]$ is less than the probability that the measurement represents a new target (β_{NT}) or false alarm (β_{FA}) and that the target was missed ($1 - P_D$). Thus, the ML test eliminates those pairings that are guaranteed to be eliminated anyway by the subsequent ML data association algorithm. If the pairing does pass the ML gate threshold, then the data association algorithm may still eliminate the pairing anyway. Yet, (11.12) describes a reasonable way to eliminate completely unlikely assignments prior to data association.

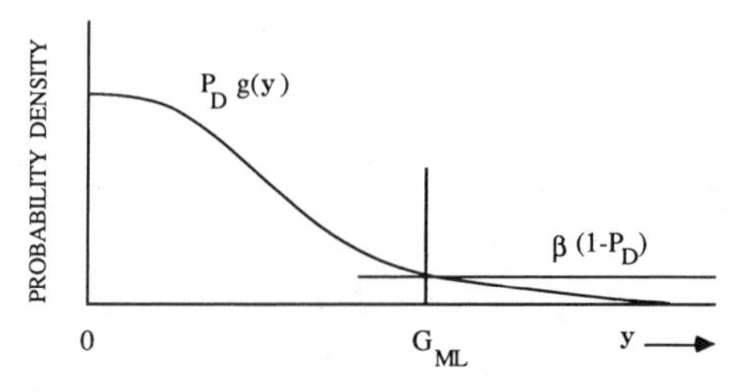

Figure 11.3 The maximum-likelihood gate G_{ML}.

Further simplifying, the following result is obtained:

$$d_{ij}^2 \gtrless 2 \ln \left[\frac{P_D}{(1 - P_D)(\beta_{\mathrm{NT}} + \beta_{\mathrm{FA}})(2\pi)^{M/2}|\mathbf{S}_{ij}|^{1/2}} \right] \tag{11.13}$$

The maximum-likelihood gate in (11.13) should replace the Neyman–Pearson gate if computational resources permit since it is based on more exact statistical considerations. In turn, the maneuver gate takes precedence over the ML gate since it is based on a more exact target dynamical model.

11.7 MULTIPLE HYPOTHESES TRACKING

All data association algorithms are hypothesis testing exercises that attempt to answer the question: "Which data associations are most likely?" This involves some manner of evaluating or ranking these hypotheses relative to each other (scoring), and some manner of storing and retrieving the best hypothesis (or best hypotheses).

With the recent advent of computers with more storage capability, it is appropriate to consider the possibility of retaining more than one hypothesis. The argument goes that a hypothesis that was originally considered to be the best may turn out, based on later evidence, not to be the correct hypothesis solution after all (Reid, 1979). In essence, this amounts to maximizing the score statistic over several scans as well as over other tracks in the vicinity.

Figure 11.4 illustrates an example where insufficient information is available in the current data to make any single assignment with a high degree of confidence. As shown, two closely spaced targets share a single measurement. The measurement lies within the large gate of both targets and is exactly midway between both tracks. Clearly, any sequential assignment algorithm will have at most a 50 percent probability of correctly assigning this measurement.

The generation of the multiple hypotheses proceeds along the same lines as outlined previously. Again, for small assignment matrices, the use of enumera-

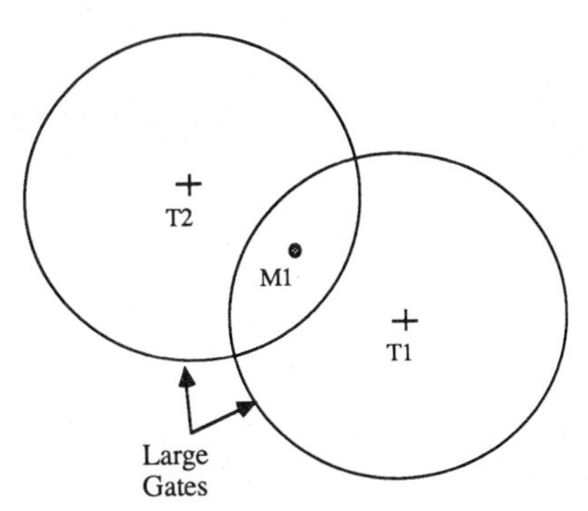

Figure 11.4 Ambiguity between tracks 1 and 2 is difficult to solve sequentially.

tion is appropriate. The technique outlined in Blackman (1986, p. 408) consists of enumerating all feasible assignments, pruning to the N best, and then considering the next measurement in turn. This logic can be shown to work well provided the intermediary pruning logic and the order of the measurement sequencing is properly chosen.

For large assignment matrices, of course, other solutions must be investigated. Unfortunately, no one solution has appeared in the literature that efficiently handles this case. The two most promising approaches are the ones by Nagarajan et al. (1987), which is basically an enumeration method with efficient cut-off rules, and the branch-and-bound method outlined in Chen et al., (1988) and Tucker (1984, p. 99).

Next a new solution is outlined that the author calls *binary search*. The algorithm can be stated as follows. The best sequential solution to the assignment matrix is first found using a search technique such as the Munkres algorithm. To find the second best solution, search through the set of assignments for a likely candidate for removal and break that connection. This can be simply done by replacing a single assigned entry by infinity (∞). The optimal solution is then recalculated using the same gradient technique as before (starting from where the old solution ended).

The set of candidate solutions is continually subdivided in this manner into binary subsets. One of the binary subsets contains the original solution, and the other binary subset does not contain the original solution. The optimal sequential solution for the second subset is found and then evaluated for possible propagation as an alternative hypothesis. Also, the optimal solution represents the subset for purposes of terminating the search routine in that direction.

The advantage of binary search is that it always proceeds from best to worst. It terminates only when the computer has exhausted its available resources, or whenever the solutions no longer appear worthy of further investigation. The optimal sequential solution is always contained in the set of solutions generated. A second advantage is that it decomposes the problem into a series of parallel subproblems with each subproblem sharing a common computational structure.

The disadvantage of binary search is that it presupposes that only a few alternate hypotheses are to be propagated forward. If the number of multiple hypotheses propagated forward is large, then this method requires too many passes through the gradient algorithm, and too much wasted effort setting up the optimal sequential solution for each binary subset.

Hypothesis Testing. The set of propagated hypotheses are initially selected based on their relative probabilities. The probability of an hypothesis is represented by its *hypothesis score*—by definition, the sum of the scores of all tracks contained in that hypothesis.

Further testing may be required to thin the number of hypotheses propagated forward. The function of these additional tests is to prevent utilization of needed computational resources in maintaining multiple hypotheses that are not sufficiently different to impact track accuracy.

For example, a simple rule is the *small gate rule*. The small gate rule states that "no branching shall occur for any target assigned a measurement in its small gate region" (Bath et al., 1980). The role of multiple hypotheses tracking, in this context, is to branch only whenever the target maneuvers or goes undetected due to a deep fade while a false alarm appears inside its large gate region. Sequential operations are performed for quiescent targets, or targets too close together to separate in an efficient manner.

Another rule in this same vein is the *single branching rule*. The single branching rule states that a newly split hypothesis (i.e., a hypothesis that splits into two hypotheses) cannot split again until one of the two hypotheses has been deleted. The function of the single branching rule is to prevent monopolization of computer time by a single track or probabilistic outcome.

11.8 ORGANIZATION OF TRACKS IN COMPUTER MEMORY

A *track file* is a computer record containing all information pertaining to the track. The set of all track files are joined together in a list, linked by pointers, as shown in Figure 11.5.

It is often convenient to order the track records according to increasing angle [called *sectoring* (see Farina and Studer, 1985a, p. 193)]. A software pointer indicates the position of the antenna beam relative to the angle-ordered list. The primary objective of sectoring is to reduce the number of gating tests required to those that lie within the antenna beamwidth. As new measurements are received, track-to-measurement associations are considered by gating the reduced list of tracks contained within the beamwidth.

Multiple hypotheses processing can be implemented by making this list two dimensional. Figure 11.6 illustrates a proposed computer structure for the case illustrated in Figure 11.4. As shown, hypotheses that are mutually contradictory are stored separately in computer memory. Each hypothesis or *branch* represents a collection of track files that altogether form a set of valid associations for a group of interacting targets. Each branch is linked by pointers to a *hypothesis record* containing the total score of the hypothesis.

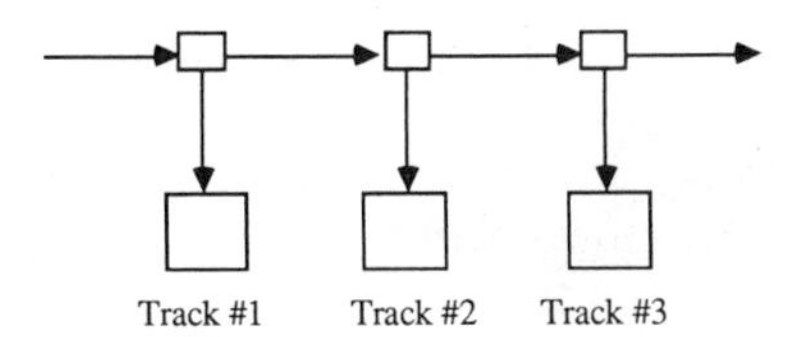

Figure 11.5 Organization of tracks in computer memory.

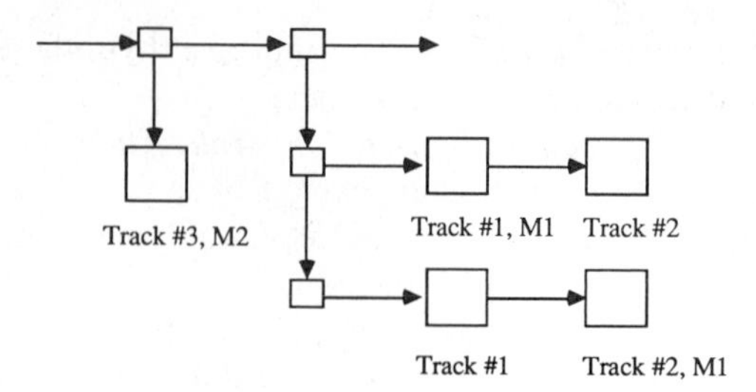

Figure 11.6 Assignment conflict between tracks 1 and 2 has generated two alternate branches or "hypotheses."

11.9 COMPARATIVE REVIEW

We are now in a position to discuss the various approaches to data correlation represented by nearest neighbor, track splitting, sequential processing, and multiple hypothesis tracking (MHT).

In multiple hypothesis tracking (MHT) the number of hypotheses is limited to some maximum value N, for N larger than unity (Reid, 1979). The value N is chosen such as to use all available computer resources without overloading. Typically, all hypotheses are deleted whose scores fall below a threshold, and then up to N best of these hypotheses are retained. The best hypotheses are chosen based on their relative probabilistic and/or kinematic properties.

Track splitting (TS) refers to the special case in which multiple hypotheses are retained, but consistency checks are not maintained (Smith and Buechler, 1975). Furthermore, track splitting does not keep a separate hypothesis record, as per Figure 11.6, and thus information contained in previous assignment decisions are not propagated forward. Track splitting results in some computational savings and performs well provided the tracks are not closely spaced and not interacting. In this special environment track splitting performs well since inconsistent tracks tend to be merged or to age and be deleted due to their own lack of inertia.

Sequential association algorithms correspond to the special case in which only a single hypothesis is retained after scoring, that is $N = 1$. The nearest-neighbor (NN) algorithm is a sequential association algorithm that utilizes a minimum distance criteria.

11.10 SUMMARY

Most data correlation methods are implemented by assigning to each track-to-measurement pair a normalized distance d^2. The assignment matrix consists of the two-dimensional matrix of all distances d^2. For each column of the assignment matrix a unique row is sought that optimizes performance by some criteria of merit.

Gating is a preliminary test that reduces the number of entries in the assignment matrix and allows the matrix to be decoupled into a number of smaller submatrices. Conventional gating tests, such as the Neyman–Pearson test, rely on probability theory and tend to be conservative. More complex gating tests can be formulated to more efficiently gate the entries. If no measurement passes the gating test for a particular track, special logic may be introduced to consider nonprobabilistic events such as a target maneuver.

Data association takes the output of the gating algorithm and makes final measurement-to-track associations. The simplest method is the so-called nearest-neighbor (NN) approach. The NN approach looks through the gated measurements and pairs the minimum distance d^2 with the track file under consideration. A more optimal algorithm would minimize the sum of the distances rather than each distance when viewed in isolation. In addition, a more general criteria would incorporate the probability of observing a particular detection sequence as given by the probability of detection and false alarm, as well as incorporating the possibility of not updating the track on that scan.

General strategies were outlined to iteratively determine the optimal solution to the assignment problem. To illustrate the concepts, three approaches were singled out for special attention: total enumeration, the auction algorithm and Munkres algorithm. The reader should be warned that this list is certainly not exhaustive and numerous other approaches are available in the literature.

Multiple hypotheses tracking (MHT) proceeds along the same lines as outlined for the single hypothesis case. The use of enumeration is appropriate for small assignment matrices, whereas other solutions should be investigated for large assignment matrices. Several alternative approaches were discussed including branch-and-bound, enumeration with pruning, and binary search. The discussion concluded with a historical overview of the various approaches to MHT implementation proposed in the past.

In general the purpose of this chapter was to provide a concise introduction, rather than a detailed description, of the principles of data correlation. The reader is referred to the following references for more information: Farina and Studer (1985a, 1985b); Mori et al. (1986); Blackman (1986); and Bar-Shalom and Fortmann (1988).

CHAPTER 12

A REPRESENTATIVE TWS SYSTEM

The traditional track-while-scan (TWS) problems are data correlation and data filtering. Therefore, after some preliminary remarks, the majority of this chapter's discussion focuses upon these two subjects.

12.1 THE RADAR HARDWARE

The basic differences between search and track are reflected in the choice of equipment. In its most conventional form a search radar uses a lower microwave radio frequency (UHF, L-band, and S-band) with a corresponding lower angular resolution. Lower RFs are preferred for search since a larger power-aperture product can be obtained, less atmospheric attenuation is present, and the detection of high-speed targets in LPRF is enhanced since the blind speeds are larger for a fixed PRF. The appendix discusses these subjects in more detail.

12.2 MEASUREMENT FORMATION

12.2.1 Range/Doppler Measurement

The first question that needs addressing is how to group detections preliminary to measurement formation. If a solitary detection is present on the range/Doppler grid, and none of the neighboring bins contain enough SNR for measurement formation, then the Kalman filter must be supplied with center-bin values (Cole et al., 1986). In the more usual case, however, a single target return will generate multiple detections on the range/Doppler grid. Thus, as a first step, it is necessary to group all detections together that belong to the same target. In a dense multiple-target environment it may not be immediately apparent how to group detections (Cole et al., 1986).

One method is to convolve a stored replica of the radar's theoretical matched-filter response against the observed pattern of target detections. The best-fit point would then correspond to the target's measured range/Doppler position according to matched filter theory. If the RMS error associated with the best fit is small enough, then the best-fit point is declared to be from a single target. Otherwise, two interfering targets are declared.

The previous method is probably the optimum approach to this problem. Nevertheless, an actual radar's ability to resolve two closely spaced targets is limited by the radar's finite A/D sampling rate and instability of the radar's front-end video response (resistors–capacitors). In those cases where these factors preclude the radar from achieving its theoretical performance, then simpler procedures suffice.

A typical pattern of detections is illustrated in Figure 12.1 along with a typical grouping. Based on this diagram, an alternate (and simpler) procedure can be immediately suggested. First, the signal peaks are located after thresholding, and each peak is assumed to correspond to a distinct target detection. The next highest amplitude cell adjacent in range comprises the next cell grouped. If Doppler is measured, then the highest amplitude cell adjacent in Doppler comprises the third cell. The remaining cell in the quadrant rounds out the set. Measurement formation then proceeds as discussed in Chapter 6 using the four cells grouped. The remaining cells in the immediate area are discarded as sidelobe suspects.

The method described can be improved by using a more judicious selection of the group. For instance, the primary detection can be located using simple peak

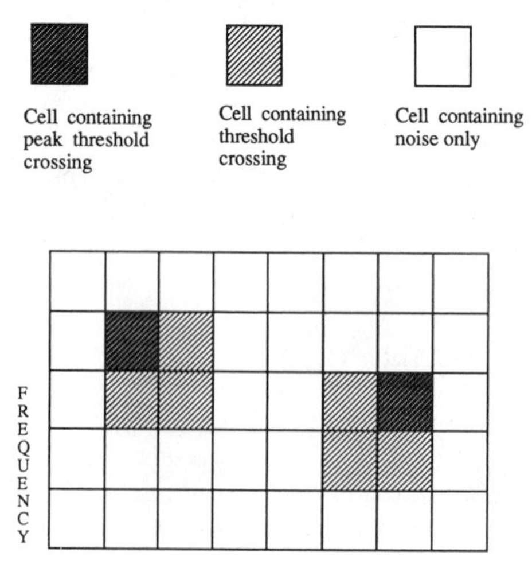

Figure 12.1 Pattern of observed threshold crossings on the range/Doppler grid. Example given is of two targets present with each target generating a block of four hits.

rules as already outlined, but the surrounding cells used by the measurement set can be more carefully selected to minimize sidelobe energy and maximize the measurement slope coefficient. Furthermore, the two-target hypothesis is more closely examined by examining the relative distribution of detections in the surrounding area (Cole et al., 1986).

12.2.2 Angle Measurement

To form an angle measurement, it is normally necessary to pair sequential detections between adjacent dwells. A reasonable test to pair detections is

$$\begin{aligned} |(r_m)_i - (r_m)_j| &< R_G \\ |(\dot{r}_m)_i - (\dot{r}_m)_j| &< V_G \end{aligned} \tag{12.1a}$$

for R_G the range gate, V_G the velocity gate, r_m the range measurement, and $\dot{r}_m$ the range rate (or Doppler filter) value. The use of rectangular gates is justified because of the small time differential between adjacent dwells. If unambiguous range rate measurements are available, a further refinement may be introduced as follows:

$$\begin{aligned} |(r_m + \dot{r}_m\Delta)_i - (r_m - \dot{r}_m\Delta)_j| &< R_G \\ |(\dot{r}_m)_i - (\dot{r}_m)_j| &< V_G \end{aligned} \tag{12.1b}$$

where Δ is one-half the difference between the i-j measurement times.

Having selected those detections to be used to form an angle measurement, the processing then proceeds as outlined in Chapter 6. If the processing cannot identify enough detections to form a valid angle measurement, then the Kalman filter may be supplied with center (boresight) values.

12.2.3 Measurement Variance

The SNR is, by definition, the total signal power divided by the total noise power. "Noise power" includes thermal noise as well as any interference from neighboring targets that enter through the range/Doppler sidelobes. Thus, in a dense target environment, the SNR fed to the Kalman filter should be adaptively modified based on the positions of the neighboring targets and the known sidelobe structure.

Data miscorrelations represent a slightly different type of error source. In this case it probably makes more sense to directly increment the r_{ij} values in the $\mathbf{R}$ covariance matrix of (9.3). The correct amount of increase should be proportional to the suspected error magnitude as weighted by the probability of miscorrelation (Blackman, 1986, p. 105).

12.3 BAR-TO-BAR ELIMINATION LOGIC

Redundant detections occur in TWS because of the manner in which the mechanically controlled antenna is scanned. Scanning is accomplished by moving the beam in a predetermined pattern in space, typically a raster scan comprised of multiple elevation bars. A certain amount of overlap between the bars is provided to minimize the beamshape loss, implying that a strong target can be detected on more than one bar. The time between these redundant detections depends on the way the search volume is scanned.

Early elimination of redundant detections can be accomplished by delaying the processing of any detections until neighboring bars have been scanned. A *pool* of new detections is thus formed. New detections are correlated against all other detections generated within the time span of the pool. The highest SNR detection (or averaged detection across the bars) is then sent to the data correlation logic to be processed as a single entity.

12.4 DATA FILTERING

In any TWS application, it is highly desirable to reference all tracks to a common inertial coordinate system. This allows the data filtering to be decoupled from the mechanical scanning and simplifies the state extrapolation of the multiple tracks.

The remaining topic is selection of state variables and target model. The optimum filter configuration would model the acceleration as a random process as per the Markov model. However, the relatively infrequent update times associated with mechanical scanning influences the performance of the resulting acceleration estimates. The acceleration state can often be eliminated in this case with negligible loss in performance. Elimination of the acceleration state is particularly desirable if the computational resources of the TWS system are limited, such that the marginal performance benefits do not offset the additional computational burden.

In order to assess the need for a separate acceleration state, the performance of the acceleration estimate should be studied. If the acceleration estimate performs poorly, then it can be argued that a reduced-state filter can be designed that performs almost as well as the full-state filter.

Such a set of performance curves was conveniently tabulated by Fitzgerald (1980) and is reproduced in Farina and Studer (1985a, p. 233). By inspection of these curves, it can be immediately seen that the circumstances under which the acceleration estimate tends to perform poorly occurs whenever

$$\alpha \Delta T > f(p_2) \tag{12.2}$$

where α is the target's correlation time, ΔT is the radar's update time, and the ratio p_2 is a dimensionless quantity introduced by Fitzgerald to parameterize the performance of his Kalman filter curves. The exact expression is given by $p_2 \triangleq \Delta T^2 \sigma_a / \sigma_x$, for σ_a the standard deviation of the target's acceleration and σ_x the standard deviation of the position measurement. As p_2 varies from 0.1 to 1000, $f(p_2)$ can be seen to vary from 0.1 to 1 (e.g., see Figure 7 of Fitzgerald, 1980; or Figure 4.9 of Farina and Studer, 1985a). States in these terms, $\alpha \Delta T$must be less than 0.1 before the acceleration estimate begins to behave well at low values of p_2. Clearly, this is outside the range of $\alpha \Delta T$ values characteristic of most TWS systems.

The previous remarks implicitly assumed that only position measurements are available to be filtered. If unambiguous Doppler measurements are available, an analogous relationship might be stated as

$$\alpha \Delta T > f(p_2, p_3) \tag{12.3}$$

for $p_3 = \Delta T \sigma_a / \sigma_{\dot{r}}$, $\sigma_{\dot{r}}$ the standard deviation of the range rate measurement, and p_2 as previously defined. Performance curves for the Doppler case are not available in the open literature (to the author's knowledge) and so the exact functional form of (12.3) is unknown. However, it is probably safe to say that the acceleration state should always be modeled if Doppler is one of the measurements.

12.4.1 Position-Velocity Filters

Given that the acceleration state is eliminated, the state variable reduces to

$$x_{\text{NED}} = \begin{vmatrix} x_N \\ v_N \\ x_E \\ v_E \\ x_D \\ v_D \end{vmatrix} \tag{12.4}$$

In the absence of Doppler measurements, the measurement vector is of the form

$$\mathbf{m}_{bae} = \begin{vmatrix} \zeta_{b_m} r_m \\ \zeta_{a_m} r_m \\ \zeta_{e_m} r_m \end{vmatrix} = \begin{vmatrix} x_{b_m} \\ x_{a_m} \\ x_{e_m} \end{vmatrix} \tag{12.5}$$

where r_m = range measurement

$(x_{b_m}, x_{a_m}, x_{e_m})$ = measured target position in antenna coordinates

$(\zeta_{a_m}, \zeta_{e_m})$ = measured direction cosines

$\zeta_{b_m} \triangleq \sqrt{1 - \zeta_{a_m}^2 - \zeta_{e_m}^2}$

Assuming $\zeta_e, \zeta_a \ll 1$, the measurement reduces to

$$\mathbf{m}_{bae} = \begin{vmatrix} r_m \\ \zeta_{a_m} r_m \\ \zeta_{e_m} r_m \end{vmatrix} \tag{12.6}$$

and the corresponding covariance matrix is

$$\mathbf{R}_{bae} = \begin{vmatrix} \sigma_r^2 & 0 & 0 \\ 0 & r^2\sigma_{\zeta_a}^2 & 0 \\ 0 & 0 & r^2\sigma_{\zeta_e}^2 \end{vmatrix} \tag{12.7}$$

where σ_r^2 = range measurement variance
$\sigma_{\zeta_a}^2$ = azimuth measurement variance
$\sigma_{\zeta_e}^2$ = elevation measurement variance

The assumption ($\zeta_e, \zeta_a \ll 1$) is valid in TWS since the target must lie close to mechanical boresight in order to receive an appreciable amount of illumination.

Next, a transformation is made to transform the above antenna-referenced measurement vector to NED coordinates:

$$\mathbf{m}_{\text{NED}} = \{\text{NED}, bae\} \begin{vmatrix} r_m \\ \zeta_{a_m} r_m \\ \zeta_{e_m} r_m \end{vmatrix} \tag{12.8}$$

where $\{\text{NED}, bae\}$ is the transformation matrix between the two coordinate systems. The associated covariance matrix is

$$\mathbf{R}_{\text{NED}} = \{\text{NED}, bae\}\mathbf{R}_{bae}\{\text{NED}, bae\}^T \tag{12.9}$$

In the specific case of a ground-based fan-beam antenna, this transform can be further reduced to a single operation:

$$\mathbf{R}_{\text{NED}} = \begin{vmatrix} \cos(v) & \sin(v) & 0 \\ -\sin(v) & \cos(v) & 0 \\ 0 & 0 & 1 \end{vmatrix} \mathbf{R}_{bae} \begin{vmatrix} \cos(v) & -\sin(v) & 0 \\ \sin(v) & \cos(v) & 0 \\ 0 & 0 & 1 \end{vmatrix}$$

for v the single gimbal rotation. In the general case, the transformation (9.18) must be taken.

The residual is then

$$\text{Residual} = \mathbf{m}_{\text{NED}} - \mathbf{H}_{\text{NED}}\hat{\mathbf{x}}_{\text{NED}} \tag{12.10}$$

for

$$\mathbf{H}_{\mathrm{NED}} = \begin{vmatrix} 1 & 0 & 0 & 0 & 0 & 0 \\ 0 & 0 & 1 & 0 & 0 & 0 \\ 0 & 0 & 0 & 0 & 1 & 0 \end{vmatrix} \tag{12.11}$$

In conclusion, the residual is given by (12.10) with $\mathbf{R}_{\mathrm{NED}}$ and $\mathbf{H}_{\mathrm{NED}}$ given by (12.9) and (12.11), respectively. The state transition matrix is given by (9.28a) with $\mathbf{\Phi}_{1\mathrm{D}}$ given by (7.10), and $\mathbf{Q}$ is given by (9.30) with $\mathbf{Q}_{1\mathrm{D}}$ given by (7.28). This completes the set of coupled TWS filtering equations.

Finally, the design of the data-filtering logic may be further simplified by decoupling the above six-state filter into three two-state filters (Brammer, 1982). This corresponds to replacing the off-diagonal submatrices with zero in $\mathbf{R}_{\mathrm{NED}}$ and $\mathbf{Q}$. Decoupling reduces the computational complexity by a factor of 9 since the computational complexity grows as the cube of the number of states modeled. Nevertheless, although decoupling in antenna coordinates or covariance coordinates is routinely done, one must be careful when decoupling the measurements in NED coordinates. In general, decoupling in NED coordinates is a good idea only if the measurement error ellipsoid can be approximated as roughly circular (equal range and cross-range variances).

12.5 TRACK INITIATION, CONFIRMATION, AND DELETION

Measurements not assigned to any existing tracks are candidates to initiate new tentative tracks. Then, once a tentative track is formed, at least one other correlating measurement on a subsequent scan is required to place the track on the list of *confirmed* tracks.

Additional restrictions may be placed on the confirmation procedure to reduce the probability of a confirmed track being produced by false alarms. The most widely used rule is that M correlating measurements should be received within N scans for successful track confirmation (M out of N). Typical values are 3-out-of-4 or 3-out-of-5, although in cases that the SNR is high only two measurements may suffice (2 out of 2). Postdetection integration (PDI) may also be employed, either in place of M out of N or in conjunction with it (Cantrell, 1986). In essence, this amounts to incorporating amplitude information as part of the track initiation process.

Other restrictions may be placed on the type of measurement used in the confirmation test. The intent is to prevent tentative tracks from being confirmed based on measurements of questionable origin. A common restriction is to prohibit measurements that fall within the large gate of an existing track file from being used in the confirming process, even if the measurements are not assigned to any existing track. In any case, if a tentative track is never confirmed, then the entire set of measurements are dismissed as false alarms. This avoids saturating the data processor with false tracks and is one means of false alarm control.

12.5.1 Track Deletion

A degraded track must be deleted. A simple rule is to delete any track after N_D consecutive scans produce no correlating measurements, where N_D is a fixed constant. Alternately, the score function may be monitored and the track deleted whenever its current score falls below a threshold relative to its previous maximum value (Blackman, 1986, pp. 261, 272). Finally, a deletion criteria may be defined based on N_D misses, but the selected N_D based on prior score analysis.

Deletion logic for a confirmed track is different than that for a tentative track. Indeed, the primary motivation for the tentative/confirm classification is track deletion. A tentative track is deleted based on the hypothesis that the track actually consists of a series of false alarms. A confirmed track is deleted based on the hypothesis that the target is lost or has moved outside the scan volume. This may imply different N_D values for confirmed and tentative tracks.

12.6 MERGE LOGIC

Merge logic is included in some systems to handle the case where the system initiates a new track file on a target already in the data base. When this happens, merge logic is needed to purge the data base of redundant tracks. The merge logic begins by occasionally comparing track files against each other two at a time. Any two track files whose states are the same (within a specified gate size) are judged to be from the same target. The two redundant tracks are merged or the weaker track is deleted.

12.7 GROUP TRACKING

Group tracking is appropriate in those situations characterized by targets moving in tight groups (Blackman, 1986, Chapter 11). In group tracking a single track file is used to store an estimate of the group centroid, extent, orientation, and count. The group count is kept to flag dropouts and to distinguish a member of the group from false alarms.

The primary motivation for the use of group tracking is reduction of computational complexity. To understand this comment, visualize a large convoy of trucks traveling along a road. The computational complexity of associating and tracking each truck individually would clearly exceed present-day computational capability. Another example in this same vein occurs when tracking a formation of closely spaced missiles or a raid of aircraft flying in formation. As long as the overall group extent is less than the illuminated area of interest, then there is no need to track each individual element in the group, and so simply tracking the centroid and extent suffices.

12.8 DOPPLER SPECTRA SUPPRESSION

Doppler spectra (DS) are a critical problem when tracking aircraft or helicopter targets in Doppler. DS returns are caused by such phenomenon as propeller blades (Nathanson, 1969, p. 175), jet engine turbine blades (Blackman, 1986, p. 139), helicopter blades (Fliss and Mensa, 1986), or tank treads (Currie and Brown, 1987, p. 271). DS possess the same azimuth, elevation, and range as the target but have a different Doppler frequency, typically some multiple (harmonic) of a base frequency to the true (skin) return. The base frequency depends on the physical characteristics of the rotating source, but is around 1 kHz for jet engines and slightly smaller for prop modulation.

The average power of the DS returns is less than that of the skin. However, at close range, the DS returns are strong enough to be consistently registered as detections. Also, the DS returns scintillate (as does the skin) and may sometimes have an instantaneous power that is higher than the skin's.

The net effect of the DS returns is to produce a pattern of false detections centered around the skin return as illustrated in Figure 12.2. This creates two related problems. First, it erroneously makes a single target appear as a group of closely spaced targets with slightly different range rates. Second, in those cases in which the true skin return has "scintillated down" and is temporarily too weak to be registered as a detection, the skin track may be incorrectly updated with a DS return. This would falsely make the target appear to have accelerated.

This last issue is particularly troublesome when tracking targets in Doppler at the higher RFs (C-band and above). At the higher RFs the Doppler gates may be 1 kHz or more given a typical TWS-like update time of one or two seconds. At the lower RFs, the Doppler gates are narrower (i.e., $f_G = 2V_G/\lambda$).

It should be emphasized that DS returns only present a problem if targets are being tracked in Doppler frequency. In LPRF systems the DS phenomenon has traditionally been viewed as beneficial since it allows the target to be tracked through a blind zone or deep fade. Indeed, some targets such as hovering helicopters can only be tracked by means of their Doppler spectra.

The correlation logic presented in this section attempts to suppress the detrimental effects of DS returns by identifying measurements that are likely to

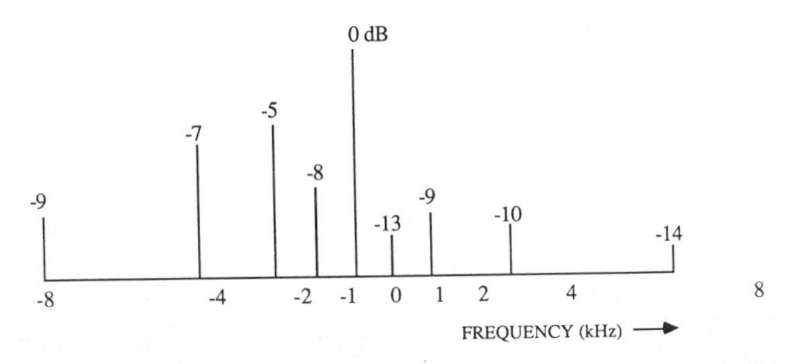

Figure 12.2 Typical DS modulation pattern from a jet engine.

be DS, and by not allowing them to initiate new tracks or update established tracks. This is mainly done through a set of ad hoc rules similar to those discussed earlier for track initiation and confirmation.

12.8.1 DS Inhibition Gates

One way to suppress DS is to simply ignore all excess returns that appear at the same range as the target being tracked. This procedure works well provided the granularity of the range bins is fine enough to separate targets. The obvious objection to this tactic is that multiple targets may be flying close together. In this case the logic would initiate a single track on one of the targets while the remaining targets would be ignored as DS suspects. The seriousness of this objection can be understood when one envisions several aircraft flying in close formation or a single aircraft plus recently launched missile.

12.8.2 DS Offset Tracking

Another option is to track the DS as a track file offset. In this scheme each DS line is tracked separately as an offset from the true skin return. As shown in Figure 12.3, measurements that do not correlate with the DS gate of an established DS offset are then treated as possible new targets.

This procedure enables the system to keep track of the DS returns given minor changes in target speed. However, if the target suddenly accelerates, the DS positions will suddenly change and a new set of lines may appear. Thus, it is necessary to set up an orderly procedure to initiate and delete DS offsets. Procedures for DS offset initiation and deletion can be performed similarly to regular trackfile initiation and deletion. The only difference is the need to associate a new DS line with an established track file of sufficient SNR to produce a DS. Any errors in DS initiation or classification can be eventually flagged by range/range-rate divergence. Then, by examining the data file, the true skin and DS combinations can be resolved as per page 181.

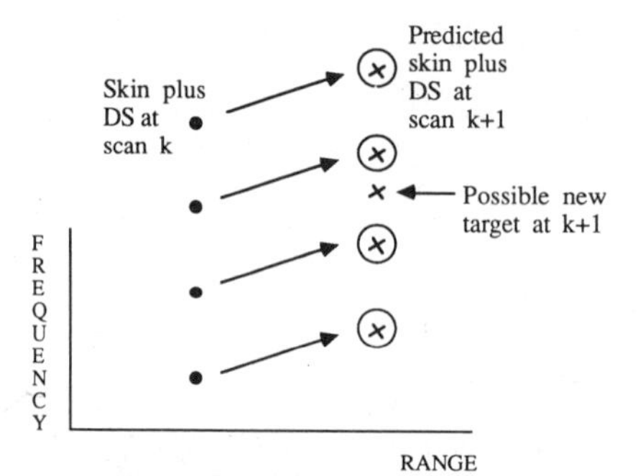

Figure 12.3 DS offset tracking.

There are potential problems to this approach. For one, the radar's update rate may be so low that the predicted DS gate regions overlap. Second, if a track is initiated on a designated target at short range, the radar will simultaneously see the target's skin plus all of its DS lines at once. Initially, there is no trackfile data to sort DS returns from the skin return. Third, there remains the problem of initiating and deleting DS trackfiles after sudden target accelerations. Despite all these drawbacks, this solution method is probably the most powerful approach to this problem.

12.8.3 DS Range Matching

Another way to recognize DS lines is to correlate them in range. Since the range of the set of DS lines are identical, an accurate measurement of range should provide the necessary information to associate DS returns.

In HPRF, implementing this idea may prove difficult since range gating is not normally provided. In one approach, incremental range is measured by transmitting a series of dwells and varying the PRF. Since the eclipsing loss is a function of both range and PRF, incremental range can be inferred from the resultant eclipsed power loss. Provided the power measurement is accurate enough, the DS can be paired to the skin by this technique.

12.8.4 DS Pattern Recognition

Another approach is to group returns based on their observed Doppler signature. This approach requires that the TWS designer postulate a "likely" pattern of DS returns. For example, it may be known that DS lines are always separated by at least Y kilohertz in Doppler, where Y is around 1 kHz for jet engine modulations and somewhat lower for prop modulations. Hence, Doppler spacings much smaller than Y can be ignored in grouping the DS suspects.

Based on the prior comments, a suggested sequence of operations might be adopted as follows:

1. For each range bin, rank the measurements by SNR.
2. Eliminate returns with SNR too low to have generated DS.
3. For each remaining measurement, search for a pattern of associated DS lines around the postulated skin.
4. Require that all lines in the DS set reside at the same range bin and have approximately the same range measurement.
5. Require that the spacing between any two lines in a proposed DS set be an integer multiple of a fundamental harmonic spacing Δf.
6. Threshold Δf to eliminate harmonic spacings that are unrealistically small, for example, $\Delta f > 800$ Hz.
7. Require that a valid DS set contain no more than one or two contiguous dropouts.

12.8.5 Interactive Processing

A fourth approach is denoted the *interactive* method. Basically, this method combines the best properties of the previous methods. Trackfile information is used to identify one or two possible skin returns, according to both kinematic and SNR measurements. Then, for each hypothetical skin return, a "feature" set is extracted that may include the harmonic spacing, Δf, and measured incremental range. This skin-and-DS information is then correlated against the trackfile's predicted values and the best correlation used to update the track. The remaining measurements are associated with other tracks or used to start tentative tracks.

12.9 SUMMARY

How detections should be grouped prior to range and Doppler measurement formation is perhaps the first question that needs addressing. The optimal solution is probably to perform a correlation test using the known shape of the radar response. Various suboptimum approaches that have been implemented in the past include peak detection or thresholding relative to peak values. Formation of angle measurements was discussed next using a correlation test similar to Chapter 11.

Further data collapsing may be required. For example, elimination of redundant detections from bar to bar can be accomplished by gating detections across bars and averaging the result. After all data collapsing has been performed, the averaged detections are sent to the data correlation logic to be processed as a single entity.

Data filtering is conventionally arranged as a reduced-state filter in inertial Cartesian coordinates. However, to provide a concrete theoretical basis to support these statements, the performance loss was examined as a function of the filter parameters. A parameterized version of the Kalman filter, introduced by Fitzgerald, was used for this purpose. Finally, a set of reduced-state filter equations was presented to illustrate a feasible approach to data filtering.

Track file initiation, confirmation, and deletion procedures typically use a simple M-out-of-N procedure to decide when a track should be reclassified. However, selection of the numbers (M, N) may be based on elaborate probabilistic reasoning. Other performance measures may also be incorporated such as target amplitude or position in the scan volume.

The last subject discussed was the important topic of Doppler spectra (DS) suppression. DS returns are caused by such phenomena as jet engine modulation (JEM), prop modulation, tread modulation, or blade modulation. DS returns are a major source of confusion when tracking any target in Doppler. Suppression techniques include inhibition gating, offset tracking, range correlation tests, pattern recognition, or a combination of the above.

APPENDIX: POWER-APERTURE PRODUCT

Equation (3.55) gives the radar range equation as

$$\mathrm{SNR} = \frac{P_{\mathrm{avg}} T G^2 \sigma_T \lambda^2 L}{(4\pi)^3 R^4 F_S k T_0} \tag{12A.1}$$

This form of the radar range equation is valid for all STT applications. During search, however, there is an additional constraint imposed that modifies this equation somewhat. This constraint is that the radar is required to search a specified volume of space within a specified amount of time.

Let us substitute in

$$T \cong \text{time-on-target}$$
$$\cong t_s \theta_E \theta_A / \Omega$$

for Ω the total angular region to be searched, t_s the total time to scan Ω, and θ_A, θ_E the azimuth and elevation beamwidths, respectively. The antenna gain can be approximately written as $G \cong 4\pi/\theta_E\theta_A$, so that with (3.43)

$$G^2 \cong (4\pi)^2 A_e/(\lambda^2 \theta_E \theta_A)$$

With the above substitutions into (12A.1), the noise-limited radar range equation becomes

$$\mathrm{SNR} = \frac{P_{\mathrm{avg}} A_e t_s \sigma_T L}{4\pi R^4 F_S k T_0 \Omega} \tag{12A.2}$$

This indicates that the important parameters in any search radar are the average power P_{avg}, the antenna aperture area A_e, the propagation loss L, the system noise figure F_S, and the target cross section σ_T. Frequency only comes into play indirectly in how it influences these parameters. Notice that these comments are true regardless of the source of the noise, whether thermal noise or jammer-induced noise, assuming a fixed noise environment and sidelobe level in dBi.

The major factor that favors the use of a lower RF in search is that it is easier to achieve a large power-aperture product, $P_{\mathrm{avg}} \cdot A_e$. The average power P_{avg} is largely limited by heat dissipation requirements, thus the larger, heavier radars at the lower RFs can achieve higher average powers. The fabrication of the antenna aperture is also simplified at the lower frequencies, allowing a larger antenna aperture A_e. In particular, the absolute mechanical design tolerance is relaxed and the dipole density becomes less.

For military applications, another factor that favors the lower RFs is the slightly higher RCS against stealth aircraft. Although exact numbers are subject to argument, roughly a 10-dB increase in RCS has been observed in going from X-band to L-band. Also, it is easier to place a sharp null in the direction of a sidelobe jammer at the lower RFs since absolute phase is easier to control, thus reducing the effectiveness of sidelobe jamming. The effectiveness of mainlobe jamming is a more complicated issue that is highly dependent upon the operational scenario.

Other factors that favor the lower frequencies include the system noise figure F_S, the clutter cross section σ_c, and the propagation loss. Above S-band, weather attenuation increasingly becomes severe (Eaves and Reedy, 1987, p. 68; Schleher, 1986, p. 285), and becomes a significant factor at Ku-band and above, restricting these frequencies to short-range applications only. Below VHF, on the other hand, external noise from graybody radiation raises the system noise figure F_S thereby reducing the detection range (Skolnik, 1980, p. 462).

Factors that favor the use of higher RFs include better angle and velocity resolution, and the larger absolute RF bandwidth for a fixed percentage RF bandwidth limitation. Also, for airborne applications, the width of mainlobe clutter is directly proportional to the antenna beamwidth, favoring a higher RF to reduce the antenna beamwidth and commitant clutter spreading.

CHAPTER 13

ESA ALLOCATION LOGIC

An electronically scanned antenna (ESA) radar system performing both search and track requires additional logic to adaptively allocate its time between these two functions. Design tradeoff issues for the required allocation logic are discussed in this chapter and a representative architecture is described in detail.

13.1 DESIGN TRADE-OFFS

The design of the logic is driven by the need to be both "fast" and "smart." A fast reaction time is desired in the detection process, in track updating, and in responding to external control inputs. This implies the ability to modify the scheduling right up to the time when a definite commitment is needed. Consequently, the logic should be able to quickly respond to incoming constraints and to alter its programming according to the information received. This imposes a task–schedule–task sequence of operational events. On the other hand, the full benefits of the ESA are not attained unless the logic can effectively plan ahead. Planning permits issues beyond the requirements for each individual dwell to be considered and allows the waveform to be tailored to the predicted track loading requirements. Furthermore, some planning is absolutely necessary if the logic must interface with an external device with a fixed timeline, such as a missile illumination at a fixed rate, or the antenna is itself in motion such that the target occasionally falls outside the antenna field-of-view.

Based on this dual requirement, Figure 13.1 shows a possible design for the required allocation manager. The proposed architecture basically consists of two queues, called the *command queue* and *planning queue*. The planning queue is used to construct a tentative list of tasks. As commands near the final

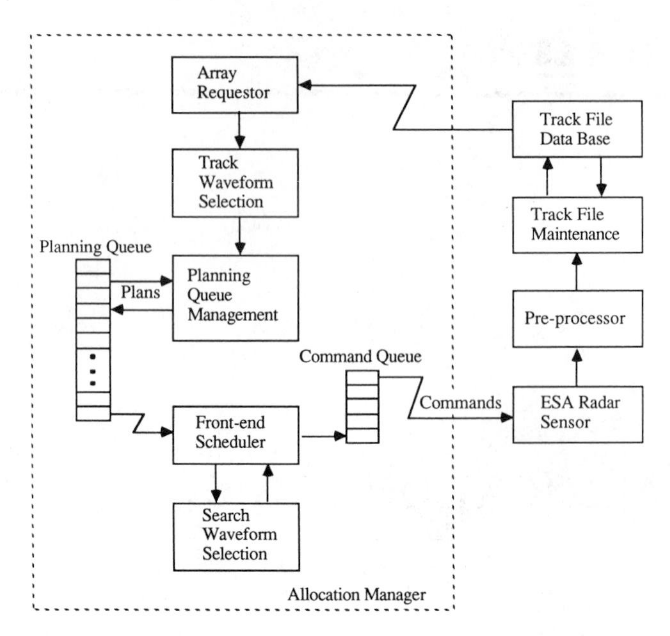

Figure 13.1 Functional elements of the proposed allocation manager.

planning, they are promoted to the command queue, which handles the detailed timing operations necessary for sequencing the front-end commands through the radar's hardware.

13.2 PLANNING QUEUE

The planning queue consists of a list of anticipated tasks that are ordered by time and priority. Each task may itself comprise a list of dwells that are to be executed concurrently. Each dwell consists of an array of transmit or receive actions. The front-end scheduler retrieves tasks sequentially from the planning queue and sends them to the command queue for execution.

New dwells are entered into the planning queue near the specified time within an allotted window. In the event a dwell cannot be entered in the queue at the time specified, the planning queue loader may bump a lower priority dwell previously scheduled. The preempted dwell is then rescheduled, if possible, within an allotted window. If an available slot cannot be found, then it may be necessary to reschedule a dwell of higher priority if possible within its allotted window. This rescheduling process may be performed immediately or by means of later logic that sorts out conflicting requests on a second pass.

13.3 COMMAND QUEUE

The command queue is a list of tasks or front-end commands that are about to be executed. No further planning or adaptation is allowed from this point on due to the system's finite response time. Hence, commands in the command queue are committed for transmission.

As shown in Figure 13.1, commands ready for execution are entered into the command queue by the front-end scheduler. Potentially other sources of input may exist; for example, a new detection in search may be immediately followed up with a confirm sequence (in effect, bypassing the planning queue). Confirmation, in this case, would be treated as an extension of search timeline.

If no commands are loaded into the command queue by any source, a default task is loaded. This type of task is typically a low-priority dwell such as long-range search or volumetric search.

13.4 WAVEFORM SELECTION

Requests for waveform parameters for a particular dwell are sent to the waveform selector, which then processes this information and assigns the appropriate parameters. These parameters include beam direction, beamwidth, PRFs, dwell time (T), pulsewidths, pulse compression ratio, amount of overlapping, number of dwells, RF, and (in a LPI environment) peak radiated power. Receiver parameters may also be tied to the dwell at this point, including detection thresholds, receiver AGC settings, clutter rejection levels, Doppler filter weights. Also, the waveform selector may specify limits (bounds) within which the subsequent logic can change the assigned waveform parameters.

13.4.1 Search Waveform Selection

The search waveform is primarily influenced by the required search coverage, update time, and detection range. Thus, these system inputs must be established beforehand by the radar designer.

To give a concrete example, consider the case where too many tracks are present to execute all track dwell requests and meet the minimum search rate requirements. In this case waveform selection could be configured as follows. First, the waveform selector accesses the planning queue (on a read-only basis) and computes the total time available to perform search assuming all track requests take priority. Next, the dwell time (T), beamwidth, and amount of overlapping are selected based on the time allocated to search. Reducing the dwell time by a factor of one-half, for instance, doubles the number of dwells that can be transmitted, but only degrades the detection range by a few percent (due to the R^4 relationship).

13.4.2 Track Waveform Selection

Track waveform selection should be initially based on track file information. The track dwell time should always be selected to provide a high probability of detection. Likewise, the PRF should be chosen to provide uneclipsed detection, with the kinematic uncertainty determining the number of PRFs necessary to guarantee a high probability of uneclipsed detection. If an early PRF should produce all the desired information, however, the remaining PRFs may be canceled.

In a dense target environment the locations of all other targets in the beam necessarily influence the waveform selection procedure. Again, the trackfiles are the best source to obtain this information. Each trackfile can be checked to determine if its target might interfere with the target to be updated. This includes checking whether its sidelobes might produce an apparent return within the selected track window after accounting for the different Doppler degradation effects. If possible, the waveform is then adjusted to minimize the amount of sidelobe energy competing with the target return.

As Figure 13.1 shows, the track waveform is initially selected prior to placing the request on the planning queue. However, the variable time delay between when a track dwell is placed on the planning queue and when it is delivered to the command queue is a potential concern. During this time, parameters such as angle, range, and Doppler can change significantly from the predicted values. Furthermore, the front-end scheduler may impose additional changes on the waveform to minimize interference with other tracks before finally placing the request in the command queue.

A possible solution to this dilemma is to perform initial waveform selection based on a nominal elapsed time, and final waveform selection prior to loading onto the command queue based on the actual elapsed time. Any stale information, such as beam direction, is corrected at the final stage. A pointer may be maintained between the dwell and the requesting track file to aid in this updating process.

13.5 PULSE PACKING

The ESA radar has the ability to update several different tracks concurrently. Since the estimated time delay of the pulse return is known from the target's range, it is possible to transmit a pulse to one target during the waiting period for the return pulse to a second target. This process is referred to as *pulse packing* (Farina and Studer, 1985a, p. 296; Robinson, 1983).

Pulse packing is implemented by making the planning queue two dimensional. One dimension corresponds to time of execution, as before, and the second dimension corresponds to a list of target dwells that are to be executed simultaneously. Each list of dwells (or pseudodwells) performed simultaneously is referred to as a *task*. A dwell is packed into a task provided it does not interfere with any dwells already present at that time tag. *Interference*, in this context, refers to any mutual eclipsing that occurs during either the pulse transmit (τ) or receive (r) periods.

Pulse packing is especially attractive for those circumstances where several targets must be tracked with a low-duty-factor waveform using a high-duty-factor transmitter. A good example is tracking multiple long-range targets using a solid-state LPRF radar.

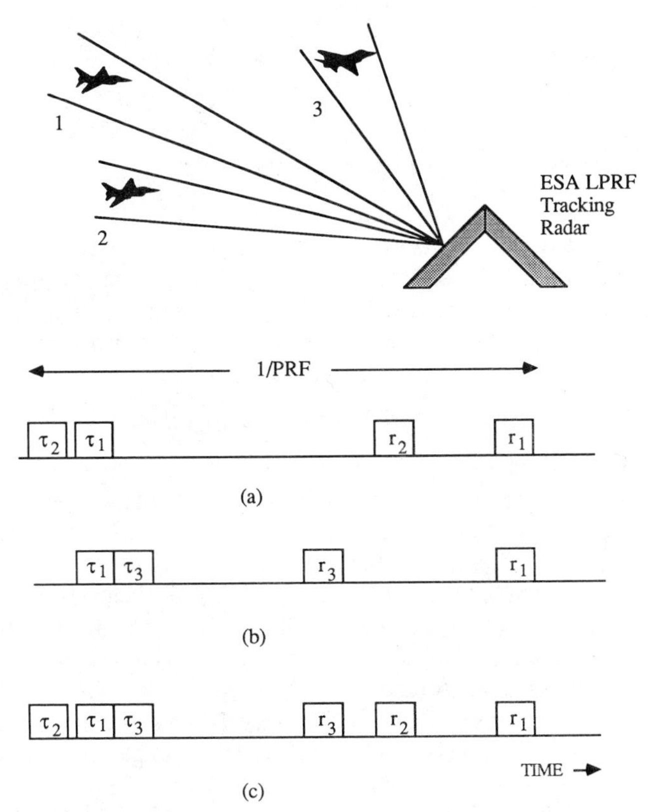

Figure 13.2 Example of ESA pulse packing: (a) Interleaved pulse packing. (b) Nested pulse packing. (c) Combined interleaved and nested pulse packing.

Pulse packing is a nontrivial problem and is known to be NP-complete (Robinson, 1983). Various ad hoc algorithms have been suggested, of which the one illustrated in Figure 13.2 closely follows that described by Farina and Studer (1985a). To begin, the scheduler first selects the set of tracks to be packed within a preset window on the planning queue. The set of tracks are sorted according to range, from the longest range to the shortest range. Pulse interleaving is employed for long-range targets, as illustrated in Figure 13.2, and pulse nesting is employed for short-range targets (Farina and Studer, 1985a, p. 297). The assigned PRF for each dwell is always a multiple of the base PRF to prevent range/Doppler walk.

13.6 PREPROCESSING

The preprocessor, as the name implies, performs the initial processing of the returned data. This includes (1) initial thresholding and detection, (2) eliminating redundant detections, (3) notifying the scheduler in the event of an incomplete data set, and (4) grouping multiple returns from the same target. After performing these tasks, the preprocessor transfers the data to the track file maintenance function for final correlation to tracks.

13.6.1 Eliminating Redundant Detections

Redundancy elimination was discussed in Section 12.3. In terms of an ESA the primary consideration is that the search time span be reasonably large to efficiently eliminate redundant detections. One way to avoid interruptions to search is to cluster the track dwells on the planning queue such that a majority of the track dwells are scheduled at the end of each search bar.

13.6.2 Redetecting an Old Target During Search

After redundancy elimination, a condensed list of detections is available. This list contains new (previously undetected) targets as well as targets already in an established trackfile. The former is of primary interest, while the latter represent detections that are unrelated to the search function.

The most straightforward approach to this problem is to confirm all detections and send all data to the correlation logic to sort out. It should be noted, however, that this approach can be time consuming, particularly if the number of tracks is large. A more efficient approach is to ignore all detections that occur in the vicinity of a tracket target. This procedure requires the preparation of a list of targets that are expected to be angle correlating. Based on this list, a set of inhibition gates are developed. Detections occurring within the inhibition gate of any established trackfile are ignored by the search logic. Detections not correlating with any inhibition gate are considered to be from a new target, and so initiate a confirm sequence.

The use of inhibition gates has the advantage of allowing search to be completely decoupled from track. In addition, it avoids wasting time and energy reconfirming targets already in an established track file. On the other hand, not inhibiting track space has an advantage in that it opens up a procedure whereby tracked targets can be updated from search. Thus, extrapolated tracks can be reacquired from normal search hits as soon as the search beam scans over the target. A timely reacquisition is also facilitated if the coverage is intentionally altered to visit the region containing the degraded target early or a special miniscan is initiated around that point.

From the foregoing, it can be argued that a more sophisticated system would incorporate both approaches. Detections are first correlated against a reduced list of angle-correlating trackfiles, containing tracks to be updated individually. If a correlation occurs, the confirmation process is terminated and the detection data is dropped. If no correlation occurs, the confirm dwell sequence is transmitted. The confirmed detections are then sent to the trackfile maintenance function to be sorted out as either (1) a new target, (2) an extrapolated target, or (3) a low-priority target for which the search rate was considered sufficient.

13.6.3 Detecting Multiple Targets During Track

As mentioned previously, a common procedure during search is to inhibit detections that occur within the gates of an established track. This tactic blinds the radar to any new target(s) that appear in the suppressed zones. Consequently, when processing track returns, the logic must account for the possibility of a new target appearing in the vicinity of the target to be updated. A simple procedure is to expand the gate size and look for any new targets appearing next to the target to be tracked. In effect, this just allows a group of interacting targets to be updated as a single entity.

13.6.4 Incomplete Data Sets

In the event of failure to detect any target during a scheduled track update, a new task must be immediately rescheduled. Likewise, if no target is detected in the small gate region, but one or more detections occur in the large gate region, an additional task may be scheduled to ensure that the target has not been missed or is not in a deep fade. Finally, it is necessary to schedule a new task in the event of failure to obtain a complete measurement set.

13.7 TRACKFILE MAINTENANCE

Trackfile maintenance is responsible for maintaining the track files. It correlates the updates to the track files, uses the updates to filter the trackfiles, deletes old trackfiles, keeps a record of track status, promotes tentative tracks to confirmed tracks, performs maneuver detection, and forms new tracks. In addition, trackfile maintenance does the "bookkeeping" necessary to know where dwell requests have been granted and where they are in the planning queue.

13.8 DWELL REQUESTER

The dwell requester establishes the proper time to update each track and makes the necessary requests to the planning queue manager. As such, the dwell scheduling must be based on some postulated criteria of merit. Reasonable criteria include:

1. Update each track at a predetermined rate, with the rate determined by the track's priority.
2. Maintain the filter accuracy of all tracks at less than an operator input threshold.
3. Minimize the total radar resources allocated to tracking.
4. Maintain the probability of losing the target at less than some threshold.
5. Proceed along the path of highest marginal utility.
6. Any combination of the preceding criteria

To formalize our discussion, it is assumed that the third criterium is our design goal. Our intent here is to pick a reasonable criteria of merit and then concentrate on its implications, rather than discuss all criteria in detail.

To minimize the track time, the updates should be relatively infrequent (i.e., the update time interval, ΔT, should be large). On the other hand, the updates should be frequent enough to ensure a high probability of a successful detection to avoid costly reacquisitions. Hence, a tradeoff exists between assigning ΔT a small value to ensure a successful detection, versus assigning ΔT a large value but running the risk of having to spend additional time reacquiring the target should the update prove unsuccessful.

An *unsuccessful update* refers to any update for which the target is outside the beamwidth or gate region. The probability of this happening is directly related to the error in the track's prediction at the time of the update. Stated in these terms, the interval ΔT should be chosen such that $\mathbf{S} < \mathbf{T}$, where $\mathbf{S} = \mathbf{H}^T\mathbf{P}\mathbf{H} + \mathbf{R}$, and $\mathbf{P}$ is an estimate of target prediction covariance. The threshold, $\mathbf{T}$, is chosen to maintain an upper bound on the probability of the update being unsuccessful. The target update task is placed in the planning queue if the selected ΔT value is less than the regular search period. Otherwise, it is placed on the list of tracks to be updated via search.

The target can become lost or ambiguous in either angle, range, or Doppler. Hence, the error $\mathbf{S}$ must be monitored in all three dimensions simultaneously. In angle, the target can be lost by escaping the antenna beamwidth or suffering significant beamshape loss. In range, possible eclipsing are the primary concerns. In Doppler, the target can be lost by miscorrelation with its own or other's Doppler spectra (DS) lines. Furthermore, there always exists the possibility of closely spaced targets interfering with each other in angle/range/Doppler (i.e., confusion between different targets).

13.8.1 Deriving the Covariance Matrix

A reasonable procedure is to derive the required prediction covariance matrix $\mathbf{P}$ from the Kalman filter. This approach closely follows that proposed in (Browne et al., 1980). The growth in error $\mathbf{P}$ with elapsed time ΔT is given by

$$\mathbf{P}(k+1) = \mathbf{\Phi}\mathbf{P}(k)\mathbf{\Phi}^T + \mathbf{Q}(k) \tag{13.1}$$

which is a function of ΔT through $\mathbf{\Phi}$. A practical simplification that may be introduced is to set the off-diagonal elements to zero during the dwell requesting process.

The previous approach assumes the Kalman filter prediction covariance gives an accurate representation of the target's true prediction error. However, the two covariances can significantly deviate upon occasion, such as during periods of severe target maneuvering or following miscorrelation.

Several approaches have been previously proposed to remedy the maneuver situation, including incrementing the Kalman filter covariance matrix whenever a maneuver is declared (Blackman, 1986). The update rate is then indirectly increased via (13.1). The amount of covariance increment may be based on a worst-case target model wherein the target is assumed to maneuver at its maximum acceleration limits at the last sampling instant.

The conservative approach is to *always* assume that the target executed a worst-case maneuver at the last sampling instant. In effect, this amounts to calculating **P** based on a worst-case maneuver model rather than to derive it from the filter.

Alternatively, the update rate may be directly increased whenever a maneuver is declared (Cohen, 1986; Gardner and Mullen, 1988). The amount of increase may be made a function of the observed residual deviation. Normal update procedures are restored whenever the end of the maneuver is declared, or whenever the target's new heading has been well established. This method is attractive if the data filtering is implemented using precomputed filter coefficients of the α-β type since a measure of covariance is not otherwise available.

13.8.2 Track Status

Besides kinematics, it is sometimes suggested that the track's *status* should also influence the update rate. Indicators of track status include target identity, perceived threat level, relative geometry, tactical scenario, and so on. The basic idea is that more important targets should be assigned a higher status and have a correspondingly higher update rate.

This idea is also consistent with our stated design philosophy. Presumably, if an important target is lost, it must be immediately reacquired *and* reidentified. The track's status, in this case, represents an investment in time that is lost once the track is lost. Conversely, if a nonimportant or nonthreatening target is lost, it may be assumed that the target's status and kinematics may be safely reestablished during the normal course of search. Hence, the process of reacquiring a less important track does not impose the same resource penalty.

To mathematically quantify these statements, consider that to guarantee

$$\mathbf{S} < \mathbf{T} \tag{13.2}$$

it is sufficient to guarantee

$$f(\mathbf{S}) < f(\mathbf{T}) \tag{13.3}$$

for $f(\,)$ any monotonically increasing function. The advantage of introducing $f(\,)$ is that it allows the preceding analysis to be generalized to any arbitrary number of variables. The test can be more generally expressed as

$$f(\mathbf{S}, I_n) < f(\mathbf{T}, 1) \tag{13.4}$$

where, for track n, I_n is the target importance as a function of target status and ranges between unity and infinity. The importance function, I_n, is normalized such that unity ($I_n = 1$) corresponds to background updating procedures. A target having a large value of importance corresponds to a target that is singled out to be updated more frequently.

Assuming multiplicative independence, the overall target importance is the product of the target's geometric importance, I_G, identification importance, I_{ID}, and the system's resource availability, I_{RA}:

$$I_n = I_G I_{\mathrm{ID}} I_{\mathrm{RA}} \tag{13.5}$$

These functions are discussed next.

I_G represents the importance of the nth track as a function of geometry and kinematics. Factors influencing I_G include geography, flight pattern, track stage, and inverse time to go, $|\dot{r}|/r$. A simple expression for I_G, taken from a fire control radar, is

$$I_G = \exp\left\{c_l \frac{\mathbf{P}_n \cdot \mathbf{I}_T |\dot{r}|/r}{c_2 + |\dot{r}|/r}\right\} \tag{13.6}$$

where $\mathbf{P}_n$ = target type probability vector for track n
$\mathbf{I}_T$ = weighting vector representing relative importance of target n in the interval (0, 1)
$r/|\dot{r}|$ = time to go

The product of the weighting vectors, $\mathbf{P}_n \cdot \mathbf{I}_T$, approach unity as the target is declared a hostile target, and approaches zero as the target is declared friendly. The time-to-go parameter is a common way to prioritize threats in a military application. It represents the time for the target to reach an undesirable location, namely, ownship position. Thus, I_G approaches its maximum value for hostile near-range targets that are rapidly closing.

I_{RA} represents the importance of all the tracks as a function of resource availability. If system resources permit, it is always preferable to assign a higher importance to all tracks and, thus, use up all available resources. As the ratio of track time to search time becomes large, however, this policy is no longer practical. An ad hoc expression relating the importance of all tracks might be proposed:

$$I_{\mathrm{RA}} = \exp\left\{c_1 \frac{N_{\mathrm{MAX}}}{N_{\mathrm{MAX}} + N}\right\} \tag{13.7}$$

where N_{MAX} = number of tracks that the system can hold without track degradation
N = number of actual tracks
c_1 = tuning parameter

The form of Eq. (13.7) was chosen empirically based on the desired limiting properties. These include (1) I_{RA} must approach some large value $\exp\{c_1\}$ as N becomes small relative to N_{MAX}, and (2) that I_{RA} must approach unity as the number of tracks N becomes large, allowing more tracks to be serviced under overload conditions.

Finally, if too many tracks exist to service properly, it may be necessary to set up an orderly procedure of dropping lower-priority tracks.

13.9 SUMMARY

There are no "standard" approaches to the design of ESA allocation logic. In the approach described here, requests for the radar's time was organized into a planning queue. Finalized requests, constrained by the physical capabilities of the system, were moved to the command queue and subsequently executed. The advantage of this structure is that it permitted the overall performance to be optimized by iteration.

Waveform selection was done in a series of steps as the dwell was moved toward execution. Search waveform selection was based on anticipated track loading requirements as characterized by the number of requests pending. Track waveform selection attempted to minimize mutual interference based on the trackfile of the target to be updated as well as the trackfile of neighboring targets.

Pulse packing is an extension of these concepts to two dimensions. Pulse packing is an attractive concept for ESA track and potentially for ESA search. To date, it is mainly used in long-range LPRF applications where the estimated time delay is accurately known relative to the PRF and TR switching times.

Initial return processing was described with emphasis placed on those procedures unique to ESA applications. Initiating a confirm sequence based on a flow of search detections, problems encountered during track updating, and track maintenance were discussed and some opposing viewpoints presented.

The subject of track dwell requesting was discussed last. Track dwell requesting was based upon the track covariance matrix such as to avoid target loss. The track covariance matrix, in turn, must be adaptively varied as a function of all available inputs. Finally, track status was related to the requesting logic in a simple fashion.

CHAPTER 14

A REPRESENTATIVE ESA RADAR SYSTEM

Electronically scanned antennas (ESAs) have become increasingly popular over this decade as a consequence of hardware and software advances as well as demands for more sophisticated functionality. This chapter presents a representative ESA radar system design and addresses some of the major design tradeoffs.

14.1 THE RADAR HARDWARE

One well-known problem with the ESA concerns the loss of effective aperture area A_e with increasing electronic scan angles. As with any antenna, the effective aperture A_e is equal to the antenna aperture area as projected in the direction of the scan angle. During ESA scanning, this area is foreshortened by the cosine of the angle between the electronic scan angle and mechanical boresight. When scanned 60° off mechanical boresight, for instance, the effective aperture is reduced by half [$\cos(60°) = 0.5$], the beamwidth is twice as broad, and the sidelobes are distorted.

Thus, by itself, the conventional ESA cannot match the ability of a mechanically scanned antenna (MSA) to scan over all angles. Additional hardware is needed if this ability is required.

A possible solution is to mechanically steer the ESA itself (as per Figure 14.1b). The advantage of this configuration is that it enables a single aperture to scan over a large volume while enabling some beam agility. The disadvantage is that two sets of steering mechanisms must be provided, and less reliable operation results as a consequence.

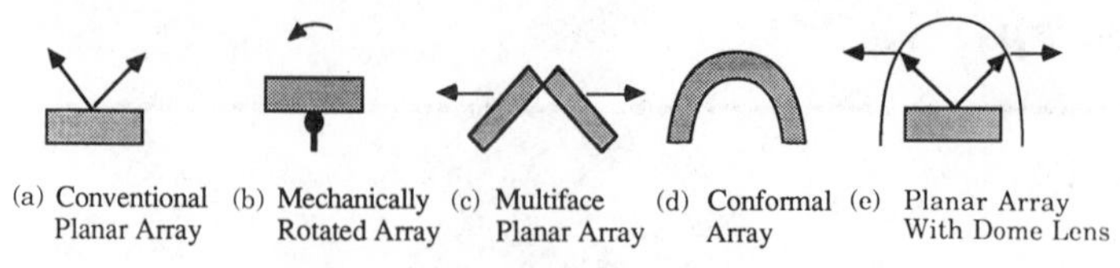

Figure 14.1 ESA antenna options.

The brute force approach is to employ a number of ESA radars facing in different directions (Figure 14.1c). Equivalently, one may use a single radar that transmits from a multitude of antenna faces, or a single "conformal" array that is shaped to the surface of the platform (Figure 14.1d). Finally, the planar array may be mounted under a lens (see Figure 14.1e) that refracts the beam thereby increasing the angular deflection (Brookner, 1981). These approaches are more reliable and eliminate mechanical gimbal errors at the expense of additional cost and antenna weight.

The previous remarks indicate that electronic steering suffers an inherent liability in terms of loss of effective aperture, beam broadening, and, possibly, weight and cost. As no clear-cut hardware solution is apparent, the question reduces to whether the overall radar system design can be improved by using an ESA to the point where these drawbacks are justified.

14.2 ADVANTAGES OF ESA TRACK

Electronic steering allows the search and track processes to be optimized independently. This permits multiple targets to be tracked with STT-like quality in the presence of maneuvers, competing clutter, and hostile interferences, while simultaneously enabling a large search volume to be scanned. Electronic steering also allows a target to be tracked as it moves outside the normal scan pattern. Furthermore, since mechanical gimbal errors are eliminated, electronic steering is more accurate than mechanical steering, further improving track quality. Finally, adaptive array techniques to be introduced to null out clutter and interference.

14.3 ADVANTAGES OF ESA SEARCH

Electronic steering is able to achieve a faster scan rate than a comparable MSA and eliminates the need for end-of-bar turnaround. Therefore, an ESA can exceed the capacity of a mechanically scanned antenna to scan a given size volume for a given size beamwidth. It also allows the possibility of more exotic scan patterns and multiple search areas. Electronic steering also allows the beam to immediately stop scanning and confirm any search detections. This opens up the possibility of detection optimization using alert/confirm logic.

14.4 ALERT/CONFIRM LOGIC

Alert/confirm logic refers to a detection strategy whereby the detection process is split into two distinct stages—an alert stage and a confirm stage. In the alert stage the scan volume is searched for potential new targets. A candidate detection in alert triggers the transmission of a confirm waveform (or waveforms). The confirm waveform screens out false target declarations, verifies the presence of a new target, and acquires the information necessary for track file initiation and confirmation.

One advantage of alert/confirm is that a lowered detection threshold can be maintained in alert. This is possible since the net probability of false alarm, P_{FA}(net), is approximately equal to the product of the probability of the alert stage producing a false alarm, P_{FA}(alert), times the probability of the confirm stage producing a false alarm, P_{FA}(confirm), that is,

$$P_{FA}(\text{net}) \cong m P_{FA}(\text{alert}) * P_{FA}(\text{confirm}) \tag{14.1}$$

where m refers to the number of cells examined for confirm acceptance. Hence, P_{FA}(alert) can be increased, while a low P_{FA}(confirm) is used to maintain a reasonable value for P_{FA}(net).

Clarification of this advantage may be in order. First, P_{FA}(alert) cannot be made arbitrarily large since the average scan time is dependent on the rate with which confirm is entered. The implication of this statement can be illustrated by means of a simple example. In the absence of targets the average scan time T_S is approximately given by:

$$T_S \cong N T_{\text{ALERT}}[1 + \text{FAR(alert)} T_{\text{CONFIRM}}] \tag{14.2}$$

where N = number of beam positions searched
T_{ALERT} = time to transmit the alert waveform
T_{CONFIRM} = average time to transmit the confirm waveform
FAR(alert) = average number of false alarms per alert time

As a baseline design let us limit the percentage of time that confirm is entered to be less than 50 percent, that is,

$$\text{FAR(alert)} T_{\text{CONFIRM}} < 50\%$$

Taking T_{CONFIRM} equal to 25 msec, the resulting FAR(alert) is

$$\text{FAR(alert)} < \frac{50\%}{25 \text{ msec}}$$

$$< 20/\text{sec}$$

In conclusion, alert/confirm logic is most appropriate whenever the desired FAR(net) is small (i.e., much less than 20/sec). To be more precise, alert/confirm logic is most appropriate whenever it is desired to eliminate most of the false alarms early (before they reach the data processor). In those instances where the desired FAR(net) is large, the benefits of alert/confirm logic are not realized. Furthermore, alert/confirm logic is appropriate only if the target density is sparse relative to a beamwidth. Only in this case is it worthwhile to decouple the search process from the track process.

Nevertheless, many applications do meet these two criteria. In this regard the true utility of the alert/confirm approach is threefold. First, from a detection perspective, it permits a lowered threshold to be maintained in alert (Dana and Moraitis, 1981). It also avoids wasting an excessive amount of time radiating energy into empty space such as with a conventional search mode using more simplistic detection logic.

Second, it enables the detection process to be decoupled from the measurement process. Decoupling the detection and measurement processes allows the alert waveform to be optimized for the sole task of detection, and the confirm waveform to be optimized for the sole task of measurement formation and trackfile confirmation.

Third and finally, alert/confirm logic directly improves confirmation of newly initiated trackfiles. In conventional TWS, 2-out-of-3 detections are typically required before successful trackfile confirmation. With alert/confirm a single (confirmed) detection initiates a sequence that may result in a confirmed trackfile. Thus, its performance is more aptly described by cumulative probabilities P_{cum}, rather than single-scan probabilities.*

14.4.1 Sequential Detection

Sequential detection is a branch of alert/confirm detection that breaks up the confirm process into a series of decisions (Nathanson, 1969; Corsini et al., 1985; Blackman, 1986, p. 152). Each decision involves comparing the output signal to two thresholds. If the signal fails to pass the lower threshold, then the confirm sequence is dropped ("no target" decision). If the signal passes the upper threshold, a valid target detection is declared ("yes" decision). If neither of these events occur, the decision is made to transmit another confirm dwell ("maybe" decision).

The interested reader is referred to Corsini et al. (1985) for more details. In general, sequential detection is known to work well under certain conditions. However, it does have two drawbacks. First, detection is only improved in the

*The use of P_{cum} has been avoided in this text, mainly because P_{cum} is an unfair criterion to use in comparing an ESA's performance relative to an MSA's performance. The P_{cum} measure is useful in evaluating the various MSA performances and/or evaluating the various ESA's performance. But, to compare an ESA's performance to a MSA's, one must adopt a more fair measure such as the probability of a firm trackfile (Ramstein and Georges, 1987).

presence of noise. If detection is limited by clutter, instead of SNR, then the length of the dwell is primarily fixed by the need to form an efficient Doppler filter. Second, if the target's SNR is lower than the design SNR, the sequential test may be unable to make a yes–no decision, and be "hung up" in a single beam position (Corsini et al., 1985, p. 143). This phenomenon is well known and is traditionally solved by truncating the sequential test after a fixed duration. If the individual dwell times are long, as needed for efficient Doppler filter formation, then realistically only a few sequential decisions may be permitted.

14.4.2 An Example Operation

A hypothetical alert/confirm timeline might be constructed as follows. The ESA radar system continuously scans a predetermined search volume according to sparse sampling concepts. During the search process, PRF switching may be used to reduce the severity of the range and Doppler blind zones. The PRF and the pulse width may be switched in synchronization so that a constant duty factor is always maintained after PRF switching.

All alert detections that occur within a single beamwidth are PDI-averaged together, and then corroborated by means of a confirm sequence. After determining the most likely spatial location of the target, the beam is repositioned to center the target in the beam. Next, the confirm sequence is transmitted. The confirm sequence may transmit several dwells before a firm decision is made (*a la* sequential detection). The confirm sequence uses any knowledge gained in alert to increase the probability of detection. For example, if the time between alert and confirm is short, each confirm dwell may use the same PRF and RF as the alert waveform. Hence, targets clear and scintillated "up" in alert, remain so in confirm.

The confirm process may also resolve the target's range and Doppler, either by FM ranging, chirping, or coincidence detection. The confirm sequence may utilize knowledge gained in alert to aid in this task. To demonstrate, consider transmitting a single HPRF dwell in alert, and in confirm transmitting one additional HPRF dwell to complete an FM ranging set. Only one additional dwell is needed to complete the FM ranging set since the initial alert detection provides the information required to round out the set (Stimson, 1983). An analogous example can be drawn from LPRF. Consider transmitting an unramped chirped LPRF pulse in alert, and in confirm transmitting a down-ramped chirped pulse to unambiguously resolve the target's range and velocity.

If the confirm process includes coincidence range/velocity resolving, then the confirm PRF is changed by some small amount from the alert PRF. Small changes in the PRF are desired for coincidence resolving since a large change could potentially translate the target into an adjacent blind zone. To distinguish these small PRF changes from the major changes used in alert, the terminology *minor PRF* is sometimes used. The total strategy of using major PRF changes for alert detection and minor PRF changes for coincidence ranging is conventionally referred to as the *major–minor* approach.

Finally, the confirm process may include measuring target acceleration by range or Doppler differentiation. Prefiltering may be employed over short time increments to provide a more statistically well-behaved acceleration estimate.

14.4.3 Nongeneric PRF Operation

In all the previous examples both alert and confirm used the same generic type of waveform. To generalize this still further, consider using different types of waveforms to perform different types of functions. The advantage of nongeneric PRF operation is that the PRFs complement each other. A case in point is using HPRF to provide high duty factor alert detection, and using LPRF to provide fine range and angle information during the trackfile initiation stage.* HPRF can also be used to provide unambiguous target radial velocity, whereas LPRF can be used to provide beam-aspect detection and to facilitate target hand-off to other LPRF radars. Finally, in any tactical scenario, perhaps the most important role of nongeneric PRFs is to improve the robustness of the target identification logic through joint range/Doppler analysis (Evans, 1986).

Using HPRF for alert and LPRF for confirm, or vice versa, is the most straightforward example of nongeneric PRF operation. As illustrated in Figure 14.2, the PRF used by the LPRF dwell should be made adaptive to the target's Doppler frequency shift as determined by the proceeding HPRF dwell, thus avoiding any degradation due to Doppler blind zones. In turn, by capitalizing on the range information supplied by the previous LPRF dwell, the subsequent HPRF dwell can be chosen to minimize eclipsing loss.

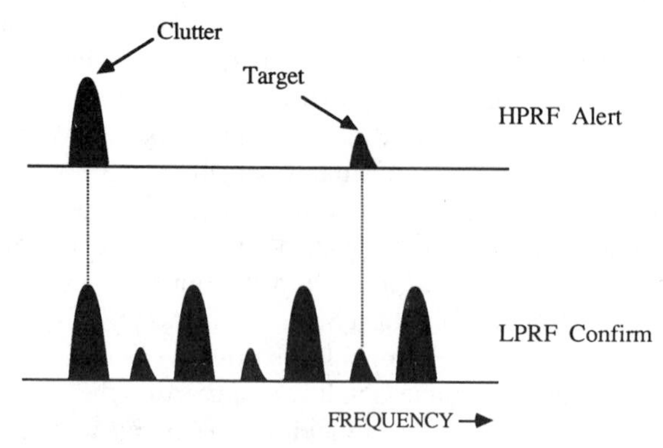

Figure 14.2 Example of using information gained on alert to optimize overall alert/confirm performance.

*The concept of HPRF/LPRF interleaving was invented by D. Goudey and J. Beuerman of Hughes Aircraft.

14.5 SCAN GENERATOR

Electronic steering allows the search scan generator to be programmed according to sparse sampling concepts. Since new targets are more likely to appear at the borders of the search region, the borders can be sampled at a higher rate (Van Keuk, 1975). The border defines a surface with the maximum target range as the forward border, the horizon as the lower border, and the azimuth scan limits as the lateral borders.

Sampling that favors the maximum range border can be implemented using the "fence" type of search procedure. Only an annulus of range is searched for new targets corresponding to the target's nominal detection range. The search rate would be commensurate with detecting the target before it penetrates through the fence (Ethington, 1977). The horizon and azimuth angle limits are also sampled at a higher update rate.

14.5.1 Horizon Search

The horizon represents an especially important border, particularly if target detection is limited by terrain masking effects. *Horizon search* refers to any search mode that follows the contours of the horizon. Depending on the specific environment, this may comprise a straight line or a more complicated contour. Horizon search is commanded more frequently than long-range search, typically every two to five seconds, due to the shorter target ranges involved and the shorter time to intercept.

Horizon search may comprise a single elevation bar (quick scan) or multiple elevation bars (extended horizon search). Multiple elevation bars are used to prevent a pop-up target at close range from climbing above the horizon scan pattern and so remain undetected. The probability of a target climbing above the scan pattern is a function of the beamwidth, the number of beams overlapped in elevation, the range to the horizon, the scan time, and the target velocity characteristics.

Conceptually, at least, horizon search can be generalized to include scanning into valleys and along ridgelines to detect terrain hugging aircraft. In this case, multiple terrain contours would be scanned instead of a single terrain contour. Each contour would be scanned individually until masked by a closer range contour. The waveform is matched to the terrain contour and expected search volume.

14.5.2 Horizon Tracking

Horizon tracking refers to the process of tracking the horizon's elevation angle. This type of tracking aids the search process since the waveform and detection apparatus can be optimized according to the local terrain profile. The track process can be enhanced as well since potential multipath can be minimized by placing a null of the difference pattern on top of the horizon's angle.

The horizon's angle can be measured and tracked by examining the returns in the zero Doppler filter. If elevation monopulse is available, an accurate horizon estimate can be immediately derived. Alternately a coarser horizon estimate can be inferred by computing the power gradient over successive elevation looks.

If disturbances such as rain or chaff are present, however, either monopulse or power gradient alone is insufficient to accurately measure the horizon's true angle. In this case a combination of approaches is required. A simple procedure follows. First, by occasionally looking up, the mean rain/chaff level can be quickly established. Next, as the beam is scanned down in elevation, the first beam position where the returned power crosses this rain threshold can be construed as containing valid terrain returns. Finally, monopulse provides a refined angular estimate as needed for scan optimization.

14.6 DATA FILTERING

Chapter 9 prevented the basic theory of data filtering. Here we consider in detail some of the more practical issues involved in filtering for an ESA radar system. Two related filtering methods are examined for reducing the filter computational complexity: (1) full decoupling, and (2) covariance decoupling. In the case of full decoupling both the state and the covariance matrices are decoupled. In covariance decoupling only the covariance matrix is decoupled. Each method has been used on practical ESA radar systems.

Decoupling reduces the computational burden by separating the full transition matrix into a number of smaller matrices. In any three-dimensional application decoupling reduces the computational complexity by roughly a factor of 9. This ninefold benefit results because the multiplication of two 9×9 matrices requires 9^3 operations, whereas the multiplication of two 3×3 matrices requires 3^3 operations for each of the three matrices. This benefit applies to those operations where decoupling is actually applied. Thus, covariance decoupling achieves an actuall overall reduction of approximately 4.3 (Daum and Fitzgerald, 1983, p. 274; Baheti, 1986).

It is sometimes argued that, as future computational resources improve, there will be less need for filter decoupling. Nevertheless, a number of other radar analysts have argued that any future increases in computational power can be more wisely spent in upgrading the ESA allocation logic and associated multiple-target processing. Hence, filter decoupling is likely to remain a valuable filter technique for some time to come.

In addition, there are a number of other advantages to decoupling, primary among these being a reduction in the numerical precision required. This reduction is a consequence of the fact that computer roundoff errors are physically separated in computer memory (i.e., decoupled along the range and angle axis), and also because square-root filtering can be more conveniently applied.

Decoupling has the additional benefit of simplifying the detection of target maneuvers. Maneuvers can be best detected by independently cross checking the range filter's estimate of acceleration with the Doppler filter's estimate of acceleration. Any divergence is taken as an indication that the target has maneuvered in range and (possibly) in angle. Maneuver detection is a concern of this chapter since the ESA allocation logic improves significantly in the presence of an efficient maneuver detector.

14.6.1 Covariance (or Partial) Decoupling

If the target is characterized by highly directional acceleration properties, it makes sense to extrapolate the state in a coordinate system well suited to modeling these directional characteristics. Given a constraint of this nature, the design of the filter then reduces to "which coordinate frame to use when propagating the covariance."

One coordinate frame found to be successful in the past is the RPCC frame (Browne et al., 1980) which in effect diagonalizes the **R** covariance matrix. If the matrices $\mathbf{HP}(k)\mathbf{H}^T$ and $\mathbf{R}(k)$ are perfectly aligned then the diagonalization of $\mathbf{R}(k)$ results in $\mathbf{P}(k)$ being diagonalized as well. Therefore, the propagation and updating of $\mathbf{P}(k)$ can be reduced to a set of decoupled submatrices.

Consider propagating **P** in RPCC:

$$\mathbf{P}(k+1) = \mathbf{\Phi}_{3D} P^{+}(k) \mathbf{\Phi}_{3D}^{T} + \mathbf{Q}(k) \tag{14.3}$$

where the covariance is *assumed* to be block diagonalized:

$$P^{+}(k) = \begin{vmatrix} P_{rr}^{+} & 0 & 0 \\ 0 & P_{\theta\theta}^{+} & 0 \\ 0 & 0 & P_{\phi\phi}^{+} \end{vmatrix}$$

and where the transition matrix is *assumed* to be block diagonalized:

$$\mathbf{\Phi}_{3D} = \begin{vmatrix} \mathbf{\Phi}_{1D} & 0 & 0 \\ 0 & \mathbf{\Phi}_{1D} & 0 \\ 0 & 0 & \mathbf{\Phi}_{1D} \end{vmatrix} \tag{14.4}$$

for

$$\mathbf{\Phi}_{1D} = \begin{vmatrix} 1 & \Delta T & \frac{1}{\alpha^2}(-1 + \alpha \Delta T + \rho) \\ 0 & 1 & \frac{1}{\alpha}(1 - \rho) \\ 0 & 0 & \rho \end{vmatrix}$$

or, for reduced-state filters,

$$\boldsymbol{\Phi}_{1\mathrm{D}} = \begin{vmatrix} 1 & \Delta T \\ 0 & 1 \end{vmatrix}$$

Unfortunately, if the matrices $\mathbf{HP}(k)\mathbf{H}^T$ and $\mathbf{R}(k)$ are not perfectly aligned, then both matrices cannot be diagonalized simultaneously. This occurs when tracking targets with a rapid rotation of the line of sight, in which case $\mathbf{P}(k)$ lags $\mathbf{R}(k)$. Decoupling can be forced in this case by assuming that $\mathbf{HP}(k)\mathbf{H}^T$ does indeed align with $\mathbf{R}(k)$. However, the Kalman gain matrix $\mathbf{K}(k)$ computed from the covariance is no longer optimum. The errors so induced have been analyzed extensively in (Rogers, 1986, 1988). One conclusion reached by Rogers was that the induced errors are insignificant if the angular change in the target's line of sight is small over the filter's memory (or filter time constant). "Small" is defined relative to the angular extent of the **R** error ellipsoid (i.e., minor axis divided by major axis). An example where this condition would *not* be satisfied is in tracking exoatmospheric targets, where the time constants involved are long due to predictable target dynamics.

One proposed solution to this problem is to keep the filter coupled in the plane of motion and decoupled in the direction orthogonal to that plane. This can be implemented by propagating the covariance **P** in range–velocity (r–v) coordinates, with the covariance coupled in the r–v plane (Daum and Fitzgerald, 1983). Page 189 discusses range-velocity coordinates in more detail.

In conclusion, the central idea is that the state $\bar{\mathbf{x}}$ and covariance **P** are updated in either RPCC or range–velocity coordinates (RVCC). The state $\bar{\mathbf{x}}$ is propagated in a second (inertial) coordinate frame, whereas the covariance is propagated in the updated coordinate frame (block diagonalized).

14.6.2 A Representative Fully Decoupled ESA Filter

If the target can be assumed to be characterized by omnidirectional acceleration properties, then it is possible to decouple both the state as well as the covariance without suffering any loss in track accuracy. The advantage of full decoupling is that the computational benefits of decoupling apply to both the state and covariance calculations. Also, the Doppler measurement can be more easily incorporated into the filter, and range data can be more easily derived from angle-only measurements.

The disadvantage of full decoupling is that it represents an inherently invalid model. The actual target dynamics are *always* coupled in some fashion. Nevertheless, these modeling errors are sometimes tolerated for the sake of the stated computational advantages.

To address the topic of full decoupling, the first and most important subject is selection of a suitable coordinate frame. The modeling errors that occur after full decoupling are strongly influenced by the choice of coordinate frame. Candidate coordinate frames include measurement coordinates, antenna coordinates, covariance coordinates, and inertial Cartesian coordinates. It can be argued that measurement coordinates represents a suitable starting point for this discussion. Recall that the primary quantities of interest in ESA tracking are the commanded (ζ_e, ζ_a) values to the predicted target position, since these quantities directly determine the antenna phase shifter settings. Therefore, the most direct filter is that one in which these physical quantities are filtered. Furthermore, for ground-based ESA systems the measurements are already in an inertial frame, hence avoiding the need for rotating the coordinate frame.

For mobile ESA radar systems, the solution is less clear-cut since the measured cosines (ζ_e, ζ_a) are not inertial. One approach is to reformulate the measurements in terms of "pseudo" inertial direction cosines $(\zeta_N, \zeta_E, \zeta_D)$. This is the approach taken here. The drawback of this method, as we shall see, is the creation of a redundant state.

To begin, start from Eq. (10.3):

$$\dot{\mathbf{i}}_r r + \mathbf{i}_r \dot{r} = \mathbf{v}_T - \mathbf{v}_A \tag{14.5}$$

and the associated second derivative

$$\ddot{\mathbf{i}}_r r + 2\dot{\mathbf{i}}_r \dot{r} + \mathbf{i}_r \ddot{r} = \mathbf{a}_T - \mathbf{a}_A \tag{14.6}$$

where r is the scalar range to the target, $\mathbf{i}_r$ is a unit vector in the line-of-sight direction, the term $\mathbf{a}_T$ is the inertial acceleration of the target, and $\mathbf{a}_A$ is the inertial acceleration of the radar's antenna.

In terms of the line-of-sight rate vector, $\boldsymbol{\omega}$, this equation is

$$\dot{\boldsymbol{\omega}} \times \mathbf{i}_r r + \boldsymbol{\omega} \times (\boldsymbol{\omega} \times \mathbf{i}_r) r + 2(\boldsymbol{\omega} \times \mathbf{i}_r)\dot{r} + \mathbf{i}_r \ddot{r} = \mathbf{a}_T - \mathbf{a}_A \tag{14.7}$$

One can verify that the second and fourth terms lie along the range vector. Combining these components gives

$$-|\boldsymbol{\omega}|^2 r + \ddot{r} = a_{T_r} - a_{A_r} \tag{14.8}$$

Equation (14.5) can be reformulated into

$$\dot{\mathbf{i}}_r = -\mathbf{i}_r \dot{r}/r + (\mathbf{v}_T - \mathbf{v}_A)/r \tag{14.9}$$

The vector $\mathbf{i}_r$ is next expressed in terms of its scalar components along the axis of the selected coordinate system. Let us denote these coefficients as $(\zeta_N, \zeta_E, \zeta_D)$, such that

$$\mathbf{i}_r = (\zeta_N, \zeta_E, \zeta_D)$$

Taking the derivative, and using the Coriolis relation,

$$\dot{\mathbf{i}}_r = (\dot{\zeta}_N, \dot{\zeta}_E, \dot{\zeta}_D) + (\boldsymbol{\omega}_{\text{NED}} \times \mathbf{i}_r) \tag{14.10}$$

where $(\boldsymbol{\omega}_{\text{NED}} \times \mathbf{i}_r)$ is the rotational rate vector of the coordinate system. When (14.10) is substituted into (14.9), this vector couples the differential equations for $\dot{\zeta}_N$, $\dot{\zeta}_E$, and $\dot{\zeta}_D$ together. By expressing all vectors in a nonrotating coordinate system ($\boldsymbol{\omega}_{\text{NED}} = 0$), the $(\zeta_N, \zeta_E, \zeta_D)$ filters can be partially decoupled.

Substituting (14.10) into (14.9), setting $\boldsymbol{\omega}_{\text{NED}}$ equal to zero, and equating components along each axis, the following three equations result:

$$\begin{aligned}
\dot{\zeta}_N &= -(\dot{r}/r)\zeta_N + (v_{T_N} - v_{A_N})/r \\
\dot{\zeta}_N &= -(\dot{r}/r)\zeta_E + (v_{T_E} - v_{A_E})/r \\
\dot{\zeta}_D &= -(\dot{r}/r)\zeta_D + (v_{T_D} - v_{A_D})/r
\end{aligned} \tag{14.11}$$

where $\mathbf{v}_T$ and $\mathbf{v}_A$ are the components of the target and the antenna velocity vectors, respectively.

The unknowns in these equations are the components of the target velocity vector $\mathbf{v}_T$. By differentiating $\mathbf{v}_T$:

$$\begin{aligned}
\dot{v}_{T_N} &= a_{T_N} \\
\dot{v}_{T_E} &= a_{T_E} \\
\dot{v}_{T_D} &= a_{T_D}
\end{aligned} \tag{14.12}$$

The net differential equation becomes

$$\frac{d}{dt}\begin{vmatrix} r \\ \dot{r} \\ \zeta_N \\ v_{T_N} \\ \zeta_E \\ v_{T_E} \\ \zeta_D \\ v_{T_D} \end{vmatrix} = \begin{vmatrix} \dot{r} \\ |\boldsymbol{\omega}|^2 r + a_{T_r} \\ -(\dot{r}/r)\zeta_N + v_{T_N}/r \\ a_{T_N} \\ -(\dot{r}/r)\zeta_E + v_{T_E}/r \\ a_{T_E} \\ -(\dot{r}/r)\zeta_D + v_{T_D}/r \\ a_{T_D} \end{vmatrix} + \begin{vmatrix} 0 \\ -a_{A_r} \\ -v_{A_N}/r \\ 0 \\ -v_{A_E}/r \\ 0 \\ -v_{A_D}/r \\ 0 \end{vmatrix}$$

Strictly speaking, the "extended" Kalman filter should be introduced to implement this data filtering, based on the nonlinear state equation. However, the quantities r, $\dot{r}$, and $\boldsymbol{\omega}$ are usually known to high accuracy. Therefore, it is *usually* legitimate to treat these quantities as known parameters in the state equation thus simplifying the state equation. Given this assumption, the form of the filter reverts to the standard linear Kalman filter:

$$\frac{d}{dt}\begin{vmatrix} r \\ \dot{r} \end{vmatrix} = \begin{vmatrix} 0 & 1 & 0 \\ |\omega|^2 & 0 & 1 \end{vmatrix} \begin{vmatrix} r \\ \dot{r} \\ a_{T_r} \end{vmatrix} + \begin{vmatrix} 0 \\ -a_{A_r} \end{vmatrix} \tag{14.14}$$

where the term $|\omega|^2$ is supplied by the angle filters. Likewise,

$$\frac{d}{dt}\begin{vmatrix} \zeta_i \\ v_{T_i} \end{vmatrix} = \begin{vmatrix} -(\dot{r}/r) & 1/r & 0 \\ 0 & 0 & 1 \end{vmatrix} \begin{vmatrix} \zeta_i \\ v_{T_i} \\ a_{T_i} \end{vmatrix} + \begin{vmatrix} -v_{A_i}/r \\ 0 \end{vmatrix}$$

where i is North, East, Down and where the terms r, $\dot{r}$ are supplied by the range filter.

Representing the acceleration as a decoupled Markov process, the acceleration equation becomes

$$\begin{aligned} \dot{a}_{T_r} &= -\alpha a_{T_r} + w_r \\ \dot{a}_{T_N} &= -\alpha a_{T_N} + w_N \\ \dot{a}_{T_E} &= -\alpha a_{T_E} + w_E \\ \dot{a}_{T_D} &= -\alpha a_{T_D} + w_D \end{aligned} \tag{14.15}$$

where w is white noise and α is the correlation time constant. Thus, the differential equations are now reduced to

$$\frac{d}{dt}\begin{vmatrix} r \\ \dot{r} \\ a_{T_r} \end{vmatrix} = \begin{vmatrix} 0 & 1 & 0 \\ |\omega|^2 & 0 & 1 \\ 0 & 0 & -\alpha \end{vmatrix} \begin{vmatrix} r \\ \dot{r} \\ a_{T_r} \end{vmatrix} + \begin{vmatrix} 0 \\ -a_{A_r} \\ 0 \end{vmatrix} \tag{14.16}$$

and

$$\frac{d}{dt}\begin{vmatrix} \zeta_i \\ v_{T_i} \\ a_{T_i} \end{vmatrix} = \begin{vmatrix} -(\dot{r}/r) & 1/r & 0 \\ 0 & 0 & 1 \\ 0 & 0 & -\alpha \end{vmatrix} \begin{vmatrix} \zeta_i \\ v_{T_i} \\ a_{T_i} \end{vmatrix} + \begin{vmatrix} -v_{A_i}/r \\ 0 \\ 0 \end{vmatrix}$$

The exact discretization of (14.16) is unnecessarily complicated. The following is a close approximation for small $|\omega|$:

$$\begin{vmatrix} \hat{r}(k+1) \\ \hat{\dot{r}}(k+1) \\ \hat{a}_{T_r}(k+1) \end{vmatrix} = \begin{vmatrix} 1 & \Delta T & \dfrac{1}{\alpha^2}(-1+\alpha\Delta T+\rho) \\ |\omega|^2\Delta T & 1 & \dfrac{1}{\alpha}(1-\rho) \\ 0 & 0 & \rho \end{vmatrix} \begin{vmatrix} \hat{r}(k) \\ \hat{\dot{r}}(k) \\ \hat{a}_{T_r}(k) \end{vmatrix} + \begin{vmatrix} 0 \\ -\Delta T\, a_{A_r}(k) \\ 0 \end{vmatrix}$$

and

$$\begin{vmatrix} \alpha\eta\gamma\hat{\zeta}_i(k+1) \\ \hat{v}_i(k+1) \\ \hat{a}_i(k+1) \end{vmatrix} = \begin{vmatrix} 1-\Delta T\dot{r}/r & \Delta T/r & \dfrac{1}{\alpha^2}(-1+\alpha\Delta T+\rho)/r \\ 0 & 1 & \dfrac{1}{\alpha}(1-\rho) \\ 0 & 0 & \rho \end{vmatrix} \begin{vmatrix} \hat{\zeta}_i(k) \\ \hat{v}_i(k) \\ \hat{a}_T(k) \end{vmatrix} + \begin{vmatrix} -\Delta T\, v_{A_i}/r \\ 0 \\ 0 \end{vmatrix}$$

For ΔT the sampling time and $\rho \triangleq e^{-\alpha\Delta T}$.

14.7 SUMMARY

The immediate advantage of the ESA is enhanced quality tracking and elimination of mechanical errors. A secondary advantage is that search can be more efficiently implemented since turnaround times are eliminated, more exotic scan patterns are possible, and alert/confirm logic can be employed to lower the FAR at no expense to the scan rate or detection performance. The ESA also allows the detection, resolution, measurement, and confirmation processes to be independently optimized and thus performance is improved. The disadvantages of the ESA include loss of effective aperture A_e, cost, weight, and (for an airborne radar) clutter broadening.

A hypothetical ESA search timeline was constructed to illustrate these remarks. The timeline was composed of a series of tasks, with the waveform and processing optimized to perform the current task subject to passing the previous task.

The remaining discussion mainly concerned methods to reduce the complexity of the data filtering for the ESA system. Included was a discussion of a full decoupled filter and a covariance decoupled filter. Suggestions to increase the effectiveness of the signal processing were also covered as needed to capture the full benefits of the ESA.

GLOSSARY

α Precomputed gain for a position–only data filter.

$\alpha\beta$ Precomputed gains for a position–velocity data filter. Also called Benedict–Bordner gains or g–h gains.

$\alpha\beta\gamma$ Precomputed gains for a three-state data filter (position–velocity–acceleration).

Active array Technology that combines an array antenna with multiple distributed solid state transmitters, thereby reducing RF losses and increasing reliability.

A/D Analog-to-digital.

Ada® Ada is a registered trademark of the U.S. government, Ada Joint Program Office.

ADC Analog-to-digital converter.

Adaptive array ESA technology in which the antenna pattern is automatically adjusted to create nulls in the directions of undesired signals. In general, the response is slower than an alternative "sidelobe canceler" type of null generation.

ADT Automatic detection and tracking.

A_e Antenna effective aperture. Weighted antenna aperture (physical aperture plus weighting) projected in the line-of-sight direction.

AGC Automatic gain control.

Alert/confirm Logic in which initial detection decisions "alert" a confirmation sequence that attempts to confirm the existence of a target and learn all relevant target properties needed for track file initiation.

Alias Ambiguity that occurs at harmonics of the sampling frequency (PRF). *See* **Nyquist rate.**

Altitude Height above a specific reference, usually referenced to above ground level (AGL) or mean sea level (MSL).

Altitude return The return from the earth's surface directly beneath the radar platform in an airborne radar. Characterized by near zero Doppler shift and range approximately equal to the radar's altitude above the ground. The power present in the altitude return can be very large, particularly when flying over water at low altitude.

Ambiguity Uncertainty in the true value of a parameter by discrete steps, e.g., a radar waveform that cannot distinguish between a possible target at frequency f_T and a possible target at frequency f_T + PRF is said to be ambiguous in frequency.

Ambiguity function Two-dimensional plot of a waveform's autocorrelation function in range and Doppler. Sometimes called Woodward ambiguity function.

Amplitude scintillation Random fluctations in the signal amplitude due to constructive and destructive interference. *See* **Fading.**

Amplitude weighting Same as **Weighting.**

A-MTI Adaptive MTI.

AMTI (1) Airborne version of MTI; (2) Adaptive MTI.

Analog device Any device that operates on analog signals.

Analog-to-digital converter Device that converts analog voltages to digital numbers.

Angels Unexplained sources of radar backscatter that are detected by the radar, but whose origin is not immediately obvious. Sometimes caused by birds, insects, or clouds, but most often attributed to reflections off of fluctuations in the atmospheric refractive index.

Angular resolution The minimum angular distance needed to separate targets. Usually taken to be equivalent to the 3-dB antenna beamwidth.

Anisotropic Not isotropic; not equal in all directions.

Antenna pattern Plot of antenna gain (in the Fraunhofer region) as a function of elevation and azimuth angle.

Aperture Area of the antenna face.

Area MTI Early type of MTI in which clutter cancellation is not based on the Doppler principle; cancellation is based on lack of position change from scan to scan.

Array antenna Antenna constructed from an array of radiating elements. The array may be arranged in a line (linear array), flat plane (planar array), or more complicated surface (conformal array).

Array time Time to collect one complete array of coherently related pulses. Same as **Coherent processing interval.**

ASR Airport surveillance radar.

ATC Air traffic control.

Automatic gain control (AGC) Controlling the gain of the radar in order to keep the average signal roughly constant. Usually implemented at the IF stage (before the A/D converter). See **Fine AGC, Digital AGC,** and **Coarse AGC.**

Azimuth Angle in the horizontal plane in some reference system.

Bandpass Pertaining to a narrow band of frequencies (not centered at dc), e.g., a *bandpass filter* passes only a narrow band of frequencies.

Bandwith (1) Width of the band of frequencies passed by a filter. (2) Width of the band of frequencies over which a system can operate effectively. (3) Band of frequencies occupied by a signal.

Bar Same as **Scan bar.**

Barker codes Binary phase codes that display the desirable property that the sidelobe amplitudes are 1 or -1, and the mainlobe is the same amplitude as the code length. Only codes of length 2, 3, 4, 5, 7, 11, and 13 are known to exist.

Baseband Lowpass signals (i.e., signals near dc). In communication links, baseband usually refers to the band of frequencies occupied by the message signal before it modulates the carrier. In radar, a baseband signal can refer to any signal near dc.

B-display Radar display presented in rectangular range-versus-azimuth coordinates.

Beam Mainlobe of an antenna.

Beam steering computer (BSC) Processing element dedicated to translating beam locations supplied by the data processor into a set of commands to be executed by an ESA.

Beamwidth Angular width of the antenna mainlobe. Generally measured between -3 dB points (one way) and denoted θ_{HP}.

Bias Constant offset, usually undesirable.

Binary integration Same as **M-out-of-N detection.**

Binary phase codes Type of digital pulse compression code that switches between only two phase shift values $(0, \pi)$.

Biphase codes Same as **Binary phase codes.**

Birdies Spurious signals created in the mixing process or generated in the receiver. May be produced by power supply surges, switching harmonics, etc.

Bistatic Type of radar in which (a) the transmitter is physically separate from the receiver by some large distance; (b) the transmitter and receiver use separate antennas.

Bit 1 or 0. A *Bi*nary uni*t*.

BIT See **Built-in test.**

Blind zones Range/Doppler regions where a significant portion of the target return is lost because of eclipsing (in range) or clutter rejection (in range and Doppler).

Blip-scan ratio Same as **Single-look probability of detection.**

BMEWS Ballistic Missile Early Warning System.

Boresight Center of the antenna beam; the direction of the peak of the antenna pattern. Also refers to the direction orthogonal to the antenna's face.

Boltzmann's constant 1.38×10^{-23} joules/K° (by difinition).

BSC Beam steering computer.

Built-in test (BIT) Any fault-detection or self-diagnosis procedure. For example, a common BIT procedure is to insert known digital data into the signal processor and ask for immediate readout to verify correct functioning. Another BIT procedure is to inject a synthetic target into the antenna and measure the resultant receiver output.

Butler array Type of ESA (electronically scanned array) that forms N beams from an N element array.

BW Bandwidth.

Byte 8 bits.

c Velocity of light in free space. $c \simeq 3.0 \times 10^8$ m/sec.

$C(\omega)$ Symbol for clutter spectrum.

CAGC Coarse AGC (automatic gain control).

Carrier The original signal upon which any information is molulated. The word "carrier" is sometimes used synonymously with "RF tone."

Cassegrain antenna Type of parabolic antenna consisting of a main reflector and a subreflector. Permits a reduction in the longitudinal dimension as compared to the conventional parabolic antenna, and reduces feed blockage by placing the antenna feeds behind the reflector.

C-band See **Radar bands.**

CCD Charge-coupled device.

CFAR Constant false alarm rate.

Chaff Small conducting strips that, when scattered in the air, can create false alarms and so obscure the real target.

Charge coupled device (CCD) Analog shift registers that transfer energy from successive capacitive storage elements at discrete times. Clutter filters constructed of CCDs are sometimes used to reduce the dynamic range required of subsequent devices, especially the A/D.

Chirp Method of pulse compression that uses a ramping frequency.

Chi-squared Probability density function (pdf) obtained by summing independent Gaussian variables after squaring. Commonly used type of probability density in radar analysis.

Closed loop Same as **Feedback.**

Clutter Unwanted returns from objects of no immediate interest, e.g., ground, buildings, birds, cars, insects, clouds, rain, chaff, and other meteorological phenomenon. These returns "clutter" the display with false target declarations that may overload the processor elements and/or desensitize the radar to the true target.

Clutter cancelation ratio Improvement in signal-to-clutter ratio (SCR) provided by the analog and digital signal processing, considering only mainlobe clutter.

Clutter filter Any filter whose sole function it is to remove clutter. Distinct from a Doppler filter in that a clutter filter does not necessarily separate targets or improve the SNR.

Clutter-referenced MTI Type of noncoherent MTI (moving target indicator) that detects moving targets by observing the cross product between the background clutter and the target. Suffers the unusual drawback that it only works well *if* clutter is present.

Coarse AGC Type of AGC (automatic gain control) in which the gain is applied in coarse steps (e.g., 10 to 20 dB steps), as opposed to fine AGC, in which the steps are much smaller. Also called "Clutter AGC."

Coherent (1) Signal whose phase and amplitude vary in a predictable fashion; (2) any processing that uses amplitude and phase information.

Coherent processing interval The amount of time necessary to collect one complete array of coherently related pulses. Same as **Array time.**

COHO *Coh*erent *O*scillator. Part of the exciter in a **MOPA** or part of a **COR.**

Coincidence resolving In this text the phrase "coincidence resolving" is used to refer to range/Doppler resolving after PRF (pulse repetition frequency) switching using **M-out-of-N processing.** This terminology is not standard but seems to be a reasonable way to refer to this process. In this type of M-out-of-N test the value of M is usually 2 or 3; two detections are used for resolving ambiguities, the third for resolving ghosts. In MPRF (**Medium PRF**) a common value for N is 8. Smaller values of N may be found in **RGHPRF** (range-gated high PRF) and MPRF in a look-up mode.

Colored noise Noise with a frequency distribution that deviates from a flat spectrum over the bandwidth of interest.

Comb filter Filter with periodic transfer function.

Comparator In a monopulse antenna, the assembly of directional couplers needed to convert the **A** and **B** signals into sum **S** and difference **D** signals.

Complementary codes Pairs of codes whose output after matched filtering are complementary, i.e., their sidelobe structures are equal and opposite so that the sidelobes cancel when added coherently.

Conical scan Early type of angle sensing technique that uses a single rotating, or nutated, beam. Amplitude modulation indicates target angle. See **Lobing.**

Constant false alarm rate (CFAR) Signal processing algorithm that continuously adjusts the detection threshold to maintain a constant false alarm rate. The algorithm adapts to changes in the mean noise level and, in some CFAR algorithms, changes in the noise correlation and noise probability density.

Control antenna Same as Guard antenna.

COR *C*oherent-*O*n-*R*eceive. System where a coherent signal is not transmitted but pulse-to-pulse phase errors are removed in the receiver. **MTAEs (Low PRF)** are not canceled in this type of system, and hence COR is used with LPRF only.

Correlated Two variables that possess nonzero correlation.

Correlation coefficient Normalized expected value of the product of two random variables.

COSRO *Co*nical *S*can on *R*eceive *O*nly. Type of conical scan where the receive beam is nutated in some manner but not the transmitted beam. Also called Silent lobing.

Covariance matrix Matrix of expected values of the product of two random vectors.

CNR Clutter-to-noise ratio.

CPI See **Coherent processing interval.**

CPU Central processing unit. A digital computing or processing element. When stated without reservation, usually implies a data processor element. For instance, a typical data processor might comprise several CPUs.

CR Cancellation ratio.

Crossing target (1) Target moving orthogonal to the radar's line-of-sight vector, such that it cannot be easily distinguished from clutter on the basis of relative Doppler frequency. (2) Targets with overlapping gates.

Cross range Orthogonal to the line of sight.

$c\tau/2$ Range resolution corresponding to the radar's pulse width τ.

Cumulative P_D Probability of having detected a target at least once over a specified region, usually before it closes to within a certain range. A common way to evaluate search mode performance in the case of approaching targets.

CW Continuous wave.

DAGC Digital AGC (automatic gain control).

Data Numbers or symbols used to represent information.

Data filter (1) Any filter implemented in the data processor; (2) any filter that smooths target data rather than analog or digitized voltages.

Data processor Computer that performs most of the higher-level data operations, including target tracking and display processing. The data processor performs a large variety of tasks at a low data rate, as opposed to the signal processor which performs a few special-purpose tasks at a high data rate. The data processor is implemented with a von Neumann type of computer architecture, whereas the signal processor is often implemented with some kind of pipeline approach.

dBi Decibels relative to an isotropic radiator.

dBm Decibels relative to one milliwatt.

dBsm Decibels relative to one square meter.

dBw Decibels relative to one watt.

dc Direct current.

Dead time (1) Time interval inserted between coherent dwell times. (2) Time the receiver is turned off in preparation for the next pulse.

Decibel (dB) Logarithmic unit used to express power ratios. A quantity P expressed in decibels is $10 \log(P/P_0)$, where P_0 is the implied power reference.

DECM (1) Deception **ECM** (electronic countermeasures); (2) defensive ECM.

Δf Incremental frequency; frequency deviation.

Delay-line canceler Filter based on differentiation.

Detection (1) The process of determining the presence of a target, usually by declaring a target present if the voltage exceeds a threshold. (2) The process of translating a bandpass signal to video signals, e.g., envelope detector, synchronous detector, crystal detector, I/Q detector, etc.

DFT Discrete Fourier transform.

Digital AGC Type of AGC (automatic gain control) where the gain is applied in small discrete steps that are adjusted by commands derived from digital signals. Note that *digital* does not refer to the actual location of the device, which usually is analog in any case. The phrase *fine AGC* is sometimes used instead of *digital AGC*, and is more descriptive of the function performed.

Digital filter Any filter implemented in a digital computer.

Discard time Time interval between arrays. During the discard time, the transmitter may be shut down (dead time) or the transmitter may be transmitting "fill" pulses in preparation for the next dwell (fill time). The discard time is used for clutter echo fading, AGC setting, filter settling, and to give the signal processor some additional time to process the previous dwell.

Discrete Quantized into distinct levels. Samples may be discrete in time (sampled) or discrete in voltage (binary words).

Discrete clutter A large specular return from a small object, e.g., a building or metallic shed. Discrete clutter typically varies between 20 and 60 dBsm RCS.

Discrete Fourier transform (DFT) Any Fourier transform that is formed digitally.

Discriminant (1) The error signal after measurement formation. For example, the product **D** ***S**/**S** ***S** is often called a *discriminant in angle*, where **D** is the difference signal and **S** is the sum signal. (2) Any algorithm whose output is proportional to target error.

Display Device that provides visual output to a human operator.

Displaced phase center antenna (DPCA) Processing that reduces the effect of platform motion on clutter spreading. Motion compensation is achieved by electronically displacing the phase center of the antenna by an amount that causes it to appear motionless to stationary clutter. DPCA is mainly used in LPRF, (**Low PRF**) since DPCA cannot correct for platform motion that is perpendicular to the antenna aperture, and hence can only correct clutter at a single unambiguous range.

DMTI Digital version of MTI (moving target indicator).

Dolph-Chebyschev weighting Weighting that produces equal sidelobes. Also called Chebyschev weighting or Tschebyscheff weighting.

Doppler blind zones Doppler zones for which the radar is blind due to rejection of areas of high clutter concentration.

Doppler filter Narrowband filter used to separate targets from other targets, clutter, and noise on the basis of sensed Doppler frequency.

Doppler frequency Frequency shift caused by relative radial velocity. Denoted as f_D (Doppler frequency) or f_T (target frequency).

Doppler spectra Modulations on the target return due to moving parts on the target, e.g., propellers, helicopter blades, treads, wheels, rotating engine blades, etc. Usually characterized by discrete lines (harmonics) offset from the base RF (radio frequency).

Doppler resolving Determining the true target Doppler frequency from a sequence of ambiguous frequency measurements.

DP Data processor.

DS Doppler spectra.

DSP Digital signal processing (or processor).

Duplexer High-speed switch or passive device that protects the receiver from transmitter leakage, while passing the low-power returns to the receiver.

Duty factor Proportion of time for which a device operates, e.g. transmit duty factor $d_T = \tau$ PRF, where τ = transmit pulse width.

Dwell time (1) Time to transmit and receive one complete array of coherently related pulses (i.e., array time plus fill time). (2) time interval between coherent data collection times (i.e., array time plus fill time plus dead time); (3) time associated with a beam position. Same as time-on-target.

Dynamic range Ratio of highest level signal to lowest level signal that can be handled in the same manner, expressed in decibels. The total dynamic range is further divided into two parts: (1) the instantaneous dynamic range, or that proportion of dynamic range that is available at any given instant, and (2) the AGC dynamic range, which part is generally not instantly available.

ECCM Electronic counter-countermeasures. Radar measures designed to combat the effect of ECM.

ECM Electronic countermeasures. Any measure designed to counter (disrupt) an enemy radar's operation. Includes jamming, chaff, and deception countermeasures.

Eclipsing Signal loss due to target return arriving back when the receiver is shut off for transmission of another pulse. Primarily associated with **HighPRF**, **RGHPRF**, and **MediumPRF**.

Elevation Angle in the vertical plane in some reference system.

Envelope detector Device that extracts the envelope (amplitude) of a bandpass signal.

ESA Electronically scanned antenna (or array). An antenna that scans its beam electronically. Also called Phase array steering, Phase scanned, Frequency scanned, etc.

ESM Electronic support measures. Branch of electronic warfare that is concerned with passive reception of signals. See **RWR**.

Estimation Assessment of the true value of a parameter in the presence of noise and other disturbances.

EW (1) Early warning; (2) electronic warfare.

Exciter Device that supplies the low level CW (continuous-wave) signal needed to maintain coherent phase relationships on transmit. Also called a radar master oscillator (RMO).

Fading Random fluctuations in the target echo amplitude due to constructive and destructive interference between scattering centers. Also known as **Amplitude scintillation.**

False alarm Incorrect target declaration.

False alarm rate (FAR) Rate (relative frequency) at which false alarms occur.

False alarm time T_{fa} (1) The inverse of the false alarm rate, i.e., $T_{fa} = 1/\text{FAR}$; (2) time for which the probability of a false alarm is 0.5, i.e., $T_{fa} = 0.69/\text{FAR}$.

Fan beam Literally, any antenna beam having a fan shape. That is, any beam whose extent in one direction is much larger than the other direction. Sometimes used in search applications since it permits a wider scan volume to be swept in a shorter amount of time, with each beam position receiving enough time on target to efficiently mechanize clutter cancelation. See **Pencil beam.**

Fast Fourier transform (FFT) A computationally efficient form of the DFT (discrete Fourier transform) invented by J. W. Cooley and J. W. Tukey.

Feedback Feeding back a portion of the output to the input of the device to help control the overall system characteristics.

Feeds (1) The radiating elements in the parabolic focus of a reflector antenna; (2) any transmission system that distributes energy to the radiating elements.

FET Field-effect transistor. As applied to radar, an FET is a common type of RF amplifier.

FIR filter Finite impulse response filter. Any filter that is formed from a single block of data (without feedback). The fast Fourier transform (FFT) is a special type of FIR filter. However, as applied to radars, it has become commonplace to use the phrase "FIR" to specify any narrowband filter that is formed directly (without the use of the Cooley–Tukey FFT algorithm). The goal is to improve performance by permitting greater control of the filter weights than that afforded by the FFT algorithm.

Fine AGC Type of AGC (automatic gain control) where the gain is applied in small discrete steps.

FM ranging (FMR) Technique for determining a target's range by modulating the frequency (usually linearly) and measuring the difference between the return frequency and the transmitter frequency. Differs from chirp in that the frequency modulation continues for a long time relative to the transit time to the target, and also in that the receive processing is performed in the frequency domain rather than the time domain.

F_N Noise figure.

Folding (1) Superposition of the positive frequencies onto the negative frequencies whenever only a single channel (either I or Q) is processed; (2) spurious noise created whenever nonperfect (I, Q) detection is employed.

Frank codes Digital version of chirp that requires (relatively) fewer phase shifter values than other versions of digital chirp.

Frame time Time to complete one radar scan.

Frequency agility Very rapid changing of the operating RF, ideally from pulse to pulse.

Frequency diversity Ability of the radar to tune to several operating RFs, not necessarily from pulse to pulse.

F_S System noise figure.

f_T Symbol for target Doppler frequency shift.

FTC See **Log-FTC.**

G Gain.

G_{SCR} Average signal-to-clutter gain of the signal processor as a function of Doppler filter index number.

G_{SNR} Average signal-to-noise gain of the signal processor as a function of Doppler filter index number.

Gain Ratio of output to input. Improvement in signal level in a specified region due to the presence of a device, relative to that attained in the absence of that device.

Gate (1) In tracking, a region that surrounds a target and within which that target may be reasonably expected to be detected; (2) in the receiver, an electronic switch that passes all returns within a specified region.

GCA Ground control approach. Monitoring and directing aircraft as they approach the runway at airports.

Ghosts Multiple false solutions that occur when two or more targets are detected simultaneously in the same range or Doppler bin during the process of **coincidence resolving** or **FM ranging.** The problem is one of too many unknowns (targets) and not enough equations (**PRFs** or FM segments).

Glint Angular scintillation. Random variation of the target line-of-sight angle due to mutual interference of the returns from different scattering centers distributed across the target's cross-range dimension.

GMT Ground moving targets, e.g., surface targets such as cars and tanks that are moving with sufficient radial velocity to be detected by the radar.

GMTI A radar mode of operation designed to detect GMTs. See **Displaced phase center antenna.**

GMTT A radar mode of operation specifically designed to track GMTs.

Grazing angle Angle between the earth and direction of signal propagation.

Guard antenna Broadbeam antenna, usually located beside the main antenna, that is used to detect energy entering through the sidelobes.

Guard processing Processing that inhibits detection of sidelobe targets by comparing the main antenna (sum) signal with an auxiliary broadbeam antenna (guard) signal. A type of sidelobe blanking technique.

Half-power beamwidth See **Beamwidth.**

Hamming weights Cosine-on-a-pedestal filter weighting.

Hanning weights Raised cosine weights. Hanning weights are frequently used because they can be simply formed after an unweighted FFT (**Fast Fourier transform**).

Harmonization errors All mechanical errors that arise in aligning the various hardware devices to a common reference direction. For example, a common type of harmonization error occurs in aligning the antenna to its pedestal. Harmonization errors are usually assumed to be constant for a particular system but random from system to system.

HF See **Radar bands.**

High PRF (HPRF) Operating at a high enough PRF that the desired target is unambiguous in Doppler frequency.

Horizon Range beyond which a target is masked by the earth's shadow. Widely used rule of thumb is $R_H = k(\sqrt{h_A} + \sqrt{h_T})$ for R_H = radar horizon in nautical miles, h_A = antenna height in feet, h_T = target height in feet, and $k = 1.23$.

HPRF High PRF.

h(t) Filter impulse response in time.

***H*(ω)** Filter impulse response in frequency.

I Improvement factor.

IF Intermediate frequency.

I. F. See **Improvement factor.**

IIR filters Infinite impulse response filters. Class of filters that use feedback to improve performance, and whose response is theoretically infinite. Feedback is convenient if the number of pulses is very large, such as in HPRF or RGHPRF, since greater control of the spectrum characteristics can be achieved for a given number of coefficients. IIF filters are rarely used in LPRF both because of the smaller number of pulses involved and the undesirable transient response.

Image frequency Ambiguity created in a superheterodyne receiver between frequency f_T and its mirror image at frequency $f_T + 2f_{IF}$, such that both frequencies are translated to the same location after mixing.

Improvement factor (I.F.) Ratio of output **SCR** to input SCR averaged over all target Doppler frequencies. Also called MTI improvement factor, Clutter improvement factor, or Reference gain. As originally stated, the improvement factor is strictly an MTI concept since it presupposes a completely random target frequency over (0, PRF) and assumes Doppler separation of targets is not attempted. Various authors have suggested generalizing the I.F. concept by averaging over all target frequencies in the radar's passband or Doppler filter.

Incoherent Not coherent.

INS Inertial navigation system (or set). System of accelerometers and gyroscopes that enables a moving platform to determine its position and velocity by integration.

Instrumentation errors Same as **Harmonization errors.**

Integration time Time span of integration.

Interference (1) Clutter plus noise; (2) clutter plus noise plus jamming.

Interlacing In reference to an array antenna, interleaving the feed connections such that each of the four ports (**A, B, C, D**) has access to the entire antenna face. Used to reduce the one-way sidelobe pattern.

Intermediate frequency (IF) Frequency intermediate between the transmitted RF and video frequency. The returns may be translated to an intermediate frequency to simplify amplification and bandpass filtering.

Interpulse period Same as **PRI** or 1/PRF.

I/O Input/output. Communications part of a module.

(*I*, *Q*) I = in phase (with the reference) and Q = quadrature phase (90° out of phase with the reference).

***I*/*Q* detector** Detecting the I (in-phase) and Q (quadrature-phase) components. Most common type of coherent detection method.

Isotropic Equal in all directions.

j Imaginary number; $\sqrt{-1}$.

Jacobian Denote f_i to be a N × 1 vector and x_j to be a M × 1 set of parameters, such that f_i is a function of x_j. Then the Jacobian of f with respect to x is the N × M matrix of partial derivatives df_i/dx_j.

JEM Jet engine modulation.

Jet engine modulation (JEM) Doppler spectra caused by rotating jet engine blades.

Jitter Change back and forth by a small amount.

Junk Generic term that refers to all sources of interference, including thermal noise (both external and internal), processing noise, clutter, jamming, and RFI (radiofrequency interference). Junk is a useful term to unambiguously denote every type of random interference that can possibly degrade performance and produce false alarms.

k (1) Boltzmann's constant. (2) FM slope factor $= \Delta f/\Delta t$.

Ka-band See **Radar bands.**

Kaiser weights Zero-order Bessel functions of the first kind.

Kalman filter Type of data filter wherein models are postulated for the filter accuracy, radar measurement, and target dynamical processes, and the data filter adaptively changes its gains according to the prescribed models.

K-band See **Radar bands.**

Klystron Type of high-power amplifier. An electron beam is passed through two or more resonant cavities, and the signal to be amplified is inserted into the first cavity and produces a modulation in the beam. In passing through the last cavity, the modulated electron beam radiates an electromagnetic field that is coupled to the output. See **Transmitter.**

Ku-band See **Radar bands.**

L Symbol for "loss".

λ Lambda. Symbol for wavelength.

L-band See **Radar bands.**

Limiter Nonlinear RF or IF device that clips high-power signals to a constant value. Limiters are used to restrict the dynamic range required of subsequent devices.

Linked list Convenient means of dynamically creating a list of computer records by linking each record with a software pointer to the next record.

Lobe Nonzero pattern between two nulls. May be a mainlobe or a sidelobe.

Lobing In general any technique for measuring the target angle by moving the mainlobe beam in angle. May be sequential or simultaneous, active or passive.

Log Logarithm.

Log-FTC Logarithmic fast time constant. Type of **CFAR** processing consisting of a logarithmic circuit followed by a highpass filter. Will provide a constant FAR at the output of the log-FTC when the input is noise with a Rayleigh probability density function.

Look Same as time on target. Time that the target is within the radar's 3 dB beamwidth.

LORO Lobe on receive only. Technique for measuring the target angle by synthetically moving the beam on receive. **COSRO** is a common type of LORO.

Low-pass filter Filter designed to pass those frequencies between dc and a maximum (cutoff) frequency.

Low PRF (LPRF) Operating at a low enough PRF that the desired target is range unambiguous.

LPRF Low PRF.

LPIR Low probability of intercept radar. Radar designed to minimize the probability of enemy **ESM** interception (detection) and tracking. Power management, adaptive scan limits, pulse-burst waveforms, and pulse compression are all features that might be used in a LPIR.

LRU Line replaceable unit. A hardware unit that is easily removed from its housing and hence is easily replaceable in the field. Also called a WRA (weapon replaceable assembly).

LSB Least significant bit.

Magnetron High-power oscillator commonly used in noncoherent microwave radars. Also used in micowave ovens.

Mainlobe Central part of an antenna or filter pattern. Used to distinguish that part of the pattern from the sidelobes.

Mainlobe clutter High-power narrowband clutter received through the antenna's mainlobe. Distinct from sidelobe clutter, which arrives through the antenna's sidelobes. The distinction is especially important in the case of an airborne radar since mainlobe clutter and sidelobe clutter possess different Doppler characteristics.

Major–minor Type of **PRF** switching scheme where major changes in the PRF are used to search for the target, and minor changes in the PRF are used to resolve range and Doppler ambiguities. Also called Major–minor–minor. Used in **ESA** applications and some older **MSA** applications.

Markov process Any random process whose probability density function (pdf) depends only on the immediate past.

Matched receiver In general, any receiver that is matched to the spectral characteristics of the received signal-plus-junk, such that a maximum of signal energy passes through the filter passband and a minimum of junk passes through. In radar terminology, however, it has become common to consider a "matched" receiver to be one matched to a signal embedded in *white noise*. In this case the matched receiver's frequency response is the same as the signal spectrum, and the time response is the reverse of the signal's.

Maximum likelihood (ML) estimate Type of estimator that maximizes the likelihood. The likelihood is the conditional probability density (or logarithm of the conditional pdf) when viewed as a function of the parameter and evaluated at the observation.

MC Mission computer.

Measurement A measurement usually refers to a physical observation of a parameter (i.e., a parameter plus noise). It differs from an *estimate* because an estimate operates on multiple measurements to extract a more accurate assessment of the parameter. Sometimes the term *state estimate* is used to refer to the smoothing part of estimation theory, and *parameter estimate* is used to refer to the process of extracting a single parameter from noise. In this case, depending on the context, there is a great deal of overlap between a *measurement* and a *parameter estimate.*

Medium PRF Operating at a PRF such that the desired target's range and Doppler frequency are both ambiguous.

MHT Multiple-hypothesis tracking.

Microwave Radio wavelengths in the centimeter regime.

Millimeter wave Radio wavelengths in the millimeter regime.

Mismatched processing Deliberately not matching the receive processing to the transmitted signal to achieve some goal. The radar may be mismatched in Doppler (e.g., Doppler filter weighting), mismatched in time (e.g., $\tau_V \neq \tau$), or mismatched in pulse compression code (e.g., ripple suppression or code weighting).

Mission computer In an airborne radar, the computer that controls an entire sensor suite, external control signals, and interfaces with the human operator(s) through displays. Also called a "central computer."

Mixer Nonlinear device used to mix two or more signals together. The output of the mixer will contain components at the desired cross-product frequency. Used in a superheterodyne receiver to translate the signal frequency to a new frequency at which it can be more conveniently processed.

ML Maximum likelihood.

MLC Mainlobe clutter.

Modes A multimode radar divides its operation into a number of specialized modes, with each mode designed to be optimized (to some degree) for the commanded mode. For example, search, acquisition, and single-target tracking are the three standard radar modes. Each of these modes may be further subdivided, e.g., aircraft search, surface target search, horizon search, velocity search, etc.

Monopulse Angle measuring technique in which the target's angle is measured by comparing the signals contained in at least two receiver channels. Since theoretically all the information is available *simultaneously*, a single received pulse is sufficient to measure the angle to the target (hence the name). Has performance advantages over sequential lobing, particularly in the presence of **ECCM** and amplitude scintillation.

Monopulse null cancellation (MNC) Processing that reduces the effect of platform motion on clutter spreading. Motion compensation is achieved by comparing the Doppler frequency of each return with its associated monopulse measurement such as to null out stationary clutter.

MNC See **Monopulse null cancellation.**

Monostatic (1) A Radar system in which the transmitter and receiver are at the same location; (2) A radar system in which the transmitter and receiver share a common antenna.

MOPA Master oscillator–power amplifier. Any transmitter system composed of an exciter followed by a power amplifier.

M*-out-of-*N A type of noncoherent detection in which M preliminary detections are required out of N tries before final target declaration. It is less efficient than PDI (postdetection integration) since it throws away the amplitude information, although its losses with respect to PDI typically do not exceed 2 dB.

Moving target indicator (MTI) Approach to Doppler filtering in which the clutter is rejected by means of a bandpass filter, but targets are not otherwise separated on the basis of Doppler frequency; i.e., narrowband resolution is not attempted. Generally refers to **LPRF** operation since Doppler resolution is not as critical in this case.

MPRF See **Medium PRF.**

MTAE Multiple-time-around echo. Echos received from beyond the radar's unambiguous range R_u.

MTI Moving target indicator.

MSA Mechanically scanned antenna.

m/sec Meters/second.

MTT Multiple-target tracking.

MTBF Mean time between failures. A common measure of hardware reliability.

MTR Missile tracking radar.

Multipath Propagation phenomenon wherein the electromagnetic energy follows multiple paths to the target and back. At microwave frequencies, multipath is often associated with reflections off large bodies of water, which degrades the ability of the radar to accurately measure the angle to low-altitude targets when flying over water or other sources of large reflection. In addition, there is a potential source of fading and hence loss of detection.

Multiple-target tracking Tracking multiple targets.

Narrowband (1) Narrow in bandwidth; (2) passing only a small bandwidth of frequencies.

Neyman–Pearson test Approach to threshold selection in which one of the probabilities is fixed, such as the P_{FA}, and the other probabilities are maximized (or minimized) subject to this constraint.

NF Noise figure.

NN Nearest neighbor. Type of correlation technique in which tracks are paired to measurements on the basis of minimum normalized distance.

N_0 Power spectral density of noise.

Noise Unwanted random signals that interfere with detection and parameter measurement. Noise can refer to any random fluctuation, but is usually taken to refer to additive white signals that are produced by the environment or the receiver. See **Thermal noise, Processor noise,** and **Receiver noise.**

Noise factor Same as **Noise figure.**

Noise figure Measure that describes the level of thermal noise in the receiver relative to that appearing across an ideal resistor at room temperature. The noise figure may or may not include the effects of the external thermal noise (system noise figure F_S versus receiver noise figure F_N, respectively).

Noncoherent Not coherent.

Noncoherent MTI Same as **Clutter-referenced MTI.**

Noncoherent integration Adding signals after phase has been removed. Same as PDI (postdetection integration).

Nonrecursive filter Same as a **FIR filter.**

NP-complete A problem is said to be NP-complete if the optimal solution requires exponential time complexity.

Nyquist criteria Rule from sampling theory stating that the sampling frequency must be at least twice the frequency of the highest frequency component of the signal in order for the sampled version of the signal to be an unambiguous version of the original signal.

Octave A 2:1 ratio. Used to indicate a doubling of the frequency, e.g., if an antenna's bandwidth extends from 5 to 10 GHz, then it operates over a full RF octave.

ω Omega. Symbol for angular frequency.

Oscillator Any device that produces an oscillating signal (e.g., a sinusoid).

Over-the-horizon (OTH) radar Radar that operates at a low enough RF (3–30 MHz) such that reflections off the ionosphere can extend its range significantly beyond the horizon.

Parabolic antenna Antenna consisting of radiating feed(s) that illuminate a parabolic reflector. See **Cassegrain antenna.**

Parameter An attribute whose value can vary.

Paramp Parametric amplifier. A low-noise RF amplifier.

Passband Band of frequencies the filter is designed to pass with minimal power degradation (no more than 50 percent).

P_{avg} Average power.

P_{cum} Cumulative probability of detection.

PD Pulse–Doppler.

P_D Single-look probability of detection.

P_d Single-dwell or single-pulse probability of detection.

pdf Probability density function.

PDI See **Postdetection integration.**

PDS Pulse–Doppler search. The same as velocity search.

PE See **Processing element.**

P_{FA} or P_{fa} Probability of false alarm.

Pencil beam Any antenna beam having a pencil shape. Any beam whose elevation extent and azimuth extent are roughly equal and relatively small. Ideally suited for tracking applications since the angle accuracy and angle resolution are good in both dimensions. See **Fan beam.**

Phase coding Pulse compression using phase modulation.

Phased array Same as **ESA** (electronically scanned antenna).

Phase shifter Device that shifts the phase of the electromagnetic field passing through it in discrete steps. Ferrite phase shifters are common at X-band; pin diode phase shifters are common at L-band. Both types are common at C-band.

Planar array Type of array antenna mounted on a flat plane.

Phasor Vector whose horizontal component represents the cosine component (I) of a sinusoidal signal and whose vertical component represents the sine component (Q). The basis of phasor notation is the observation that sinusoid signals add and subtract *vectorially*, and so representing the signal as a vector (phasor) facilitates any mathematical operation involving summation or integration.

Pointer Address or memory location of a computer record.

Polarization Orientation of the electric and magnetic field components of an electromagnetic field.

Polyphase codes Type of digital pulse compression code that switches among more than two phase shift values. A four-phase code, for example, uses four phase shifts including 0°, 90°, 180°, and 270°.

Postdetection integration (PDI) Adding signals after phase has been removed by envelope detection (hence the name).

Power aperture product The product of the antenna's effective aperture, A_e, and the average transmitted power, P_{avg}. Useful concept for search mode design.

Power spectral density Power present in each spectral component as a function of frequency for a random power signal. The inverse Fourier transform of the power spectral density is equal to the autocorrelation function of the random signal (according to the Wiener–Khinchine relation) assuming wide sense stationary signals.

Preamplifier An IF amplifier, usually placed immediately after the first mixing stage and before the bandpass filter, with large enough gain to ensure that the bandpass filter insertion loss does not cause appreciable degradation to the receiver noise figure.

Predetection integration The opposite of postdetection integration, viz., using phase information during the integration process.

Prefiltering Process used to reduce the amount of storage required by separating the filtering into stages.

Preselection filter An RF bandpass filter used to minimize the effects of image frequency responses and other interference sources.

Processor noise Noise created in the digital signal processor, usually associated with computer roundoff errors. Normally does not include A/D quantization noise.

PPI display Plan position indicator display. Range versus azimuth display in a polar coordinate frame centered on the radar's position. Provides a plan view (i.e., view from above). May be 360° PPI or sector.

PRF Pulse repetition frequency.

PRF switching Switching the PRF, either to decrease the blind zones after switching or to spot and resolve **MTAEs** by observing the change in apparent range after switching.

PRI Pulse repetition interval (PRI = 1/PRF).

Probability Relative rate of occurrence.

Processing element (PE) Generic name for a digital computing module of any kind, comprising an assembly of logical units and registers. Processing elements may be networked together with other elements (such as data memories or program memories) to form more complex processors. A data processor or signal processor typically is composed of several PEs.

Process synch Regular clock cycles that mark the transfer of digital data. The process synch is the fundamental quantization level that governs how promptly data can be transferred from the signal processor to the data processor and vice versa. Contemporary process synchs are on the order of a millisecond, although future process synchs are likely to be much less.

PSP Programmable signal processor. A digital signal processor whose instructions can be programmed (rather than hardwired). Programmability is desired if instructions must be changed in real time, or if later upgrades are to be implemented on the same processor element, or if the same type of processing element is to be used in multiple radars.

Pulse-burst waveform Type of **LPRF** waveform for which Doppler processing is performed on the basis of a single pulse (or single burst of pulses). Pulse-burst waveform is distinct from pure LPRF in that range is sampled at the HPRF rate, rather than the pulse (range resolution) rate. Pulse-burst has often been described as a "train of pulses," although the actual formation of pulses at the HPRF rate is immaterial to the primary goal of Doppler filtering.

Pulse coding Same as **Pulse compression.**

Pulse compression Method of achieving a long transmitted pulse while simultaneously achieving a short receive range resolution. The transmitted signal is coded (modulated) to possess a large bandwidth B and the receive pulse is compressed in time to approximately $1/B$ by means of a matched filter.

Pulse-delay ranging Determining the range to a target by measuring the time delay between transmission of a pulse and reception of its echo.

Pulse(d) Doppler Pulsed radar that sorts the returns on the basis of sensed Doppler frequency.

Pulse repetition frequency (PRF) Rate at which pulses repeat.

Quadrature 90° out of phase; e.g., a cosine signal and sine signal are said to be "in quadrature" with each other.

Quantization noise Noise that occurs when converting a continuous signal into digital format with finite word length. The difference between an analog signal and its quantized equivalent.

R Range to an object.

R_0 Range at which the SNR is unity (0 dB).

R_{50} Range at which the detection probability is 50 percent, usually in reference to the single-look probability of detection.

R_{90} Range at which the detection probability is 90 percent, usually in reference to the cumulative probability of detection.

R_{85} Range at which the detection probability is 85 percent, usually in reference to the cumulative probability of detection.

Radar *Ra*dio *d*etection *a*nd *r*anging.

Radar bands Standard nomenclature denoting frequency bands:

Band Designation	Nominal Frequencies	Assigned Frequencies*
HF	3–30 MHz	
VHF	30–300 Mz	138–144 MHz
		216–225 MHz
UHF	0.3–1 GHz	420–450 MHz
		890–942 MHz
L	1–2 GHz	1.215–1.4 GHz
S	2–4 GHz	2.3–2.5 GHz
		2.7–3.7 GHz
C	4–8 GHz	5.25–5.925 GHz
X	8–12 GHz	8.5–10.68 GHz
Ku	12–18 GHz	13.4–14.0 GHz
		15.7–17.7 GHz
K	18–27 GHz	24.05–24.25
Ka	27–40 GHz	33.4–36.0 GHz

*Frequencies assigned by International Telecommunications Union (ITU).

Radar cross section (RCS) An equivalent reflection area that would give the same received power as the target if scattered isotropically. Because the radar cross section incorporates the gain of the scattering mechanism as viewed in the direction of the radar, it may be much larger than the actual physical dimensions of the scatterer.

Radar data processor Same as **Data processor.**

Radar signal processor Same as **Signal processor.**

Radar master oscillator (RMO) Same as **Exciter.**

Radial velocity Component of velocity projected on the line-of-sight (radial) direction.

Radiation Electromagnetic energy.

Radio frequency interference (RFI) Interference (usually unintentional) from another radar operating in the vicinity. Sometimes called "friendly interference."

Radome Protective dielectric cover for the antenna face.

Range bin Discrete location where all energy returned from a given time of arrival is entered and stored.

Range gating (1) In analog systems, "range gating" refers to opening and closing a switch (gate) that passes all the energy within a given time increment. (2) In radars employing digital signal processing, range gating may refer to sampling the signal in time. See **Gate** for alternative definitions.

Range rate Radial velocity. Rate of change of target range.

Range resolving Determining the true range from a sequence of ambiguous range measurements.

Rayleigh scattering Scattering that occurs when the reflector is much smaller than a wavelength. The backscatter in this case is inversely proportional to the fourth power of the wavelength ($1/\lambda^4$).

RCS Radar cross section.

Receiver Device that bandpass-filters the signal, amplifies it, and translates it down to a lower frequency for more convenient processing.

Receiver noise Internal thermal noise generated by the radar receiver. Excludes digital processor noise, A/D quantization noise, and external environmental noise. Since most of the receiver noise is introduced just prior to IF amplification, the receiver noise is often the dominant noise component in the radar.

Rectifier Nonlinear device that rectifies the voltage, that is, converts voltages that may be both positive and negative into purely positive voltages (or purely negative voltages, depending upon the ground polarity).

Recursive filter Same as an **IIR filter.**

Reference range Range at which the signal-to-noise ratio is unity. Also called R_0.

Register (1) (n.) Computer memory location used to store data; (2) (v.) Align to a common reference direction.

Resolution Resolution relates to the ability to separate targets from other targets and clutter: on the order of a pulse width in range, a beamwidth in angle, and a Doppler filter width in frequency. Different from *measurement* in that a measurement can provide a much more accurate indication of the position to the target *provided the desired target signal is the dominant signal in the resolution cell.* Since an accurate measurement is predicated upon the assumption that at most a single target exists in each resolution cell, the radar's resolution plays an important role in radar tracking.

RF Radio frequency.

RFI Radio frequency interference.

RGHPRF Range-gated high PRF.

RGT Range-gated track. Usually used to denote a MPRF STT mode of operation.

Ripple suppression Processing that attempts to suppress spurious time sidelobes (ripples) by inverse filtering on receive.

ρ Rho. Symbol for correlation coefficient.

RMO Radar master oscillator.

RMS Root mean square.

Round-trip transit time Time to travel to an object and back ($2R/c$).

RSS Root sum square.

R_u Unambiguous range, $R_u = c/(2 * \mathrm{PRF})$.

RWR Radar warning receiver. Receiver that intercepts and interprets radiations coming from radars in the vicinity. A police radar detector is a form of RWR.

Saturation Condition that occurs whenever the magnitude of the input exceeds the capability of the device to process linearly. In this case the device will saturate, that is, its output will not increase linearly in response to linear increases in the input. Because saturation creates spurious signals in an uncontrolled manner, it is generally considered less desirable than quantization errors.

SAW device Surface acoustic wave device.

SB Sidelobe blanking.

S-band See **Radar bands.**

Scan To move the radar beam in an orderly fashion in search of new targets.

Scan bar A continuous sweep of the radar beam, usually at a constant elevation angle. A typical search pattern may comprise multiple scan bars.

Scan generator Part of the data processor that determines the scan pattern.

Scan pattern Pattern the radar makes as it sweeps a given volume of space for targets. May be visualized as the trace made by the antenna boresight as it is moved in angle.

Scintillation Fluctuations in the signal due to constructive and destructive interference between discrete scattering centers distributed over the target. Scintillation occurs in amplitude (fading), angle (glint), range, and Doppler.

SCR Signal-to-clutter ratio.

SCV See **Subclutter visibility.**

Sensitivity time control (STC) Type of AGC (automatic gain control) where the gain is adjusted in an open-loop fashion, usually based upon an assumed clutter model (such as clutter falls off with an assumed $1/R^4$ or $1/R^3$ profile). This type of AGC is applicable only to LPRF or pulse-burst waveforms.

Sequential lobing Lobing the beams sequentially. The opposite of monopulse.

Servo-mechanism Literally, "slave machine." Any mechanism that is adjusted in response to error signals. In radars, servo-mechanisms are used to keep the target centered in the radar's passband in angle, range, or Doppler.

Shift register Type of register that operates on data by shifting its contents to the right or to the left.

Sidelobe blanking Any feature that inhibits the classification of sidelobe signals as a target in the mainlobe. Comprises two steps: detection (as a sidelobe) and desensitization (usually by rejection of that resolution element).

Sidelobe canceler (SLC) Technology that attempts to cancel sidelobe signals by adaptively placing a null in the direction of the unwanted signal. Differs from **Sidelobe blanking** in that desensitization is not used. Differs from an adaptive array in that it requires N antenna elements and N processing channels to cancel $N-1$ sidelobe signals, whereas an adaptive array requires only a single receiver channel.

Sidelobe clutter (SLC) Clutter received through the antenna sidelobes.

Sidelobes Undesirable but unavoidable spurious lobes that occur on either side of the mainlobe.

Sidelobe target Unwanted return, due to a target or clutter, that enters through the sidelobes and may induce a false alarm.

σ Sigma. Symbol for square root of the variance.

σ_T Symbol for **Radar cross section (RCS).**

Signal Term applied to any voltage of interest. When used without reservation, usually refers to the desired target return.

Signal-to-noise ratio (SNR) Ratio of signal power to noise power. The SNR can in fact be referenced to any stage in the receive chain and based on any combination of noise sources (usually assumed to be white).

Signal processor (SP) The signal processor is characterized by a few special-purpose tasks that are performed at a high data rate, such as Fourier transformation, thresholding, sidelobe blanking, range/Doppler ambiguity resolving and, in some radars, measurement formation. May be analog or digital, hardwired or programmable.

Signal processor complex Comprises individual processing elements that perform the data processing and signal processing functions, but the actual processing is dynamically reconfigurable to support fault tolerance and situation responsiveness.

Simultaneous lobing Same as **Monopulse.**

Single-look probability of detection (P_D) Probability of detection based on a single time on target or observation.

Single-target tracking Tracking a single target (or cluster of targets closely spaced in angle). A radar mode of operation.

Sinusoid Sine, cosine, or sine with arbitrary phase.

SLC (1) **Sidelobe clutter;** (2) **sidelobe canceler.**

SNR See **Signal-to-noise ratio.**

SP See **Signal processor.**

Space stabilization (1) In airborne TWS, compensating the scan pattern to be independent of ownship angle (roll, pitch, and sometimes yaw); (2) in airborne or missile STT, mechanically compensating the antenna itself to be independent of ownship angle.

Spectrum Signal-versus-frequency distribution.

Spillover (1) Energy that leaks out (spills over) when a radiator distributes energy to a parabolic reflector; (2) energy that leaks out from the transmitter through the duplexer into the receiver.

Split-gate measurement Method of measurement comprising the normalized difference between the magnitudes of two voltage samples from adjacent bins. Commonly used method of range measurement.

Spread spectrum Spreading the spectral bandwidth beyond that required for range resolution, usually for military purposes. Common spread spectrum features include pulse compression, frequency agility, and coherent RF hopping.

SPRT Sequential Probability Ratio Test. Branch of alert/confirm logic where all detection decisions are separated into three classes: yes, no, and maybe. Only the "yes" and "no" decisions result in termination of the test.

STAE Second-time-around echos. See **MTAE.**

Staggered PRF Varying the **PRF** on a pulse to pulse basis, as opposed to varying the PRF on a dwell to dwell basis. Used to combat Doppler blind zones. PRF staggering suffers the drawback in that all **MTAEs** are smeared in range, and hence cannot be canceled using conventional Doppler techniques.

STALO *Sta*ble *l*ocal *o*scillator. A local oscillator. May be part of an exciter in a **MOPA.**

STC See **Sensitivity time control.**

Stepped chirp Type of chirp in which the frequency is stepped in discrete increments rather than varied continuously.

Stopband The band of frequencies that the filter is designed to attenuate. Opposite of **Passband.**

Straddling loss Reduction in signal power because the signal is not centered in the range or Doppler bin.

STT See **Single-target tracking.**

Subclutter visibility (SCV) Minimum SCR (signal-to-clutter ratio) at the input to the radar at which a target is detectible. An interesting concept that presently has limited utility because of the lack of a universal agreement on a quantitative definition.

Superhet See **Superheterodyne.**

Superheterodyne From Greek; literally means two (*hetero*) power (*dyne*) levels. A common type of radar receiver that uses two or more frequencies in the process of translating the signals down to video. See **Intermediate frequency.**

Surface acoustic wave (SAW) device Type of analog–analog device used in clutter filtering and pulse compression. The signal is propagated on the surface of the device (as opposed to bulk quartz crystals).

Surface targets Surface targets, e.g., cars and ships, that are moving with sufficient radial velocity to be detected by the radar.

Surveillance Searching and monitoring a given region.

SWC Scan with compensation. A type of conical scan that uses two rotating beams. SWC can achieve many of the performance benefits of monopulse.

Swerling models Models for target RCS (radar cross section) scintillation widely used for search and TWS (track-while-scan) modes. A brief description of the Swerling RCS models follows:

Swerling Case 0: Steady target (no scintillation).
Swerling Case 1: Gaussian target with slow scintillation.
Swerling Case 2: Gaussian target with fast scintillation.
Swerling Case 3: Chi-squared with two duo-degrees of freedom and slow scintillation.
Swerling Case 4: Chi-squared with two duo-degrees of freedom and fast scintillation.
Swerling Case 5: Steady target (same as Case 0).

Note: RCS relates to power instead of amplitude, so a "Gaussian target" is actually exponetially distributed. The exponential density is also called "chi-squared with two degrees of freedom" (or a single duo-degree). A "chi-squared target with two duo-degrees of freedom" refers to a target whose density is chi-squared with four degrees of freedom. "Slow scintillation" refers to scintillation that is constant over a look, but independent from scan to scan. "Fast scintillation" refers to scintillation that varies from pulse to pulse, which implies RF agility.

Synchronous detection (1) Same as ***I/Q*** detection. (2) Any detection in which phase is detected as well as amplitude. Includes *I/Q* detection and amplitude/phase detection as special cases.

System noise Refers to internal thermal noise due to the resistors, as well as external thermal noise due to graybody radiation from the earth, sun, atmosphere, ionosphere, and galactic noise.

T Symbol for **Dwell time.**

Tangential target Same as **Crossing target.**

Tapering Same as **Weighting.**

Target Object of interest.

τ Tau. Symbol for *pulse width.* If **Pulse compression** is used, refers to the compressed pulse width.

τ_u Symbol for uncompressed (transmitted) pulse width.

τ_V Symbol for length of the receiver's video filter response.

TB Terrain bounce. Type of angle deception ECM involving reflections from the ground.

T_H or T_h See **Threshold.**

Thermal noise White Gaussian noise due to thermal agitation of free electrons. Thermal noise is a physical representation of the fundamental randomness of our universe that states that freely moving electrons always have a finite probability of creating a nonzero current. The power of thermal noise is directly proportional to the absolute temperature (hence the name) and typically arises in the front-end conductors (before the first amplifier).

Threshold (T_H or T_h) (1) Level used to establish the presence or absence of a signal. (2) Any fixed level above which a response is generated.

Time-on-target (TOT) Time for the radar mainlobe to sweep over the target, usually measured between half power points (one-way).

TOT See **Time-on-target.**

TR Transmit/receive. Adjective used to describe any device that switches between the two modes of operation.

Track file A record (or set of records) that stores data pertaining to a single target in the data processor.

Tracking Following a target in angle, range, and Doppler. Usually involves measuring the target's position, smoothing the position measurements to obtain a more accurate assessment of the target's position, predicting the target position ahead in time, and using that prediction to gather the next sample measurement.

Tracking radar (1) A radar whose beam follows (tracks) a target; (2) any radar that performs the tracking function. The actual definition of a tracking radar is subject to confusion. The original definition, which pertains to following the target in a closed-loop fashion, is largely meaningless in today's complex radar environment where everything is closed loop, e.g., a TWS radar may shift its nominal scan pattern to center on the expected target density.

Track while scan (TWS) Tracking targets directly from data supplied by a mechanically scanning antenna.

Transmitter Device that supplies the signal for transmission. It might comprise a high-power oscillator (such as a magnetron) or an amplifier (such as a klystron, traveling wave tube, or solid state amplifier) that amplifies a reference signal supplied by a low-power exciter. The radar may use a single transmitter or multiple distributed transmitters (active array).

Traveling wave tube (TWT) Type of power amplifier. The TWT typically has a wider bandwidth but lower efficiency than an equivalent klystron. See **Transmitter.**

TTR Target tracking radar.

TWS See **Track while scan.**

TWT See **Traveling wave tube.**

Turnaround time Time required for an MSA to stop scanning a bar and start the next bar.

UHF See **Radar bands.**

V_A Antenna (or ownship) velocity.

V_R Radar velocity. Same as V_A.

Variance The second central moment of a random variable's probability density.

VB Velocity blind zones.

Velocity blind zones Zones for which the radar is blind due to the need to reject areas of high clutter concentration.

Vertical return Same as altitude return.

VHF See **Radar bands.**

Video Lowpass signals that occur after envelope detection. Video signals are represented by a single scalar (either I or Q) rather than a phasor (I, Q). Distinct from baseband because baseband refers to signals that were always in lowpass form, whereas video refers to signals that may originally be in bandpass form but are converted into lowpass form for further processing.

Video integration Same as postdetection integration.

V_T Target velocity.

VS Velocity search. Usually implies a HPRF search mode. VS measures target radial velocity but not target range.

Waveform Overall form of the radio waves radiated by the transmitter.

Waveguide Hollow pipe used to carry microwave energy from one place to another. A waveguide has less loss per unit length than an equivalent coaxial cable because of the larger surface area, and so are commonly used when feasible (microwave RFs and above). Since the cutoff frequency is proportional to the physical dimensions of the waveguide, the dimensions of the waveguide are typically matched to the nominal operating wavelength of the system.

Weighting Tapering (or reducing in amplitude) some of the samples on the edges of an array to reduce the level of sidelobes after taking the Fourier transform. In the case of pulse arrays, the cost of weighting is poorer Doppler resolution and reduced SNR.

White Adjective describing voltages with a flat power spectral density.

Windowing Same as **Weighting.**

Word A set of bits, specific to a device, used to quantize or store data.

WRA Weapon replaceable assembly. See **LRU.**

X-band See **Radar bands.**

REFERENCES

Aranciba, P. O., 1978. A Sidelobe Blanking System Design and Demonstration, *Microwave Journal,* **21** (3): 69–73; reprinted in *Radar Electronic Counter-countermeasures,* S. L. Johnston, Ed., Dedham, MA: Artech House, 1979.

Aronoff, E., and N. M. Greenblatt, 1974. Medium PRF Radar Design and Performance, 20th Tri-Service Radar Symposium Record, 53–67.

Athans, M., and C. B. Chang, 1976. Adaptive Estimation and Parameter Identification using Multiple Model Estimation Algorithm, MIT Lincoln Lab. Tech. Note 1976-28, ESD-TR-76-184.

Baheti, R. S., 1986. Efficient Approximation of Kalman Filter for Target Tracking, *IEEE Transactions on Aerospace and Electronics Systems,* AES-22, 8–14.

Barton, D. K., 1964. *Radar System Analysis,* Englewood Cliffs; NJ: Prentice-Hall.

Bar-Shalom, Y., and K. Birmiwal, 1982. Variable Dimension Filter for Maneuvering Target Tracking, *IEEE Transactions on Aerospace and Electronics Systems,* AES-18, 621–629.

Bar-Shalom, Y. and T. Fortmann, 1988. *Tracking and Data Association,* Orlando, FL: Academic Press.

Bath, W. G., F. R. Castella, and S. F. Haase, 1980. Techniques for Filtering Range and Angle Measurements from Colocated Surveillance Radars, Proceedings of the 1980 IEEE International Radar Conference, Arlington VA., 335–360.

Benedict, R. T., and G. W. Bordner, 1962. Synthesis of an Optimal Set of Radar Track-While-Scan Smoothing Equations, *IRE Transactions on Automatic Control,* AC-7, 27–32.

Berg, R. F., 1983. Estimation and Prediction for Maneuvering Target Trajectories, *IEEE Transactions on Automatic Control, AC*-28, 303–313.

Bertsekas, D. P., 1979. "A Distributed Algorithm for the Assignment Problem," Laboratory for Information and Decision Systems, Working Paper, M.I.T.

Bertsekas, D. P., 1988. The Auction Algorithm: A Distributed Relaxation Method for the Assignment Problem, *Annals of Operations Research*, **14**: 105–123.

Bertsekas, D. P., 1985. A Distributed Asynchronous Relaxation Algorithm for the Assignment Problem, Proceedings of the 24th IEEE Conference on Decision and Control, FT.Lauderdale, FL, 1703–1704.

Bierman, G. J., 1977. *Factorization Methods for Discrete Sequential Estimation*, New York: Academic Press.

Billan, E. R., Eclipsing Effects with High-duty-factor Waveforms in Long-range Radar, *IEEE Proceedings Part F*, **132**: 598–603.

Blackman, S. S., 1986. *Multiple Target Tracking with Radar Applications*, Dedham, MA: Artech House.

Blake, L. V., 1986. *Radar Range Performance Analysis*, Dedham, MA: Artech House.

Bogler, P., 1986. Detecting the Presence of Target Multiplicity, *IEEE Transactions on Aerospace and Electronic Systems*, AES-22, 197–202.

Bourgeois, F., and J. C. LaSalle, 1971. An Extension of the Munkres Algorithm for the Assignment Problem to Rectangular Matrices, *Communications of the ACM*, **14**; 802–806.

Braasch, R. H., and A. Erteza, 1966. A Recursion for Determining Feedback Formulas for Maximal Length Linear Psuedo-random Sequences, *Proceedings of the IEEE*, **54**: 999.

Brammer, K. G., 1982. Stochastic Filtering Problems in Multiradar Tracking, Nonlinear Stochastic Problems, Proceedings of the NATO Advanced Study Institute on Nonlinear Stochastic Problems, 533–552.

Brookner, E., 1977. *Radar Technology*, Dedham, MA: Artech House.

Brookner, E., 1981. A Review of Array Radars, *Microwave Journal*, **24**: 25–42.

Brookner, E., and T. F. Mahoney, 1983. Derivation of Satellite Radar Architecture for Air Surveillance, Proceedings of the 1983 IEEE International Radar Conference, 465–475.

Brown, R. G., 1983. *Introduction to Random Signal Analysis and Kalman Filtering*, New York: Wiley.

Brown, K. R., A. O. Cohen, E. F. Harrold, and G. W. Johnson, 1977. Covariance Coordinates—A Key to Efficient Radar Tracking, Presented at 1977 IEEE EASCON.

Browne, B. H., L. Ekchian, and L. J. Lawdermilt, 1980. Adaptive Features and Measurement Requirements for Advanced Surveillance Radars, Proceedings of the 1980 IEEE International Radar Conference, Arlington VA., 190–194.

Bucciarelli, T., U. Carletti, M. D'Avanzo, and G. Picardi, 1982. A New Family of Selenia Tracking Radars; System Solutions and Experimental Results, RADAR-82, IEE Conference publication #216, 434–438.

Cantrell, B. H., 1984. Comparison of 6 Integraters Used in Long-term Radar, NRL Report 8796.

Cantrell, B. H., 1986. Power Density and Threshold Control Strategies for Radar Track Initiation, Proceedings of the 1986 IEEE National Radar Conference, 71–75.

Capon, J., 1964. Optimum Weighting Functions for the Detection of Sampled Signals in Noise, *IRE Transactions on Information Theory*, IT-10, 152–159.

Carew, B., and P. R. Belanger, 1973. Identification of Optimum Filter Steady-state Gain for Systems with Unknown Noise Covariances, *IEEE Transactions on Automatic Control*, AC-18, 582–588.

Carlson, E. J., 1988. Low Probability of Intercept (LPI) Techniques and Implementations for Radar Systems, Proceedings of the 1988 IEEE National Radar Conference, 56–60.

Cartledge, L., and R. M. O'Donnell, 1977. Description and Performance Evaluation of the Moving Target Detector, MIT Lincoln Lab. Rept. No FAA-RD-76-190; reprinted in *Radar Electronic Counter-countermeasures*, S. L. Johnston, Ed., Dedham, MA: Artech House, 1979.

Castella, F. R., 1980. An Adaptive Two-dimensional Kalam Tracking Filter, *IEEE Transactions on Aerospace and Electronics Systems*, AES-16, 822–829.

Chan, Y. T., A. G. C. Hu and J. B. Plant; 1979. A Kalman Filter Based Tracking Scheme with Input Estimation, *IEEE Transactions on Aerospace and Electronics Systems*, AES-15, 237–244.

Chang, C. B., and J. A. Tabaczynski, 1984. Application of State Estimation to Target Tracking, *IEEE Transactions on Automatic Control*, AC-29, 98–108.

Chen, C. W., R. A. Walker, and C. H. Feng, 1988. A Branch-and-bound Algorithm for Multiple-target Tracking and its Parallel Implementation, 1988 American Control Conference, Atlanta Georgia.

Chin, L., 1979. "Advances in Adaptive Filtering," in *Advances in Control and Dynamic Systems*, Vol. 15, C. T. Leondes, Ed., New York: Academic Press.

Cohen, S. A., 1986. Adaptive Variable Update Rate Algorithm for Tracking Targets with a Phased Array Radar, *IEEE Proceedings Part F*, **133**: 277–280.

Cole, E. L., M. J. Hodges, R. G. Oliver, and A. C. Sullivan, 1986. Novel Accuracy and Resolution Algorithms for the Third Generation MTD, Proceedings of the 1986 IEEE Radar Conference, 41–47.

Cook, C. E., and M. Bernfeld, 1967. *Radar Signals*, New York: Academic Press.

Corsini, G., E. Dalle Mese, G. Marchetti, and L. Verrazzani, 1985. Design of the SPRT for Radar Target Detection, *IEE Proceedings Part F, 132*: 139–148.

CRC Standard Mathematical Tables, 1970. 18th edition, Cleveland: The Chemical Rubber Co.

Currie, N. C., and C. E. Brown, 1987. *Principles and Applications of Millimeter-wave Radar*, Dedham, MA. Artech House.

D'Addio, E., and G. Galati, 1985. Adaptivity and Design Criteria of a Latest-generation MTD Processor, *IEE Proceedings*, 132: 58–66.

Dana, R. A., and D. Moraitis, 1981. Probability of Detecting a Swerling I Target on Two Correlated Observations, *IEEE Transactions on Aerospace and Electronic Systems*, AES-17, 727–730.

Daum, F. E., and R. J. Fitzgerald, 1983. Decoupled Kalman Filters for Phased Array Tracking, *IEEE Transactions on Automatic Control*, AC-28, 269–283.

Davenport, W., and W. L. Root, 1958. *Random Signals and Noise*, New York: McGraw-Hill.

DiFranco, J. V., and W. L. Rubin, 1968. *Radar Detection*, Englewood Cliffs, NJ: Prentice-Hall.

Di Lazzaro, M., G. Fedele, and S. Strappaveccia, 1983. Theoretical and Experimental Results of Height Measurement in a Monopulse Hard-limited 3D Radar, IRSI 1983 Proceedings, Bangalore (India), 93–98.

Dillard, B. M., 1967. A Moving Window Detector for Binary Integration, *IEEE Transactions on Information Theory*, IT-13, 2–6.

Di Vito, A., A. Farina, G. Fedele, G. Galati, and F. Studer, 1985. Synthesis and Evaluation of Phase Codes for Pulse Compression Radar, *Rivista Tecnica Selenia, 9*: 12–24; reprinted in *Optimised Radar Processing*, A. Farina, Ed., London: Peter Peregrinus Ltd., 1987.

Dixon, R. C., 1976. *Spread Spectrum Techniques*, New York: Wiley.

Dolph, C. L., 1946. A Current Distribution for Broadside Arrays Which Optimizes the Relationship Between Beam Width and Sidelobe Level, *IRE Proceedings, 34*: 335–348.

Eaves, J. L., and E. K. Reedy, 1987. *Principles of Modern Radar*, New York: Van Nostrand Reinhold.

Emerson, R. C., 1954. Some Pulsed-doppler MTI and AMTI Techniques, Rand Corp report R-274-PR; reprinted in *MTI Radar*, 77–142, D. C. Schleher, Ed., Dedham, MA: Artech House, 1979.

Escobal, P. R., 1965. *Methods of Orbital Determination*, New York: Wiley.

Ethington, D., 1977. The AN/TPQ-36 and AN/TPQ-37 Firefinder Radars, IEE International Radar Convention, London.

Evans, R., 1986. Pattern Recognition Techniques for Radar Target Recognition, IEE Colloquium on Discrimination and Identification Methods in Radar and Sonar Systems, Digest #69, 6/1-4.

Farina, A., 1977. Single Sidelobe Canceller: Theory and Evaluation, *IEEE Transactions on Aerospace and Electronics Systems*, AES-13, 690–699.

Farina, A., and G. Galati, 1985. An Overview of Current and Advanced Processing Techniques for Surveillance Radar, Proceedings of the 1985 IEEE International Radar Conference, 175–183.

Farina, A., and S. Pardini, 1980. Survey of Radar Data-Processing Techniques in Air-Traffic-Control and Surveillance Systems, *IEE Proceedings Part F, 127*: 190–204.

Farina, A., and F. A. Studer, 1985a. *Radar Data Processing*, Vol. I, Letchworth, England: Research Studies Press. (distributed by Wiley, NY).

Farina, A., and F. A. Studer, 1985b. *Radar Data Processing*, Vol. II, Letchworth, England: Research Studies Press (distributed by Wiley, NY).

Fitzgerald, R. J., 1974a. Effects of Range-Doppler Coupling on Chirp Radar Tracking Accuracy, *IEEE Transactions on Aerospace and Electronics Systems*, AES-10, 528–532.

Fitzgerald, R. J., 1974b. On Reentry Vehicle Tracking in Various Coordinate Systems, *IEEE Transactions on Automatic Control*, AC-19, 581–582.

Fitzgerald, R. J., 1980. Simple Tracking Filters: Steady-State Filtering and Smoothing Performance, *IEEE Transactions on Aerospace and Electronics Systems*, AES-16, 860–863.

Fleskes, W., and G. Van Keuk, 1980. Adaptive Control and Tracking with the ELRA Phased-array Radar Experimental System, Proceedings of the 1980 IEEE International Radar Conference, Washington DC, 8–13.

Fliss, G. G., and D. L. Mensa, (1986) Instrumentation for RCS Measurements of Modulation Spectra of Aircraft Blades, Proceedings of the 1986 IEEE National Radar Conference, Los Angeles, CA, 95–99.

Frank, R. L., 1963. Polyphase Codes with Good Nonperiodic Correlation Properties, *IRE Transactions on Information Theory*, IT-9, 43–45.

Friedland, B., 1969. Treatment of Bias in Recursive Filtering, *IEEE Transactions on Automatic Control*, AC-14, 359–367.

Gardner, L. A., and R. J. Mullen, 1988. Constant Gain Tracker with Variable Frame Time, *IEEE Transactions on Aerospace and Electronics Systems*, AES-24, 322–326.

Gelb, A., 1974. *Applied Optimal Estimation*, Cambridge, MA: MIT Press.

Golay, M. J., 1961. Complementary Series, *IRE Transactions on Information Theory*, IT-7, 82–87.

Hamming, R. W., 1983. *Digital Filters*, Englewood Cliffs, NJ: Prentice-Hall.

Hannan, P. W., 1961. Optimum Feeds for all Three Modes of a Monopulse Antenna, *IRE Trans. on Antennas and Wave Propagation*, AP-9, 444–461.

Hansen, V. G., and D. Michelson, 1980. A Comparison of the Performance Against Clutter of Optimum, Pulsed-Doppler, and MTI Processors, Proceedings of the 1980 IEEE International Radar Conference, Washington DC, 211–218.

Hansen, V. G., and J. H. Sawyers, 1980. Detectability Loss due to Greatest of Selection in a Cell Averaging CFAR, *IEEE Transactions on Aerospace and Electronics Systems*, AES-16, 115.

Hansen, V. G., R. B. Cambell, N. Freedman, and W. W. Shrader, 1973. Adaptive Digital MTI Signal Processing, EASCON 1973 Convention Record.

Harris, F. J., 1978. On the use of Windows for Harmonic Analysis with the Discrete Fourier Transform, *Proceedings of the IEEE*, *66*(1): 51–83.

Hayt, W., 1974. *Engineering Electromagnetics*, New York: McGraw-Hill.

Hoelzer, H. D., G. W. Johnson, and A. O. Cohen, 1978. Modified Polar Coordinates—the Key to Well Behaved Bearings Only Ranging, IR&D report 78-M19-0001A, IBM Federal Systems Division, Shipboard Defense Systems, Manassas, VA.

Hovanessian, S. A., 1973. *Radar Detection and Tracking Systems*, Dedham, MA: Artech House.

Hsiao, J. K., 1974. On the Optimization of MTI Clutter Rejection, *IEEE Transactions on Aerospace and Electronic Systems*, AES-10, 622–629.

Hynes, R., and R. E. Gardner, 1967. Doppler Spectra of S-band and X-band Signals, *Supplement to IEEE Transactions on Aerospace and Electronic Systems*, AES-3, 356–365.

Iglehart, S. C., 1978. Some Results on Digital Chirp, *IEEE Transactions on Aerospace and Electronics Systems*, AES-14, 118–127.

Jacovitti, G., 1983. Performance Analysis of Monopulse Receivers for Secondary Surveillance Radar, *IEEE Transactions on Aerospace and Electronic Systems*, AES-19, 884–897.

Jazwinski, A. H., 1970. *Stochastic Processes and Filtering Theory*, New York: Academic Press.

Johnson, G. R., 1977, Jamming Low Power Spread Spectrum Radar,*Defence Electronics*, 103–112.

Johnston, J. A., and A. C. Fairhead, 1986. Waveform Design and Doppler Sensitivity Analysis for Nonlinear FM Chirp Pulses, *IEE Proceedings Part F, 133*: 163–175.

Johnson, M. A. and D. C. Stoner, 1976. ECCM from the Designer's Viewpoint, *IEEE Electro-76 Record*; reprinted in *Radar Electronic Counter-countermeasures*, S. L. Johnston, Ed., Dedham, MA: Artech House, 1979.

Key, E. L., E. N. Fowle, and R. D. Haggarty, 1959a. *A Method of Pulse Compression Employing Nonlinear Frequency Modulation*, MIT Lincoln Lab. Rept. No 207, Lexington, MA.

Key, E. L., E. N. Fowle, and R. D. Haggarty, 1959b. A Method of Sidelobe Suppression in Phase Coded Pulse Compression Systems, MIT Lincoln Lab. Rept. No 209, Lexington, MA.

Kirkpatrick, G. M., 1952. Final Engineering Report on Angular Measurement Accuracy Improvement, reprinted in *Radars, Volume 1*, D. Barton, Ed., Dedham, MA: Artech House, 1977.

Klauder, J. R., A. C. Price, S. Darlington, and W. J. Albersheim, 1960. The Theory and Design of Chirp Radars, *Bell System Technical J.*, *39*(4): 745–808.

Kretschmer, F. F., and B. L. Lewis, 1983. Doppler Properties of Polyphase Coded Pulse Compression Waveforms, *IEEE Transactions on Aerospace and Electronics Systems*, AES-19, 521–531.

Kuhn, H. W., 1955. The Hungarian Method for the Assignment Problem, *Naval Research Logistics Quarterly*, *2*: 83–97.

Kurniawan, Z., and P. J. McLane, 1985. Parameter Optimisation for an Integrated Radar Detection and Tracking System, *IEE Proceedings Part F*, *132*: 36–44.

Lathi, B., 1983. *Modern Digital and Analog Communication Systems*, New York: Holt, Rinehart, and Winston.

Leonov, A. I., and K. I. Fomichev, 1986. *Monopulse Radars*, Dedham, MA: Artech House.

Levanon, N., 1988. *Radar Principles*, New York: Wiley.

Lewis, B. L., and F. F. Kretschmer, 1982. Linear Frequency Modulation Derived Polyphase Pulse Compression Codes, *IEEE Transactions on Aerospace and Electronics Systems*, AES-18, 637–641.

Long, J. J., and A. Ivanov, 1974. Radar Guidance of Missiles, *Electronic Progress*, *16*: 20–28.

Loppnov, D. H., 1983. Low Level Surveillance Radar Design, IRSI 1983 Proceedings, Bangalore (India), 424–429.

Magill, D. T., 1965. Optimal Adaptive Estimation of Sampled Stochastic Processes, *IEEE Transactions on Automatic Control*, AC-10, 434–439.

Maisel, L., 1968. Performance of Sidelobe Blanking Systems, *IEEE Transactions on Aerospace and Electronics Systems*, AES-4, 174–180.

Manassewitch, V., 1987. *Frequency Synthesizers*, New York: Wiley.

Matthews, H., 1977. *Surface Wave Filters*, New York: Wiley.

Maybeck, P. S., 1979. *Stochastic Models, Estimation, and Control*, New York: Academic Press.

Maybeck, P. S., 1985. Adaptive Tracker Field-of-View Variation via Multiple Model Filtering, *IEEE Transactions on Aerospace and Electronics Systems*, AES-21, 529–539.

Mehra, R. K., 1970. On the Identification of Variances and Adaptive Kalman Filtering, *IEEE Transactions on Automatic Control*, AC-15, 175–184.

Mehra, R. K., 1971. A Comparison of Several Nonlinear Filters for Reentry Vehicle Tracking, *IEEE Transactions on Automatic Control*, AC-16, 307–319.

Mehra, R. K., 1972. Approaches to Adaptive Kalman Filtering, *IEEE Transactions on Automatic Control*, AC-17, 693–698.

Miller, K. S., and D. M. Leskiw, 1982. Nonlinear Estimation with Radar Observations, *IEEE Transactions on Aerospace and Electronics Systems*, AES-18, 192–200.

Mitchell, R. L., and J. F. Walker, 1971. Recursive Methods for Computing Detection Probabilities, *IEEE Transactions on Aerospace and Electronic Systems*, AES-7, 671–676.

Monzingo, R. A., and T. W. Miller, 1980. *Introduction to Adaptive Arrays*, New York: Wiley.

Mori, S., C. Y. Chong, and R. P. Wishner, 1986. Tracking and Classifying Multiple Targets Without A Priori Identification, *IEEE Transactions on Automatic Control*, AC-31, 401–409.

Mulholland, R. G., and D. W. Stout, 1979. Numerical Studies of Conversion and Transformation in a Surveillance System Employing a Multitude of Radars, ADA 072085.

Mulholland, R. G., and D. W. Stout, 1982. Stereographic Projection in the National Airspace System, *IEEE Transactions on Aerospace and Electronic Systems*, AES-18, 48–57.

Munkres, J., 1957. Algorithm for the Assignment and Transportation Problems, *J. Siam*, 5, 32–38.

Nagarajan, V., M. R. Chidambara, and R. N. Sharma, 1987. Combinatorial Problems in Multitarget Tracking—a Comprehensive Solution, *IEE Proceedings Part F*, *134*: 113–118.

Nathanson, F. E., 1969. *Radar Design Principles*, New York: McGraw-Hill.

Nevin, R. L., 1988. Waveform Tradeoffs for Medium PRF Air-to-air Radar, Proceedings of the 1988 IEEE National Radar Conference, 140–145.

Oppenheim, A. V., and R. W. Schafer, 1975. *Digital Signal Processing*, Englewood Cliffs, NJ: Prentice-Hall.

Pachares, J., 1958. A Table of Bias Levels Useful in Radar Detection Problems, *IRE Transactions on Information Theory*, IT-4, 38–45.

Page, R. M., 1955. Accurate Angle Tracking by Radar, NRL Report RA 3A222A; (1955); reprinted in *Monopulse Radars*, D. Barton, Ed., Dedham, MA: Artech House, 1974.

Pearson, J. B., 1970. Basic Studies in Airborne Radar Tracking Systems, Ph.D. Dissertation, University of California, Los Angeles.

Pearson, J. B., and E. B. Stear, 1974. Kalman Filter Applications in Airborne Radar Tracking, *IEEE Transactions on Aerospace and Electronic Systems*, AES-10, 319–329.

Petrocchi, G., S. Rampazzo, and G. Rodriguez, 1978. Anti-clutter and ECCM Design Criteria for a Low-coverage Radar, Proceedings of the 1978 International Conference on Radar, Paris, 194–200; reprinted in *Radar Electronic Counter-countermeasures*, S. L. Johnston, Ed., Dedham, MA: Artech House, 1979.

Postema, G. B., 1985. Range Ambiguity Resolution for High PRF Pulse-Doppler Radar, Proceedings of the 1985 IEEE International Radar Conference, 113–118.

Potter, J. E., 1963. *New Statistical Formulas*, Space Guidance Analysis memo #40, Instrumentation Lab., Cambridge, MA: MIT.

Rabiner, L. R., and B. Gold, 1965. *The Theory and Application of Digital Signal Processing*, Englewood Cliffs, NJ: Prentice-Hall.

Ramstein, Ph., and Ph. Georges, 1987. A New Criteria to Assess Radar Performance, Proceedings of the 1987 IEE International Radar Conference, London, 409–413.

Reid, D. B., 1979. An Algorithm for Tracking Multiple Targets, *IEEE Transactions on Automatic Control*, AC-24, 843–854.

Richard, J. T., and B. M. Dillard, 1977. Adaptive Detection Algorithms for Multiple-Target Situations, *IEEE Transactions on Aerospace and Electronic Systems*, AES-13, 338–343.

Riggs, D. D., 1975. Target Scintillation Fades and their Impact on Tracking and Detection, Proceedings of the 1975 IEEE International Radar Conference, Arlington VA, 446–451.

Rihaczek, A. W., and R. M. Golden, 1971. Range Sidelobe Suppression for Barker Codes, *IEEE Transactions on Aerospace and Electronic Systems*, AES-7, 214–218.

Ringel, M. B., D. H. Mooney, and W. H. Long, 1983. F-16 Pulse Doppler Radar (AN/APG-66) Performance, *IEEE Transactions on Aerospace and Electronics Systems*, AES-19, 147–158.

Ritcey, J. A., 1986. Performance Analysis of the Censored Mean-level Detector, *IEEE Transactions on Aerospace and Electronics Systems*, AES-22, 443–453.

Rivers, D. D., 1977. Advances in Digital Signal Processing with Applications to Airborne Pulse Doppler Radar, Ph.D. Dissertation, University of California, Los Angeles.

Robinson, T. H. 1983. Active Sensor Measurement Scheduling Problems in the Context of NP-completeness, *unpublished* Ph.D. Dissertation, University of California, Los Angeles.

Rockafellar, R. T., 1984. *Network Flows and Monotropic Optimization*, New York: Wiley.

Rogers, S. R., 1986. Accuracy of the Decoupled Kalman Filter, *IEEE Transactions on Automatic Control*, AC-31, 274–275.

Rogers, S. R., 1988. Steady-state Performance of the Decoupled Kalman Filter, *NAECON 1988*, *1*: 334–339.

Scheder, R. A., 1976. A Self-adapting Target State Estimator, Army Material Systems Analysis Activity, NTIS Document #ADA 026 152.

Schimidt, H., 1970. *Analog/Digital Conversion*, New York: Van Nostrand Reinhold.

Schleher, D. C., 1982. Performance Comparison of MTI and Coherent Doppler Processors, Proceedings of the 1982 IEE International Radar Conference, London, 154–158.

Schleher, D. C., 1986. *Electronic Warfare*, Dedham, MA: Artech House.

Shank, E. M., 1986, A Coordinate Conversion Algorithm for Multisensor Data Processing, DTC AD-A176-368.

Sherman, S. M., 1984. *Monopulse Principles and Techniques*, Dedham, MA: Artech House.

Shnidman, D. A., 1975. Evaluation of Probability of Detection for Several Target Fluctuation Models, MIT Lincoln Lab. Technical Note 1975-35, ADA013733. Lexington, MA.

Singer, R. A., 1970. Estimating Optimal Tracking Filter Performance for Manned Maneuvering Targets, *IEEE Transactions on Aerospace and Electronic Systems*, AES-6, 473–483.

Skolnik, M. I., 1970. *Radar Handbook*, New York: McGraw-Hill.

Skolnik, M. I., 1980. *Introduction to Radar Systems*, New York: McGraw-Hill.

Smith, P., and G. Buechler, 1975. A Branching Algorithm for Discriminating and Tracking Multiple Targets, *IEEE Transactions on Automatic Control*, AC-20, 101–104.

Spafford, L. 1968. Optimum Radar Signal Processing in Clutter, *IEEE Transactions on Information Theory*, IT-14, 734–743.

Stark, H., F. B. Tuteur, and J. B. Anderson, 1988. *Modern Electrical Communications*, Englewood Cliffs, NJ: Prentice-Hall.

Stephens, P. A., and N. R. Krupa, 1979. A Sparse Matrix Technique for the Munkres Algorithm, 1979 Summer Computer Simulation Conference, Toronto, Canada, 44–48.

Stimson, G. W., 1983. *Introduction to Airborne Radar*, El Segundo, CA: Hughes Aircraft Company.

Stoker, A., 1987. Dwelltime Minimization for Radar Target Detection in the Clear, *IEEE Transactions on Aerospace and Electronics Systems*, AES-23, 130–137.

Stone, M. L., and W. J. Ince, 1980. Air-to-ground MTI Radar Using a Displaced Phase Center Phased Array, Proceedings of the 1980 IEEE International Radar Conference, Arlington VA, 225–230.

Strappaveccia, S., 1987. Spatial Jammer Suppression by Means of an Automatic Frequency Selection Device, Proceedings of the 1987 IEE International Radar Conference, London, 582–587.

Swerling, P., 1957. Detection of Fluctuating Pulsed Signals in the Presence of Noise, *IRE Transactions on Information Theory*, IT-3, 175–178.

Taylor, J. W., 1982. Sacrifices in Radar Clutter Suppression due to Compromises in Implementation of Digital Doppler Filters, Proceedings of the 1982 IEE International Radar Conference, London, 46–50.

Taylor, J. W., L. W. Pardoe, and S. L. Johnston, 1980. A Monopulse Radar with MTI for Tracking Through Fixed Clutter, Proceedings of the 1980 IEEE International Radar Conference, 200–205.

Toomay, J. C., 1982. *Radar Principles for the Non-specialist*, Belmont, CA: Lifetime Learning Publications.

Trunk, G. V., 1983. Survey of Radar ADT, *Microwave Journal*, 77–88.

Tucker, A., 1984. *Applied Combinatorics*, New York: Wiley.

Van Keuk, G., 1975. Adaptive Computer-controlled Target Tracking with a Phased-array Radar, Proceedings of the 1975 IEEE International Radar Conference, Washington DC, 429–434.

Van Trees, H., 1968. *Detection, Estimation, and Modulation Theory*, New York: Wiley.

Walters, C. M., 1958. A Quantitative Analysis of Automatic Target Detection, IRE National Convention Record, Part 5, 107–119.

Ward, H. R., 1969. The Effect of Bandpass Limiting on Noise with a Guassian Spectrum, *Proceeding of the IEEE*, *57*: 2089–2090.

Ward, H. R., 1973. Properties of Dolph-Chebyshev Weighting Functions, *IEEE Transactions on Aerospace and Electronic Systems*, AES-9, 785–786.

Wehner, D. R., 1987. *High Resolution Radar*, Dedham, MA: Artech House.

Wirth, W. D., 1978. Array Antennas for Electronic Scanning, Proceedings of the 1978 IEE International Radar Conference, Paris, 423–430.

Wishner, R. P., R. E. Larson, and M. Athans, 1970. Status of Radar Tracking Algorithms, presented at the Symp. on Nonlinear Estimation Theory and Applications, San Diego, CA, 32–54.

Zeoli, G. W., 1971a. IF versus Video Limiting for Two Channel Coherent Signal Processing, *IEEE Trans. Information Theory*, AES-7 579–586.

Zeoli, G. W., 1971b. Generalized Pulse-burst Radar Design, *IEEE Transactions on Aerospace and Electronic Systems*, AES-7, 486–498.

Ziemer, R., and W. Tranter, 1976. *Principles of Communication*, Boston: Houghton Mifflin.

Zoraster, S., 1980. Minimum Peak Range Sidelobe Filters for Binary Phase-coded Waveforms, *IEEE Transactions on Aerospace and Electronic Systems*, AES-16, 112–115.

INDEX